SLIDING MODE CONTROL OF FRACTIONAL-ORDER SYSTEMS

SLIDING MODE CONTROL OF FRACTIONAL-ORDER SYSTEMS

XIN MENG
School of Physical and Mathematical Sciences
Nanjing Tech University
Nanjing, China

BAOPING JIANG
Department of Electronic and Information Engineering
Suzhou University of Science and Technology
Suzhou, China

HAMID REZA KARIMI
Department of Mechanical Engineering
Politecnico di Milano
Milan, Italy

ACADEMIC PRESS
An imprint of Elsevier

Working together
to grow libraries in
developing countries
www.elsevier.com • www.bookaid.org

Contents

8. An event-triggered mechanism to observer-based sliding mode control of fractional-order uncertain switched systems

9. Nonfragile sliding mode control of fractional-order complex networked systems via combination event-triggered approach

Acknowledgments

The chapters in this title are supported by the Natural Science Foundation of the Jiangsu Higher Education Institutions of China (No. 24KJB120007), by the Natural Science Foundation of Jiangsu Excellent Youth Fund Project (No. BK20240159), and in part by the Horizon Europe's Marie Skłodowska-Curie Actions Fund (No. 101073037).

Introduction of sliding mode control and fractional-order control systems

1.1 Overview of sliding mode control and its applications

Sliding mode control, originating in the 1950s, is a control method proposed by Emelyanov [1], and further investigated by Utkin and Itkin [2,3]. The theoretical study of variable structure control was conducted by Utkin in 1977 through the paper *Variable Structure Systems with Sliding Modes*. This theoretical study received widespread attention and underwent profound development, gradually forming an independent research branch [4]. Subsequently, with the rapid advancement of computers, robots, and electronic devices in the 1980s, this theory's research scope expanded to encompass complex fields such as nonlinear large systems, hysteretic systems, distributed parameter systems, and discrete systems. Concurrently, methods like neural networks and genetic algorithms were employed in variable structure control research and found extensive application within the field of control. Additionally, advanced control strategies including neural networks, genetic algorithms, and adaptive controls were also utilized in variable structure design [5].

Variable structure control involves the presence of one or more switching functions such that, when the system state reaches a specific value, it automatically transitions from one defined structure to another. In this context, the system's structure refers to its model or mathematical expression, typically represented by a set of mathematical equations. This control method is distinguished by the continuous adaptation of the system's structure based on its current state in a dynamic process, allowing for the maintenance of a predetermined sliding mode. As a result, it is

Sliding Mode Control of Fractional-order Systems
https://doi.org/10.1016/B978-0-44-331696-8.00005-7

commonly known as sliding mode variable structure control (or sliding mode control), often abbreviated as SMC.

Definition 1.1 ([1]). For the given ordinary differential equations,

$$\dot{x}(t) = f(x(t), u(t), t), \tag{1.1}$$

where $x(t) \in \mathbb{R}^n$, $u(t) \in \mathbb{R}^m$, $t \in \mathbb{R}$, the objective is to find a suitable switching function $s(x)$ to define the control function

$$u = \begin{cases} u^+(x), & s(x) > 0, \\ u^-(x), & s(x) < 0, \end{cases} \tag{1.2}$$

where $u^+(x) \neq u^-(x)$, such that the system satisfies the following conditions:

(1) Sliding modes are present;
(2) The reachability condition is satisfied, i.e., all moving points outside the switching surface $s(x) = 0$ will reach the switching surface in a finite time;
(3) The stability of sliding mode motion is guaranteed;
(4) The dynamic quality requirements of the control system are achieved.

Among them, the former three points must be met to have the sliding mode variable structure control.

Since its inception, sliding mode variable structure control has undergone over half a century of development and has evolved into an independent and dynamic research field within control theory. Its developmental trajectory can be categorized into three stages:

(1) The period from 1957 to 1962 represents the initial phase of variable structure control research, during which Utkin and Emelyanov introduced the concept, with the focus limited to second-order linear systems;
(2) From 1962 to 1970, research expanded to encompass higher-order linear systems, but remained confined to single-input and single-output systems;
(3) Post-1970, investigations into variable structure control extended to explore the system's invariance amidst disturbances and perturbations within a linear space.

Sliding mode control, as an effective nonlinear control approach, demonstrates its nonlinearity through the utilization of nonsmooth control signals to modify the dynamic characteristics of the system and compel the state trajectory to slide along a predetermined sliding mode surface. In 1974, Utkin elucidated in his research the advantages of sliding mode control over alternative control methodologies, including its simple, yet effective, algorithm, ease of implementation, robustness against external environmental disturbances and system parameter variations, complete adaptivity, low sensitivity to system model errors and parameter fluctuations, and high precision in control.

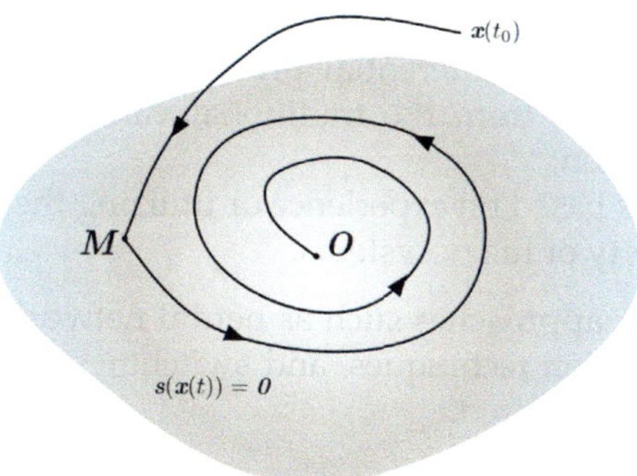

FIGURE 1.1 Schematic diagram of sliding mode variable structure control process.

The design of the sliding mode control strategy can be divided into two steps. The first involves constructing the sliding mode surface to ensure stable dynamic characteristics when the system state reaches this surface. The second entails designing the sliding mode control law to facilitate finite-time convergence of the system state to the sliding mode surface, aiming for its arrival within a specified time frame; see Fig. 1.1 for details. It is evident that this control strategy is discontinuous, reflecting its inherent nonlinearity. Its advantages include rapid response speed, resilience against system uncertainties, and robustness in handling external noise disturbances and parameter variations. Its main disadvantage is that in practical applications, the hysteresis of the switching device makes the actual sliding mode not always maintained in the switching flow shape, thus causing chattering. The causes of chattering can be categorized into: (1) spatial hysteresis switching; (2) temporal hysteresis switching; (3) the role of the physical inertia of the system; and (4) the chattering generated by the characteristics of the discrete system itself.

As is well known, chattering cannot be completely eliminated; therefore, when designing the sliding mode controller, it is necessary to select appropriate gain parameters to mitigate its effects. Currently, there are numerous domestic and international studies on the chattering phenomenon in sliding mode control. The main conclusions include:

(1) Quasi-sliding mode methods, encompassing continuous function approximation and boundary layer design;
(2) Convergence law methods such as linear and exponential convergence laws;

(3) Filtering methods that involve passing the system's control signal through a filter for smoothing and filtering;

(4) Interference observer method which utilizes an interference observer to estimate external disturbances and uncertainties followed by compensation;

(5) Dynamic sliding mode method seeking a continuous time dynamic control law using a differential process to generate an alternative switching function from the traditional variable structure control's switching function;

(6) Fuzzy methods based on experience or utilizing the universal approximation property of fuzzy systems.

Additionally, other approaches such as neural network methods, genetic algorithm optimization techniques, and switching gain reduction strategies have also been explored.

1.2 Overview of fractional-order control systems and their applications

At the end of the 17th century, when introducing the differential notation $\frac{d^n x}{dt^n}$, Professor L'Hôpital raised a question to Leibnitz: *Is it possible for order n to be expressed as a noninteger?* At that time, Professor Leibnitz did not have a satisfactory answer to this query. It was not until 1819 that Lacroix finally addressed this issue in [6], demonstrating that the $\frac{1}{2}$-order derivative of x is $2\sqrt{x/\pi}$. However, due to the absence of a unified definition of fractional calculus operators and practical physical background support, as well as contradictions with classical physical theoretical systems established by traditional calculus theory, fractional calculus had long been studied solely as part of theoretical research by mathematicians with slow progress. It was not until the end of the 20th century when Professor Mandelbrot, an American mathematician, first proposed the concept of fractals and suggested that there was a certain self-similarity between fractional-order systems and traditional systems shown in [7]. Subsequent studies by Professor Mandelbrot supported this view – namely that the fractional-order system is a generalization and extension of the traditional system in terms of its order. With further exploration into fractional calculus theory, it has been applied in research fields such as viscoelastic media [8], heat conduction equations [9], electrolytic processes [10], and digital circuits [11].

In the field of control, research on fractional-order systems can be primarily divided into two branches: system modeling and controller design [12–15]. When it comes to system modeling, more complex models are often required to meet the high demands for accuracy in practical engineering applications. The fractional-order model is favored by scholars due to its dual degrees of freedom – derivative and integral – providing an effective solution to these challenges; see [16,17] and the references therein. On the other hand, in controller design, the introduction of fractional-order calculus operators allows for the development of suitable controllers such as fractional-order PID controllers, fractional-order feedback controllers, and fractional-order event-triggered controllers. Implementing these controllers into the original system results in a closed-loop system that operates as a fractional-order system. In recent years, stability theory research on various types of fractional control systems has become a prominent topic within the field of control with notable scientific advancements emerging; see [18–20] and the references therein.

Recently, with increasing comprehension of fractional-order calculus, the introduction of fractional-order calculus operators into the original sliding mode controller has led to the development of the fractional-order sliding mode controller. This new controller preserves the fundamental theoretical characteristics of sliding mode control while incorporating novel application features that enable a more precise description of the physical attributes of actual systems. Currently, research on fractional-order sliding mode control methods has yielded promising outcomes and found widespread applications in switching systems [21], chaotic systems [22,23], electromechanical systems [24], power systems [25], and multi-intelligence systems [26]. For instance, considering a class of integer- and fractional-order controlled systems with unmatched disturbances, a fractional-order disturbance observer was designed as outlined by Wang et al. in [24]. Building upon this observer system, a corresponding sliding mode control strategy was also proposed and validated through simulation experiments involving a four-wing UAV and magnetic levitation system. Additionally, for a category of control systems affected by nonlinear perturbations, Fei et al. explored adaptive fractional sliding mode controllers based on neural estimators in [27]. Furthermore, targeting another class of nonlinear systems, Delacari et al. investigated the design problem associated with a robust fractional-order sliding mode controller in [28]. The mitigation of system chattering was achieved by substituting the symbolic function in the controller with a fuzzy logic controller.

On the other hand, as communication technology continues to advance, there is a growing consideration of developing a more rational and viable control strategy to ensure the stable operation of controlled systems within limited communication resources. In 1999, Professor Arzen [29] from Sweden introduced the concept of an event-based PID controller for the first time in his book, *A Simple Event-based PID Controller*, demonstrating its effectiveness and feasibility through application in a PID control system. Subsequently, event-triggered control has garnered significant attention and extensive research efforts from scholars globally [30–35]. When designing the event-triggering mechanism, aside from meeting the dynamic characteristics of controlled systems, an important focus lies in studying methods to prevent Zeno behavior, that is, avoiding infinite sampling within a finite time frame. Generally speaking, preventing Zeno behavior requires establishing a minimum event triggering interval $\Delta T = \min\limits_{k}\{t_{k+1} - t_k | k = 0, 1, \ldots\} > 0$. However, it is noted that existing literature lacks comprehensive exploration and discussion on fractional-order event triggering control due to the nonlocal characteristics of calculus operators in fractional-order controlled systems [35–37], with some flawed criteria being presented. Henceforth, there remains ample research space for analyzing Zeno behavior in fractional-order systems. Furthermore, unlike periodic trigger control strategies where task execution depends on specific conditions at regular intervals; event trigger control strategies rely solely on given trigger conditions. In the existing research results [36–42], there are two commonly used trigger condition design forms, namely

(1) Norm-type trigger condition, $(x(t) - x(t_k))^{\mathrm{T}}\Omega(x(t) - x(t_k)) \leq \varrho x^{\mathrm{T}}(t)\Omega x(t)$, $t \in [t_k, t_{k+1})$, where $x(t)$ denotes the state of the physical model which can be obtained, $\Omega > 0$ is a given matrix, and $\varrho > 0$ is a threshold constant;

(2) Exponential-type trigger condition, $\|x(t) - x(t_k)\|^2 \leq \varrho e^{-\varepsilon t} + \rho_0$, $t \in [t_k, t_{k+1})$, where $\varrho > 0$, $\rho_0 \geq 0$, and $\varepsilon > 0$.

Building upon the aforementioned design methods, a novel combined event-triggered mechanism was introduced for the first time by Li et al. in [43]. This mechanism effectively integrates the strengths of the previously mentioned forms, achieving a reduction in sample numbers while preserving system stability. Subsequently, in [44], Chang et al. have presented an intermittent combined event triggering control strategy and investigated exponential boundedness within a category of fractional-order multi-agent systems. The findings from these studies offer valuable theoretical insights for this research work.

1.2.1 Variable-order fractional systems

Research results show that there are some complex physical systems or phenomena in nature, with their memory and nonlocal properties changing with time, space or other conditions, so it may be more reasonable and accurate to use variable-order fractional differential equation to describe these complex physical phenomena. In 1995, Samko first gave the definition of time-varying Riemann–Liouville fractional calculus operators and their related properties in [45], and then Lorenzo and Hartley et al. conducted in-depth research on such problems; the definition and calculation methods of some other types of variable-order fractional calculus operators were presented in [46]. Since then, the theory of variable-order fractional calculus has been applied to various scientific and engineering fields; see [47–50] and the references therein. For example, the controller design problem for a class of nonlinear viscoelastic oscillators was studied by using variable-order fractional calculus operators in [47]. In [48], based on Arzela–Ascoli theory, Jiang et al. studied the existence, uniqueness, and stability of solutions for a class of variable-order fractional differential equations. In [49], the controller design problem of a class of variable-order fractional-order chaotic systems was studied, and the adequacy criterion of Mittag-Leffler stability of variable-order fractional systems was given by using the fractional-order comparison principle; the stability theory of constant fractional-order systems has been extended to the theory of variable-order fractional calculus. On the other hand, since Pecora and Carroll [51] put forward the chaos synchronization problem, chaos synchronization has also attracted extensive attention from scholars at home and abroad and has been applied to communication security, image processing, and other fields; see [52–54] and the references therein. For example, for a class of fractional-order chaotic systems with uncertain and time-delay terms, the adaptive fuzzy sliding mode control method was presented by Lin et al. to explore the system state synchronization control problem in [53]. In [55], Tolba et al. studied the synchronization control problem for a class of fractal-order chaotic systems based on the Izhikevich neuron model. In [56], Li et al. explored the exponential synchronization control problem for a class of variable-order fractional piecewise-smooth systems with short-term memory characteristics based on the event-triggered mechanism. However, the theory of variable-order fractional calculus as a new branch of fractional-order calculus theory and its theoretical system are not mature, therefore there are few academic achievements in this area. So how to explore the stability problem of variable-order fractional calculus system is still a new field.

1.2.2 Fractional-order coupled systems

In daily life, some complex problems can be described by various coupled differential equations, and the stability of coupled systems has always been one of the hot issues studied by scholars. For example, in [57], Li et al. used graph theory to design a new vertex Lyapunov function to study the stability problem of coupled systems built on strongly connected networks. In [58], Chen et al. studied the sliding mode synchronization problem of two types of multi-chaotic systems with unknown parameters and external disturbances, and proposed a suitable adaptive law to estimate the unknown parameters contained in the system, so as to realize sliding mode synchronization control between each drive system and response system state. In recent years, fractional-order coupled systems, which are generated by combining fractional-order calculus theory with coupled systems, have also been widely investigated and discussed by outstanding scholars at home and abroad; see [59,60] and the references therein. For example, Li et al. explored the global Mittag-Leffler stability problem for a class of fractional-order coupled systems by designing a novel vertex Lyapunov function in [59]. In addition, some scholars have also studied the stability problem for various coupling systems by using other methods, such as pulse control, periodic intermittent control, and feedback control, etc.; see [61–63] and the references therein. However, it is worth noting that all above-mentioned research results were based on the assumption of strong connectivity of the network, while in practical engineering, non-strongly connected networks are more common, so it is of certain scientific research value to study the stability of coupled system solutions on non-strongly connected networks.

1.2.3 Fractional-order multi-agent systems

The multi-agent system, as one of the important research objects of distributed artificial intelligence systems, has attracted great attention from the academic community due to its advantages of high efficiency, low cost, flexibility, and reliability, and has become one of the effective solutions to various complex tasks; see [64–70] and the references therein. It has been also applied in the research on computer networks [71], robot cooperative control [72], smart grid [73], etc. In recent years, fractional-order multi-agent systems, obtained by some scholars by introducing fractional-order calculus theory into multi-agent systems, have also received extensive attention and excellent research results were obtained. In this book, the consensus control and formation control problem of fraction-order multi-agent systems are studied based on the event-triggered mechanism.

(1) Consensus Control Problem

Consensus, as one of the challenges in the coordinated control of multi-agent systems, aims at reaching a global consensus goal on a specific feature [74]. For example, Rehan et al. studied the distributed consensus control problem for a class of nonlinear multi-agent systems with one-sided Lipschitz conditions by designing a state feedback controller in [75]. Meanwhile, in [76,77], the problem of fuzzy adaptive consensus control for a class of nonlinear multi-agent systems was also studied, and some excellent results have been achieved. For a class of high-order nonlinear multi-agent systems with unknown dynamics, an adaptive fuzzy finite-time consensus control strategy was proposed in [78]. For descriptor multi-agent systems, the necessary and sufficient conditions of admissible consensus were given in [79]. The admissible output consensus problem for a class of high-order descriptor linear multi-agent systems based on the dynamic feedback control was studied in [80]. A distributed controller was designed in [81] to explore the admissible consensus problem for homogeneous descriptor multi-agent systems with undirected graphs and disturbances. These excellent research results provide the theoretical basis for the study of descriptor multi-agent systems in this work.

For fractional-order multi-agent systems, reference [82] studied the problem of consistency control with input delay. At the same time, the pinning control method was used to study the robust consensus problem with external disturbances in [83]. In [84], for the leader system described by a second-order system, a fractional-order observer was designed to study the consensus control problem of system state tracking in the leader–follower mode. In addition, in many practical systems, such as power networks with supercoils and supercapacitors, robot systems running on dirt roads, etc., fractional-order descriptor linear multi-agent systems can be used to describe these models [85]. Therefore, some other excellent results have been obtained on the consensus control problem of fractional-order descriptor linear multi-agent systems; see [86–88] and the references therein. For example, in [86], a distributed consensus control strategy was proposed to study the leader–follower admissible consensus control problem for a class of fractional-order descriptor linear multi-agent systems with order $\alpha \in (0, 2)$. However, in the research of fractional-order descriptor nonlinear multi-agent systems, the consensus problem has not been fully studied, so it is still a challenging topic. The above literature results provide a necessary reference value for the study of this work.

Recently, some research results have been obtained based on the event-triggered mechanism, including adaptive fuzzy consistency, finite-time output consistency, and leader–follower consistency; see [36,89–91] and the references therein. For example, the authors studied the design problem of adaptive fuzzy consensus tracking control for fractional order multi-agent systems based on an event-triggered mechanism, and gave the bounds of tracking error in [90,91]. In [92], Gao et al. proposed a distributed adaptive finite-time observer to study the event-triggered finite-time output consistency control for a class of heterogeneous fractional-order multi-agent systems. For fractional-order descriptor linear multi-agent systems, Zhang et al. used the event-triggered mechanism to explore the leader–follower exponential consensus control problem in [36]. By designing two fractional-order event-triggered control strategies, the sufficiency criterion of leader–follower exponential consensus in the sense of Mittag-Leffler stability of fractional-order systems was obtained. It is worth pointing out that all the above research results treat every agent as a descriptor linear system and, in some special cases, it is reasonable to treat every agent as a descriptor nonlinear system. Based on the above analysis, there is still a lot of space for the study of the consistency control of fractional-order descriptor nonlinear multi-agent systems.

(2) Formation Control Problem

The formation control problem [93], as one of the hot topics in the research of multi-agent system cooperative control, has been widely used in practical engineering systems, such as UAV formation control, satellite formation control, and multi-robot cooperative formation control. Its core purpose is to design a suitable controller so that each agent system state can be driven to its desired state, such as position, displacement, speed, etc., to achieve the formation purpose. In the existing literature, formation control strategies can be divided into three categories according to the information of the leader: leaderless control strategy [94–96], virtual leader control strategy [97,98], and leader–follower control strategy [99–101]. In [94], Kang et al. presented a state-feedback control method to study the robust time-varying formation control of UAV swarm systems with a directional topology for tracking guidance. By using a special matrix decomposition method, the problem of leaderless time-varying formation control can be transformed into the problem of asymptotic stability of a low-order closed-loop control system, and the corresponding criterion of adequacy was also obtained. In [100], He et al. designed a distributed adaptive formation controller to explore the leader–follower formation control problem of a UAV surface vehicle with specific performance and an anticollision function. More excellent research results on multi-agent systems for formation control can be found in [102,103] and the references therein. In recent years, the research on formation control of fractional-order multi-agent systems has obtained some excellent research results; see [104–108] and the references therein. As shown in [104], based on the observer system, the time-varying formation distributed control problem of a class of fractional multi-agent systems was studied. In [106], Liu et al. considered the formation control problem under the conditions of relative damping and nonuniform time delay, wherein the nonuniform delay includes symmetric and asymmetric cases, and proposes two fractal-order multi-agent system formation control algorithms based on undirected network topology and directed network topology respectively. In addition, based on the event-triggered mechanism, some research achievements on formation control of multi-agent systems have been obtained, such as time-varying formation control, robust formation control, circular formation control, etc.; see [109–113] and the references therein. For example, in [113], Li et al. studied the problem of state formation control of multi-agent systems in a general network topology. By adopting a distributed asynchronous mixed event-triggered control mechanism, the system can realize the given formation target. Compared with previous research results, the control strategy proposed in [113] can save more communication resources within the same event-triggering cycle. On the other hand, with the continuous development and improvement of the event-triggered mechanism, it has become one of the important issues in the study of event-triggered mechanisms to consider how to strike a better balance between further reducing the number of event-triggering times and ensuring system performance. At present, one of the most effective solutions to this problem is to introduce a nonnegative threshold parameter that can be dynamically adjusted according to dynamic rules at the right end of the event-triggering condition, so as to obtain the dynamic event-triggered mechanism. For more research results on the dynamic event-triggered mechanism, refer to the review article of Han et al. [114–116] and the references therein. In [117], the problem of distributed dynamic event-triggered formation control for networked multi-agent systems constrained by communication resources was studied, in which dynamic parameters dependent on neighbor information of each agent system were introduced into the trigger condition. In [118], time-varying output formation control for a class of heterogeneous multi-agent systems was studied based on dynamic event-triggered mechanisms. At the same time, in [119], Zhang et al. proposed an observer-based distributed dynamic event-triggered control strategy, and the time-varying output formation tracking control problem for a class of isomorphic/heterogeneous multi-agent systems in switched communication networks was studied. The research objects of the above results were concentrating on integer multi-agent systems, and fractional multi-agent systems have not been fully studied. Therefore, it is challenging and innovative to use the dynamic event-triggered mechanism to study formation control problems.

1.2.4 Fractional-order switched systems

Switched systems, as hybrid systems composed of a family of subsystems and switching rules guiding these subsystems, have received great attention from scholars at home and abroad in the past decades; see [120–122] and the references therein. In recent years, with the continuous development and improvement of fractional calculus theory [123–126], some scholars began to apply the traditional switched system modeling idea to the modeling of fractional switched systems, that is, if there is at least one system described by a fractional calculus equation in these subsystems, such a system is called a fractional switched system. Research on the stability of such fractional-order switched systems has also made some breakthroughs and progress [127–130]. For example, in [131], Yang et al. presented the mean residence time method to give the adequacy criterion of finite-time stability for a class of fractional-order pulse systems. Subsequently, Zhang et al. studied the finite-time stability of fractional-order positive switched systems by using the Lyapunov function design and average dwell time method in [132]. For fractional-order uncertain switched systems, the problem of pulse stability was studied in [133]. In [134], Wu et al. provided the adequacy criteria of Lyapunov stability and external stability for fractional-order switched systems based on the definition of Caputo type. In [135], Zhan et al. studied the exponential stability of a class of fractional switched systems by using the model-dependent average pulse method, and the effectiveness and feasibility of the proposed method were verified by taking fractional switched circuit systems as an example. It can be seen from the existing research results that, compared with the traditional switched system, the fractional switched system has more advantages in simulating some complex phenomena in engineering. However, although great achievements have been made in practical applications, especially in circuit systems, considering the special definition of fractional calculus, the memory and genetic characteristics of its operators, how to better combine the event-triggering mechanism to avoid the occurrence of Zeno behavior in the system is still an urgent problem to be solved.

1.2.5 Fractional-order complex networked systems

At the end of the 20th century, Watts and Strogatz [136] published their seminal paper on small-world networks, thus laying the foundation for the study of complex networks. In the past few decades, with the continuous in-depth research on network systems, complex network systems have attracted the attention of many scholars because of their characteristics such as connectivity diversity, node diversity and dynamic complexity, and they have been applied to various fields of science and engineering, such as social networks, Internet, highway networks, etc.; see [137–140] and the references therein. At present, the research on complex network systems has obtained abundant results [141–146]. For example, in [141], Liu et al. explored the nonfragile exponential synchronization control problem for a class of complex network systems with time-varying coupled delays for the first time by using the sampled data feedback control technology. In addition, in the actual network system, considering that data transmission may be compromised by network attacks, such as eavesdropping attacks [142], spoofing attacks [144], etc., the stability performance of network system nodes may change, which may make the system enter an unstable state and destroy the system security. In order to reduce the impact of network attacks, Liu et al. studied the distributed fusion estimation problem for a class of multi-sensor nonlinear network systems with random access protocols by constructing multi-channel transmission network structures in [145].

In recent years, the theory of fractional complex network systems, which is formed by combining complex network systems with fractional calculus and Lyapunov stability theory, has been widely developed by scholars at home and abroad and has seen some excellent research results; see [147–152] and the references therein. For example, in [153], Wang et al. designed an adaptive containment control strategy to discuss the problem of projective cluster synchronization control for a class of fractional-order complex networked systems with coupled delay. Then, based on the Lyapunov stability theory and Barbalat lemma, the adequacy criterion of projective cluster synchronization for fractal-order complex networks was obtained. In [154], Bao et al. investigated the nonfragile state estimation problem for a class of fractional memristor BAM neural networks for the first time. In [155], Jia et al. discussed the problem of robust nonfragile finite-time synchronization control for a class of fractional-order coupled complex networks with multiple weights and uncertainties. In [156], Zhang et al. studied the finite-time synchronization control problem of fractional-order complex network systems by using the intermittent control method. As for the study of fractional-order complex network systems, based on the event-triggered mechanism, some research results have been obtained for the external network attack. For example, in [33], Shen et al. investigated the adaptive event-triggered synchronization control problem of fractional-order neural networks with uncertain variable delay subjected to dual network attacks. In [35], Xiong et al. studied the event-triggered output feedback control problem of fractional order network systems with random network attacks based on observer systems. Based on nonfragile state estimation, Tan et al. studied the adaptive event-triggered random mean square stability of fractious-order complex network systems subjected to random nonlinear network attacks in [38]. However, in the above research results, the consideration of Zeno behavior of fractional order system has not been further studied, so there is still a lot of research space for this problem.

1.3 Preview of the book

Based on the above research background and research status analysis of several fractional-order systems, this book primarily integrates sliding mode variable structure control theory, event-triggered mechanism, observer design, and other theoretical methods to investigate the controller design of fractional-order systems. It also conducts in-depth research on Zeno behavior problems existing in fractional-order event-triggered control. The content of the book is divided into nine chapters:

Chapter 1 Introduction of sliding mode control and fractional-order control systems

This chapter provides a comprehensive overview of the research background and the significance of this book, laying the foundation for understanding the subsequent discussions. It delves into the current state of research on several advanced topics in the field of fractional calculus and its applications. Specifically, it examines the existing literature on fraction-order coupled systems, elucidating their dynamic properties and potential applications. Furthermore, the chapter explores variable-order fractional chaotic systems, highlighting their unique characteristics and the complexities involved in their analysis. It also addresses fraction-order switched systems, discussing their relevance in control theory and their implications for stability and performance. In addition, the chapter reviews the research progress on fraction-order complex network systems, emphasizing their role in modeling and understanding the behavior of interconnected systems. Lastly, it considers fraction-order multi-agent systems, focusing on their coordination and control mechanisms in various contexts. Through this detailed examination, the chapter aims to provide a solid theoretical foundation and identify key research gaps that this book seeks to address.

Chapter 2   Theoretical foundations of fractional-order systems

This chapter introduces some basic knowledge of fractional calculus and stability theory used in this book. Firstly, some functions commonly used in fractional calculus theory are given, such as the Gamma function, Mittag-Leffler function, etc. Secondly, the definitions and methods commonly used in current research are given, along with some properties and fractional Laplace transformation. Then, based on traditional systems, the definition and properties of fractional-order systems and their related stability are briefly summarized. Finally, some important lemmas that will be used in this book are given.

Chapter 3   Finite-time projective synchronization control of variable-order fractional chaotic systems via sliding mode approach

First, for two kinds of fraction-order chaotic systems with different time-variable orders and based on the theory of variable-order fractional calculus, two kinds of sliding mode surface functional of integral and differential type are respectively constructed. Also the finite-time projective synchronous sliding mode control problem of the state trajectory of the driving and responding systems is investigated and transformed into the asymptotic stability problem of the projection error system. Second, by using the properties of the variable-order fractional calculus operators, the corresponding variable-order fractional sliding mode control strategy is given, and the adequacy criterion of the asymptotic stability of the new system is obtained. Then, the variable-order fractional sliding mode switching law is designed to make the system state trajectory reach the sliding mode surface in a finite time. Finally, the effectiveness and feasibility of the proposed variable-order fractional sliding mode control strategy are verified by numerical simulation.

Chapter 4 Finite-time synchronization of variable-order fractional uncertain coupled systems via adaptive sliding mode control

First, an innovative variable-order fractional integral-type sliding surface, incorporating nonlinear coupling terms, is designed utilizing the principles of variable-order fractional calculus. This approach leverages the unique properties of variable-order derivatives to enhance the adaptability and performance of the sliding surface in dynamic systems. Second, leveraging the foundational concepts of graph theory, new asymptotical stability criteria are derived for the sliding mode dynamics associated with the designed surface. These criteria provide robust theoretical underpinnings that ensure the long-term stability of the system under variable-order fractional dynamics, addressing both convergence and stability. Moreover, the finite-time reachability of the predefined variable-order fractional integral-type sliding surface is rigorously ensured through the development of a novel adaptive variable-order fractional controller. This controller is specifically engineered to adapt in real-time to the changing dynamics of the system, ensuring rapid and reliable convergence to the sliding surface within a finite time frame. Finally, two numerical studies are presented to verify the validity and superiority of the proposed control strategy.

Chapter 5 Sliding mode projective synchronization for fractional-order coupled systems based on network without strong connectedness

First, for two types of fractional-order coupled systems with time delay and external disturbance built on the network without strong connectedness, the fractional-order integral sliding mode surface function is designed. Based on hierarchical algorithm theory, the projected synchronous sliding mode control problem of the drive and response systems is transformed into the Mittag-Leffler stability problem of the new system. The adequacy criterion of Mittag-Leffler stability of the projection error system is obtained. Second, the fraction-order sliding mode switching law is designed to make the system state trajectory reach the sliding mode surface in a finite time. Finally, the effectiveness and feasibility of the proposed fractional-order sliding mode control strategy are verified by a numerical simulation.

Chapter 6 An event-triggered sliding mode control mechanism to exponential consensus of fractional-order descriptor nonlinear multi-agent systems

First, for a class of fractional-order descriptor nonlinear multi-agent systems, by sampling the states of each agent system, the event-triggered mechanism and the fractional-order integral sliding mode surface, including distributed state feedback controller, are designed, and then the corresponding closed-loop system is obtained. Second, using the definition of Mittag-Leffler stability, the adequacy conditions for the leader–follower exponential consensus of the fractional-order system are obtained. Then, according to the proposed event-triggered mechanism, a new condition is obtained to avoid Zeno behavior in the system. Finally, the effectiveness and feasibility of the proposed event-triggered sliding mode control strategy are verified by three numerical simulations.

Chapter 7   Leader–follower sliding mode formation control of fractional-order multi-agent systems: a dynamic event-triggered mechanism

First, for a class of fractional-order multi-agent systems, a dynamic event-triggered mechanism based on the combined measurement vector is designed. Different from the threshold parameter design of the existing dynamic event-triggered mechanism, the introduced auxiliary variables can be dynamically adjusted according to the fractional-order dynamic rules. Second, based on the proposed dynamic event-triggered sliding mode control strategy, the leader–follower formation control problem is transformed into the asymptotic stability problem of the new system, to realize the fractional multi-agent leader–follower formation control. Furthermore, it is theoretically proved that the proposed dynamic event-triggered control strategy can avoid the occurrence of Zeno behavior. Finally, the effectiveness and feasibility of the proposed dynamic event-triggered sliding mode control strategy are verified by numerical simulation, and the numerical results show that the proposed dynamic event-triggered mechanism can achieve a better balance between reducing sampling frequency and improving formation performance by selecting appropriate threshold parameters.

Chapter 8   An event-triggered mechanism to observer-based sliding mode control of fractional-order uncertain switched systems

First, for a class of fraction-order uncertain switched systems, a suitable state observer and a fraction-order integral sliding mode surface function are designed based on the event-triggered mechanism. Second, the event-triggered mechanism is designed by using the observer state sampling error and sliding mode surface sampling error vector. By using the average dwell time method, an adequacy criterion of the finite time stability of the closed-loop augmented system is given. Then, the event-triggered sliding mode switching law is designed, and the finite-time accessibility of the sliding mode surface is guaranteed under the proposed event-triggered mechanism. Based on the proposed event-triggered mechanism, a self-triggering mechanism is also presented, which can also ensure the finite-time stability of the system. At the same time, considering the memory and genetic properties of fractional calculus operators, combined with fractional Laplace transform and Gronwall–Bellman inequality, it is theoretically proved that the event-triggered sliding mode control strategy proposed in this chapter can avoid Zeno behavior. That is, it excludes the case that infinite trigger sampling occurs in the system in a limited time. Finally, the effectiveness and feasibility of the proposed event-triggered sliding mode control strategy are verified by three numerical simulations.

Chapter 9　Nonfragile sliding mode control of fractional-order complex networked systems via combination event-triggered approach

First, a fractional-order nonfragile state observer and a suitable fractional-order integral sliding mode surface are designed for fractional-order complex network systems under random network attacks. Second, a combined event-triggered mechanism is designed by using the observer state sampling error and sliding mode surface sampling error vector, and an adequacy criterion of closed-loop augmented system stability is given. This conclusion is also applicable to the norm- and exponential-type event-triggered mechanisms. Then, the event-triggered sliding mode switching law is designed, and the finite-time accessibility of the sliding mode surface is guaranteed under the proposed event-triggered mechanism. At the same time, a positive lower bound of the event-triggered interval is given, which proves that the proposed combined event-triggered mechanism can avoid Zeno behavior. Finally, the effectiveness and feasibility of the proposed combined event-triggered sliding mode control strategy are verified by numerical simulation.

1.4 Notations

$\mathbb{R}^n$ $(\mathbb{R}^{n \times m})$	The n-dimensional $(n \times m$-dimensional) Euclidean space
$X > 0$	$X \in \mathbb{R}^{n \times n}$ is a symmetric positive definite matrix
$X \geq 0$	$X \in \mathbb{R}^{n \times n}$ is a symmetric semi-positive definite matrix
$\lambda_{\max}(X)$ $(\lambda_{\min}(X))$	The maximum (minimum) eigenvalue of matrix X
$\lambda_{\mathrm{m}}(X)$	The minimum nonzero eigenvalue of matrix X
X^{T}	The transpose of a matrix X
X^{-1}	The inverse of a matrix X
$(X)_j$	The jth row of matrix $X \in \mathbb{R}^{n \times m}$
$\mathrm{sym}\{X\}$	$X^{\mathrm{T}} + X$
$\|X\|$	The Euclidean norm of matrix X
$\vee$ $(\wedge)$	The OR (AND) operator
$T_{\min} = \min\limits_{k=0,1,2,\dots} \{t_{k+1} - t_k\}$	The minimal interevent time

$\otimes$	The Kronecker product operator
$\text{sgn}(*)$	The signum function
$\mathcal{L}[*]$	The Laplace operator

References

[1] S.V. Emelyanov, Control of first order delay systems by means of an astatic controller and nonlinear correction, Automation and Remote Control 8 (1959) 983–991.

[2] V. Utkin, Variable structure systems with sliding modes, IEEE Transactions on Automatic Control 22 (2) (1977) 212–222.

[3] U. Itkis, Control systems of variable structure, 1976.

[4] V.I. Utkin, Sliding mode and their application in discontinuous systems, Automation and Remote Control 1 (1974) 1898–1907.

[5] X. Yu, Y. Feng, Z. Man, Terminal sliding mode control – an overview, IEEE Open Journal of the Industrial Electronics Society 2 (2020) 36–52.

[6] I. Podlubny, Fractional Differential Equations. An Introduction to Fractional Derivatives, Fractional Differential Equations, to Methods of Their Solution and Some of Their Applications, vol. 198, Academic Press, 1998.

[7] B.B. Mandelbrot, The Fractal Geometry of Nature, vol. 1, W. H. Freeman and Co, San Francisco, 1982.

[8] R.L. Bagley, P.J. Torvik, Fractional calculus in the transient analysis of viscoelastically damped structures, AIAA Journal 23 (6) (1985) 918–925.

[9] J.-D. Gabano, T. Poinot, Fractional modelling and identification of thermal systems, Signal Processing 91 (3) (2011) 531–541.

[10] K.B. Oldham, Interrelation of current and concentration at electrodes, Journal of Applied Electrochemistry 21 (12) (1991) 1068–1072.

[11] S. Westerlund, L. Ekstam, Capacitor theory, IEEE Transactions on Dielectrics and Electrical Insulation 1 (5) (1994) 826–839.

[12] J. Tenreiro Machado, V. Kiryakova, F. Mainardi, Communications in Nonlinear Science and Numerical Simulation 16 (3) (2011) 1140–1153.

[13] Y.Q. Chen, I. Petras, D. Xue, Fractional order control – a tutorial, in: 2009 American Control Conference, IEEE, 2009, pp. 1397–1411.

[14] Z. Li, L. Liu, S. Dehghan, Y.Q. Chen, D. Xue, A review and evaluation of numerical tools for fractional calculus and fractional order controls, International Journal of Control 90 (6) (2017) 1165–1181.

[15] I. Birs, C. Muresan, I. Nascu, C. Ionescu, A survey of recent advances in fractional order control for time delay systems, IEEE Access 7 (2019) 30951–30965.

[16] R. Metzler, J. Klafter, The random walk's guide to anomalous diffusion: a fractional dynamics approach, Physics Reports 339 (1) (2000) 1–77.

[17] J. Klafter, S.C. Lim, R. Metzler, Fractional dynamics: Recent advances, 2012.

[18] I. Podlubny, Fractional-order systems and $PI^\lambda D^\mu$-controllers, IEEE Transactions on Automatic Control 44 (1) (1999) 208–214.

[19] S. Dadras, H.R. Momeni, Fractional terminal sliding mode control design for a class of dynamical systems with uncertainty, Communications in Nonlinear Science and Numerical Simulation 17 (1) (2012) 367–377.

[20] I. Birs, I. Nascu, C. Ionescu, C. Muresan, Event-based fractional order control, Journal of Advanced Research 25 (2020) 191–203.

[21] C. Yin, X. Huang, Y. Chen, S. Dadras, S. Zhong, Y. Cheng, Fractional-order exponential switching technique to enhance sliding mode control, Applied Mathematical Modelling 44 (2017) 705–726.

[22] C. Yin, S. Zhong, W. Chen, Design of sliding mode controller for a class of fractional-order chaotic systems, Communications in Nonlinear Science and Numerical Simulation 17 (1) (2012) 356–366.

[23] X. Meng, Z. Wu, C. Gao, B. Jiang, H.R. Karimi, Finite-time projective synchronization control of variable-order fractional chaotic systems via sliding mode approach, IEEE Transactions on Circuits and Systems. II, Express Briefs 68 (7) (2021) 2503–2507.

[24] J. Wang, C. Shao, Y.-Q. Chen, Fractional order sliding mode control via disturbance observer for a class of fractional order systems with mismatched disturbance, Mechatronics 53 (2018) 8–19.

[25] S. Huang, L. Xiong, J. Wang, P. Li, Z. Wang, M. Ma, Fixed-time fractional-order sliding mode controller for multimachine power systems, IEEE Transactions on Power Systems 36 (4) (2020) 2866–2876.

[26] Z. Yaghoubi, Robust cluster consensus of general fractional-order nonlinear multi agent systems via adaptive sliding mode controller, Mathematics and Computers in Simulation 172 (2020) 15–32.

[27] J. Fei, C. Lu, Adaptive fractional order sliding mode controller with neural estimator, Journal of the Franklin Institute 355 (5) (2018) 2369–2391.

[28] H. Delavari, R. Ghaderi, A. Ranjbar, S. Momani, Fuzzy fractional order sliding mode controller for nonlinear systems, Communications in Nonlinear Science and Numerical Simulation 15 (4) (2010) 963–978.

[29] K.-E. Årzén, A simple event-based PID controller, IFAC Proceedings Volumes 32 (2) (1999) 8687–8692.

[30] C. Peng, F. Li, A survey on recent advances in event-triggered communication and control, Information Sciences 457 (2018) 113–125.

[31] X. Ge, Q.-L. Han, X.-M. Zhang, L. Ding, F. Yang, Distributed event-triggered estimation over sensor networks: a survey, IEEE Transactions on Cybernetics 50 (3) (2019) 1306–1320.

[32] L. Zha, R. Liao, J. Liu, X. Xie, E. Tian, J. Cao, Dynamic event-triggered output feedback control for networked systems subject to multiple cyber attacks, IEEE Transactions on Cybernetics 52 (12) (2021) 13800–13808.

[33] Z. Shen, F. Yang, J. Chen, J. Zhang, A. Hu, M. Hu, Adaptive event-triggered synchronization of uncertain fractional order neural networks with double deception attacks and time-varying delay, Entropy 23 (10) (2021) 1291.

[34] Y. Ye, H. Su, Y. Sun, Event-triggered consensus tracking for fractional-order multiagent systems with general linear models, Neurocomputing 315 (2018) 292–298.

[35] M. Xiong, Y. Tan, D. Du, B. Zhang, S. Fei, Observer-based event-triggered output feedback control for fractional-order cyber–physical systems subject to stochastic network attacks, ISA Transactions 104 (2020) 15–25.

[36] H. Zhang, Z. Gao, Y. Wang, Y. Cai, Leader-following exponential consensus of fractional-order descriptor multiagent systems with distributed event-triggered strategy, IEEE Transactions on Systems, Man and Cybernetics: Systems 52 (6) (2021) 3967–3979.

[37] T. Feng, Y.-E. Wang, L. Liu, B. Wu, Observer-based event-triggered control for uncertain fractional-order systems, Journal of the Franklin Institute 357 (14) (2020) 9423–9441.

[38] Y. Tan, M. Xiong, B. Zhang, S. Fei, Adaptive event-triggered nonfragile state estimation for fractional-order complex networked systems with cyber attacks, IEEE Transactions on Systems, Man and Cybernetics: Systems 52 (4) (2021) 2121–2133.

[39] M. Li, Y. Chen, Y. Liu, Adaptive disturbance observer-based event-triggered fuzzy control for nonlinear system, Information Sciences 575 (2021) 485–498.

[40] J. Chen, B. Chen, Z. Zeng, P. Jiang, Event-triggered synchronization strategy for multiple neural networks with time delay, IEEE Transactions on Cybernetics 50 (7) (2019) 3271–3280.

[41] Y. Zhang, H. Wu, J. Cao, Global Mittag-Leffler consensus for fractional singularly perturbed multi-agent systems with discontinuous inherent dynamics via event-triggered control strategy, Journal of the Franklin Institute 358 (3) (2021) 2086–2114.

[42] X. Li, Z. Sun, Y. Tang, H.R. Karimi, Adaptive event-triggered consensus of multiagent systems on directed graphs, IEEE Transactions on Automatic Control 66 (4) (2020) 1670–1685.

[43] Q. Li, S. Liu, Y. Chen, Combination event-triggered adaptive networked synchronization communication for nonlinear uncertain fractional-order chaotic systems, Applied Mathematics and Computation 333 (2018) 521–535.

[44] Q. Chang, A. Hu, Y. Yang, L. Li, Pinning exponential boundedness of fractional-order multi-agent systems with intermittent combination event-triggered protocol, International Journal of Systems Science 52 (4) (2021) 874–888.

[45] S.G. Samko, Fractional integration and differentiation of variable order, Analysis Mathematica 21 (3) (1995) 213–236.

[46] C.F. Lorenzo, T.T. Hartley, Variable order and distributed order fractional operators, Nonlinear Dynamics 29 (2002) 57–98.

[47] G. Diaz, C.F.M. Coimbra, Nonlinear dynamics and control of a variable order oscillator with application to the van der Pol equation, Nonlinear Dynamics 56 (2009) 145–157.

[48] J. Jiang, H. Chen, J.L.G. Guirao, D. Cao, Existence of the solution and stability for a class of variable fractional order differential systems, Chaos, Solitons and Fractals 128 (2019) 269–274.

[49] J. Jiang, D. Cao, H. Chen, Sliding mode control for a class of variable-order fractional chaotic systems, Journal of the Franklin Institute 357 (15) (2020) 10127–10158.

[50] Z. Hammouch, M. Yavuz, N. Özdemir, Numerical solutions and synchronization of a variable-order fractional chaotic system, Mathematical Modelling and Numerical Simulation with Applications 1 (1) (2021) 11–23.

[51] L.M. Pecora, T.L. Carroll, Synchronization in chaotic systems, Physical Review Letters 64 (8) (1990) 821.

[52] D. Deepika, S. Kaur, S. Narayan, Uncertainty and disturbance estimator based robust synchronization for a class of uncertain fractional chaotic system via fractional order sliding mode control, Chaos, Solitons and Fractals 115 (2018) 196–203.

[53] T.-C. Lin, T.-Y. Lee, Chaos synchronization of uncertain fractional-order chaotic systems with time delay based on adaptive fuzzy sliding mode control, IEEE Transactions on Fuzzy Systems 19 (4) (2011) 623–635.

[54] Y. Xu, Z. He, Synchronization of variable-order fractional financial system via active control method, Open Physics 11 (6) (2013) 824–835.

[55] M.F. Tolba, A.H. Elsafty, M. Armanyos, L.A. Said, A.H. Madian, A.G. Radwan, Synchronization and FPGA realization of fractional-order Izhikevich neuron model, Microelectronics Journal 89 (2019) 56–69.

[56] R. Li, H. Wu, J. Cao, Event-triggered synchronization in networks of variable-order fractional piecewise-smooth systems with short memory, IEEE Transactions on Systems, Man and Cybernetics: Systems 53 (1) (2022) 588–598.

[57] M.Y. Li, Z. Shuai, Global-stability problem for coupled systems of differential equations on networks, Journal of Differential Equations 248 (1) (2010) 1–20.

[58] X. Chen, J.H. Park, J. Cao, J. Qiu, Adaptive synchronization of multiple uncertain coupled chaotic systems via sliding mode control, Neurocomputing 273 (2018) 9–21.

[59] H.-L. Li, Y.-L. Jiang, Z. Wang, L. Zhang, Z. Teng, Global Mittag-Leffler stability of coupled system of fractional-order differential equations on network, Applied Mathematics and Computation 270 (2015) 269–277.

[60] V. Daftardar-Gejji, A. Babakhani, Analysis of a system of fractional differential equations, Journal of Mathematical Analysis and Applications 293 (2) (2004) 511–522.

[61] M. Liu, L. Zhang, P. Shi, Y. Zhao, Fault estimation sliding-mode observer with digital communication constraints, IEEE Transactions on Automatic Control 63 (10) (2018) 3434–3441.

[62] A.-J. Muñoz-Vázquez, V. Parra-Vega, A. Sánchez-Orta, Non-smooth convex Lyapunov functions for stability analysis of fractional-order systems, Transactions of the Institute of Measurement and Control 41 (6) (2019) 1627–1639.

[63] I. Stamova, G. Stamov, Mittag-Leffler synchronization of fractional neural networks with time-varying delays and reaction–diffusion terms using impulsive and linear controllers, Neural Networks 96 (2017) 22–32.

[64] S.D.J. McArthur, E.M. Davidson, V.M. Catterson, A.L. Dimeas, N.D. Hatziargyriou, F. Ponci, T. Funabashi, Multi-agent systems for power engineering applications—Part I: concepts, approaches, and technical challenges, IEEE Transactions on Power Systems 22 (4) (2007) 1743–1752.

[65] S.D.J. McArthur, E.M. Davidson, V.M. Catterson, A.L. Dimeas, N.D. Hatziargyriou, F. Ponci, T. Funabashi, Multi-agent systems for power engineering applications—Part II: technologies, standards, and tools for building multi-agent systems, IEEE Transactions on Power Systems 22 (4) (2007) 1753–1759.

[66] W. Dong, J.A. Farrell, Cooperative control of multiple nonholonomic mobile agents, IEEE Transactions on Automatic Control 53 (6) (2008) 1434–1448.

[67] A. Dorri, S.S. Kanhere, R. Jurdak, Multi-agent systems: a survey, IEEE Access 6 (2018) 28573–28593.

[68] D. Ye, M. Zhang, A.V. Vasilakos, A survey of self-organization mechanisms in multiagent systems, IEEE Transactions on Systems, Man and Cybernetics: Systems 47 (3) (2016) 441–461.

[69] P. Gokulan Balaji, D. Srinivasan, An introduction to multi-agent systems, in: Innovations in Multi-Agent Systems and Applications – 1, 2010, pp. 1–27.

[70] L. Zhao, X. Chen, J. Yu, P. Shi, Output feedback-based neural adaptive finite-time containment control of non-strict feedback nonlinear multi-agent systems, IEEE Transactions on Circuits and Systems. I, Regular Papers 69 (2) (2021) 847–858.

[71] A. Nedic, A. Ozdaglar, P.A. Parrilo, Constrained consensus and optimization in multi-agent networks, IEEE Transactions on Automatic Control 55 (4) (2010) 922–938.

[72] G. Dudek, M.R.M. Jenkin, E. Milios, D. Wilkes, A taxonomy for multi-agent robotics, Autonomous Robots 3 (1996) 375–397.

[73] C.P. Nguyen, A.J. Flueck, Agent based restoration with distributed energy storage support in smart grids, IEEE Transactions on Smart Grid 3 (2) (2012) 1029–1038.

[74] J. Qin, Q. Ma, Y. Shi, L. Wang, Recent advances in consensus of multi-agent systems: a brief survey, IEEE Transactions on Industrial Electronics 64 (6) (2016) 4972–4983.

[75] M. Rehan, A. Jameel, C.K. Ahn, Distributed consensus control of one-sided Lipschitz nonlinear multiagent systems, IEEE Transactions on Systems, Man and Cybernetics: Systems 48 (8) (2017) 1297–1308.

[76] D. Chen, X. Liu, W. Yu, Finite-time fuzzy adaptive consensus for heterogeneous nonlinear multi-agent systems, IEEE Transactions on Network Science and Engineering 7 (4) (2020) 3057–3066.

[77] Y. Li, K. Li, S. Tong, An observer-based fuzzy adaptive consensus control method for nonlinear multiagent systems, IEEE Transactions on Fuzzy Systems 30 (11) (2022) 4667–4678.

[78] H. Zhou, S. Sui, S. Tong, Fuzzy adaptive finite-time consensus control for high-order nonlinear multiagent systems based on event-triggered, IEEE Transactions on Fuzzy Systems 30 (11) (2022) 4891–4904.

[79] X.-R. Yang, G.-P. Liu, Necessary and sufficient consensus conditions of descriptor multi-agent systems, IEEE Transactions on Circuits and Systems. I, Regular Papers 59 (11) (2012) 2669–2677.

[80] J. Xi, M. He, H. Liu, J. Zheng, Admissible output consensualization control for singular multi-agent systems with time delays, Journal of the Franklin Institute 353 (16) (2016) 4074–4090.

[81] X. Liu, Y. Xie, F. Li, W. Gui, Admissible consensus for homogeneous descriptor multiagent systems, IEEE Transactions on Systems, Man and Cybernetics: Systems 51 (2) (2019) 965–974.

[82] J. Shen, J. Cao, Necessary and sufficient conditions for consensus of delayed fractional-order systems, Asian Journal of Control 14 (6) (2012) 1690–1697.

[83] G. Ren, Y. Yu, Robust consensus of fractional multi-agent systems with external disturbances, Neurocomputing 218 (2016) 339–345.

[84] W. Yu, Y. Li, G. Wen, X. Yu, J. Cao, Observer design for tracking consensus in second-order multi-agent systems: fractional order less than two, IEEE Transactions on Automatic Control 62 (2) (2016) 894–900.

[85] Z. Gao, H. Zhang, Y. Wang, Y. Mu, Time-varying output formation-containment control for homogeneous/heterogeneous descriptor fractional-order multi-agent systems, Information Sciences 567 (2021) 146–166.

[86] H. Pan, X. Yu, L. Guo, Admissible leader-following consensus of fractional-order singular multiagent system via observer-based protocol, IEEE Transactions on Circuits and Systems. II, Express Briefs 66 (8) (2018) 1406–1410.

[87] H. Pan, X. Yu, G. Yang, L. Xue, Robust consensus of fractional-order singular uncertain multi-agent systems, Asian Journal of Control 22 (6) (2020) 2377–2387.

[88] Z. Gao, H. Zhang, Y. Wang, K. Zhang, Leader-following consensus conditions for fractional-order descriptor uncertain multi-agent systems with $0 < \alpha < 2$ via output feedback control, Journal of the Franklin Institute 357 (4) (2020) 2263–2281.

[89] P. Xiao, Z. Gu, Adaptive event-triggered consensus of fractional-order nonlinear multi-agent systems, IEEE Access 10 (2021) 213–220.

[90] L. Wang, J. Dong, Adaptive fuzzy consensus tracking control for uncertain fractional-order multi-agent systems with event-triggered input, IEEE Transactions on Fuzzy Systems (2020).

[91] L. Wang, J. Dong, Event-based distributed adaptive fuzzy consensus for nonlinear fractional-order multiagent systems, IEEE Transactions on Systems, Man and Cybernetics: Systems 52 (9) (2021) 5901–5912.

[92] Z. Gao, H. Zhang, Y. Cai, Y. Mu, Finite-time event-triggered output consensus of heterogeneous fractional-order multiagent systems with intermittent communication, IEEE Transactions on Cybernetics 53 (4) (2021) 2164–2176.

[93] K.-K. Oh, M.-C. Park, H.-S. Ahn, A survey of multi-agent formation control, Automatica 53 (2015) 424–440.

[94] Y. Kang, D. Luo, B. Xin, J. Cheng, T. Yang, S. Zhou, Robust leaderless time-varying formation control for nonlinear unmanned aerial vehicle swarm system with communication delays, IEEE Transactions on Cybernetics (2022) 1–14.

[95] X. Dong, B. Yu, Z. Shi, Y. Zhong, Time-varying formation control for unmanned aerial vehicles: theories and applications, IEEE Transactions on Control Systems Technology 23 (1) (2014) 340–348.

[96] H. Rezaee, F. Abdollahi, Pursuit formation of double-integrator dynamics using consensus control approach, IEEE Transactions on Industrial Electronics 62 (7) (2014) 4249–4256.

[97] C. Yuan, H. He, C. Wang, Cooperative deterministic learning-based formation control for a group of nonlinear uncertain mechanical systems, IEEE Transactions on Industrial Informatics 15 (1) (2018) 319–333.

[98] Y. Yu, J. Guo, C.K. Ahn, Z. Xiang, Neural adaptive distributed formation control of nonlinear multi-UAVs with unmodeled dynamics, IEEE Transactions on Neural Networks and Learning Systems (2022).

[99] D. Yu, C.L.P. Chen, Automatic leader–follower persistent formation generation with minimum agent-movement in various switching topologies, IEEE Transactions on Cybernetics 50 (4) (2018) 1569–1581.

[100] S. He, M. Wang, S.-L. Dai, F. Luo, Leader–follower formation control of USVs with prescribed performance and collision avoidance, IEEE Transactions on Industrial Informatics 15 (1) (2018) 572–581.

[101] H. Zhi, L. Chen, C. Li, Y. Guo, Leader–follower affine formation control of second-order nonlinear uncertain multi-agent systems, IEEE Transactions on Circuits and Systems. II, Express Briefs 68 (12) (2021) 3547–3551.

[102] G.-P. Liu, S. Zhang, A survey on formation control of small satellites, Proceedings of the IEEE 106 (3) (2018) 440–457.

[103] Q. Ouyang, Z. Wu, Y. Cong, Z. Wang, Formation control of unmanned aerial vehicle swarms: a comprehensive review, Asian Journal of Control 25 (1) (2023) 570–593.

[104] Y. Gong, G. Wen, Z. Peng, T. Huang, Y. Chen, Observer-based time-varying formation control of fractional-order multi-agent systems with general linear dynamics, IEEE Transactions on Circuits and Systems. II, Express Briefs 67 (1) (2019) 82–86.

[105] J. Bai, G. Wen, A. Rahmani, Y. Yu, Distributed formation control of fractional-order multi-agent systems with absolute damping and communication delay, International Journal of Systems Science 46 (13) (2015) 2380–2392.

[106] J. Liu, P. Li, L. Qi, W. Chen, K. Qin, Distributed formation control of double-integrator fractional-order multi-agent systems with relative damping and nonuniform time-delays, Journal of the Franklin Institute 356 (10) (2019) 5122–5150.

[107] Z. Yu, Y. Zhang, B. Jiang, C.-Y. Su, J. Fu, Y. Jin, T. Chai, Distributed fractional-order intelligent adaptive fault-tolerant formation-containment control of two-layer networked unmanned airships for safe observation of a smart city, IEEE Transactions on Cybernetics 52 (9) (2021) 9132–9144.

[108] J. Bai, G. Wen, Y. Song, A. Rahmani, Y. Yu, Distributed formation control of fractional-order multi-agent systems with relative damping and communication delay, International Journal of Control, Automation, and Systems 15 (1) (2017) 85–94.

[109] X. Li, X. Dong, Q. Li, Z. Ren, Event-triggered time-varying formation control for general linear multi-agent systems, Journal of the Franklin Institute 356 (17) (2019) 10179–10195.

[110] B. Yan, P. Shi, C.-C. Lim, Robust formation control for nonlinear heterogeneous multiagent systems based on adaptive event-triggered strategy, IEEE Transactions on Automation Science and Engineering 19 (4) (2022) 2788–2800.

[111] M. Yu, H. Wang, G. Xie, K. Jin, Event-triggered circle formation control for second-order-agent system, Neurocomputing 275 (2018) 462–469.

[112] W. Zhu, W. Cao, Z.-P. Jiang, Distributed event-triggered formation control of multiagent systems via complex-valued Laplacian, IEEE Transactions on Cybernetics 51 (4) (2019) 2178–2187.

[113] W. Li, H. Zhang, Y. Cai, Y. Wang, Fully distributed formation control of general linear multiagent systems using a novel mixed self- and event-triggered strategy, IEEE Transactions on Systems, Man and Cybernetics: Systems 52 (9) (2022) 5736–5745.

[114] X. Ge, Q.-L. Han, X.-M. Zhang, D. Ding, Dynamic event-triggered control and estimation: a survey, International Journal of Automation and Computing 18 (6) (2021) 857–886.

[115] X. Ge, Q.-L. Han, L. Ding, Y.-L. Wang, X.-M. Zhang, Dynamic event-triggered distributed coordination control and its applications: a survey of trends and techniques, IEEE Transactions on Systems, Man and Cybernetics: Systems 50 (9) (2020) 3112–3125.

[116] X. Yi, K. Liu, D.V. Dimarogonas, K.H. Johansson, Dynamic event-triggered and self-triggered control for multi-agent systems, IEEE Transactions on Automatic Control 64 (8) (2018) 3300–3307.

[117] X. Ge, Q.-L. Han, Distributed formation control of networked multi-agent systems using a dynamic event-triggered communication mechanism, IEEE Transactions on Industrial Electronics 64 (10) (2017) 8118–8127.

[118] W. Song, J. Feng, H. Zhang, W. Wang, Dynamic event-triggered formation control for heterogeneous multiagent systems with nonautonomous leader agent, IEEE Transactions on Neural Networks and Learning Systems (2022).

[119] H. Zhang, W. Li, J. Zhang, Y. Wang, J. Sun, Fully distributed dynamic event-triggered bipartite formation tracking for multiagent systems with multiple nonautonomous leaders, IEEE Transactions on Neural Networks and Learning Systems (2022).

[120] D. Liberzon, A.S. Morse, Basic problems in stability and design of switched systems, IEEE Control Systems Magazine 19 (5) (1999) 59–70.

[121] F. Blanchini, S. Miani, Switching and switched systems, in: Set-Theoretic Methods in Control, Springer, 2015, pp. 405–466.

[122] H. Lin, P.J. Antsaklis, Stability and stabilizability of switched linear systems: a survey of recent results, IEEE Transactions on Automatic Control 54 (2) (2009) 308–322.

[123] X. Chen, Y. Chen, B. Zhang, D. Qiu, A modeling and analysis method for fractional-order DC–DC converters, IEEE Transactions on Power Electronics 32 (9) (2016) 7034–7044.

[124] L. Liu, X. Cao, Z. Fu, S. Song, H. Xing, Finite-time control of uncertain fractional-order positive impulsive switched systems with mode-dependent average dwell time, Circuits, Systems, and Signal Processing 37 (9) (2018) 3739–3755.

[125] C. Zhang, B.-S. Han, Stability analysis of stochastic delayed complex networks with multi-weights based on Razumikhin technique and graph theory, Physica A. Statistical Mechanics and Its Applications 538 (2020) 122827.

[126] S. Sui, C.L.P. Chen, S. Tong, Neural-network-based adaptive DSC design for switched fractional-order nonlinear systems, IEEE Transactions on Neural Networks and Learning Systems 32 (10) (2020) 4703–4712.

[127] S. Balochian, A.K. Sedigh, Sufficient condition for stabilization of linear time invariant fractional order switched systems and variable structure control stabilizers, ISA Transactions 51 (1) (2012) 65–73.

[128] R. Sakthivel, S. Mohanapriya, C.K. Ahn, H.R. Karimi, Output tracking control for fractional-order positive switched systems with input time delay, IEEE Transactions on Circuits and Systems. II, Express Briefs 66 (6) (2018) 1013–1017.

[129] S.H. Nabavi, S. Balochian, The stability of a class of fractional order switching system with time-delay actuator, International Journal of System Dynamics Applications 7 (1) (2018) 85–96.

[130] P. Warrier, P. Shah, Fractional order control of power electronic converters in industrial drives and renewable energy systems: a review, IEEE Access 9 (2021) 58982–59009.

[131] Y. Yang, G. Chen, Finite-time stability of fractional order impulsive switched systems, International Journal of Robust and Nonlinear Control 25 (13) (2015) 2207–2222.

[132] J. Zhang, X. Zhao, Y. Chen, Finite-time stability and stabilization of fractional order positive switched systems, Circuits, Systems, and Signal Processing 35 (2016) 2450–2470.

[133] J.-E. Zhang, Stabilization of uncertain fractional-order complex switched networks via impulsive control and its application to blind source separation, IEEE Access 6 (2018) 32780–32789.

[134] C. Wu, X. Liu, Lyapunov and external stability of Caputo fractional order switching systems, Nonlinear Analysis: Hybrid Systems 34 (2019) 131–146.

[135] T. Zhan, S. Ma, W. Li, W. Pedrycz, Exponential stability of fractional-order switched systems with mode-dependent impulses and its application, IEEE Transactions on Cybernetics 52 (11) (2021) 11516–11525.

[136] D.J. Watts, S.H. Strogatz, Collective dynamics of 'small-world' networks, Nature 393 (6684) (1998) 440–442.

[137] G. Wen, X. Yu, W. Yu, J. Lu, Coordination and control of complex network systems with switching topologies: a survey, IEEE Transactions on Systems, Man and Cybernetics: Systems 51 (10) (2020) 6342–6357.

[138] Y. Yang, Z. Gao, Y. Li, Q. Cai, N. Marwan, J. Kurths, A complex network-based broad learning system for detecting driver fatigue from EEG signals, IEEE Transactions on Systems, Man and Cybernetics: Systems 51 (9) (2019) 5800–5808.

[139] X. Liu, Synchronization and control for multiweighted and directed complex networks, IEEE Transactions on Neural Networks and Learning Systems 34 (6) (2023) 3226–3233.

[140] J. Hu, C. Jia, H. Liu, X. Yi, Y. Liu, A survey on state estimation of complex dynamical networks, International Journal of Systems Science 52 (16) (2021) 3351–3367.

[141] Y. Liu, B.-Z. Guo, J.H. Park, S.-M. Lee, Nonfragile exponential synchronization of delayed complex dynamical networks with memory sampled-data control, IEEE Transactions on Neural Networks and Learning Systems 29 (1) (2016) 118–128.

[142] W. Yang, Z. Zheng, G. Chen, Y. Tang, X. Wang, Security analysis of a distributed networked system under eavesdropping attacks, IEEE Transactions on Circuits and Systems. II, Express Briefs 67 (7) (2019) 1254–1258.

[143] J.V. Milanović, W. Zhu, Modeling of interconnected critical infrastructure systems using complex network theory, IEEE Transactions on Smart Grid 9 (5) (2017) 4637–4648.

[144] J. Wu, C. Peng, J. Zhang, B.-L. Zhang, Event-triggered finite-time H_∞ filtering for networked systems under deception attacks, Journal of the Franklin Institute 357 (6) (2020) 3792–3808.

[145] J. Lu, W. Wang, L. Li, Y. Guo, Distributed fusion estimation for non-linear networked systems with random access protocol and cyber attacks, IET Control Theory & Applications 14 (17) (2020) 2491–2498.

[146] L. Kuan, S. Qixin, Research progress of complex network synchronization control, in: 2020 International Conference on Intelligent Design (ICID), IEEE, 2020, pp. 23–28.

[147] J. Bai, H. Wu, J. Cao, Topology identification for fractional complex networks with synchronization in finite time based on adaptive observers and event-triggered control, Neurocomputing 505 (2022) 166–177.

[148] R. Li, H. Wu, J. Cao, Exponential synchronization for variable-order fractional discontinuous complex dynamical networks with short memory via impulsive control, Neural Networks 148 (2022) 13–22.

[149] H.-L. Li, C. Hu, L. Zhang, H. Jiang, J. Cao, Complete and finite-time synchronization of fractional-order fuzzy neural networks via nonlinear feedback control, Fuzzy Sets and Systems 443 (2022) 50–69.

[150] J. Bai, H. Wu, J. Cao, Secure synchronization and identification for fractional complex networks with multiple weight couplings under DoS attacks, Computational & Applied Mathematics 41 (4) (2022) 187.

[151] Y. Xu, Y. Li, W. Li, Adaptive finite-time synchronization control for fractional-order complex-valued dynamical networks with multiple weights, Communications in Nonlinear Science and Numerical Simulation 85 (2020) 105239.

[152] Y. Xu, S. Gao, W. Li, Exponential stability of fractional-order complex multi-links networks with aperiodically intermittent control, IEEE Transactions on Neural Networks and Learning Systems 32 (9) (2020) 4063–4074.

[153] F. Wang, Y. Yang, M. Hu, X. Xu, Projective cluster synchronization of fractional-order coupled-delay complex network via adaptive pinning control, Physica A. Statistical Mechanics and Its Applications 434 (2015) 134–143.

[154] H. Bao, J.H. Park, J. Cao, Non-fragile state estimation for fractional-order delayed memristive BAM neural networks, Neural Networks 119 (2019) 190–199.

[155] Y. Jia, H. Wu, J. Cao, Non-fragile robust finite-time synchronization for fractional-order discontinuous complex networks with multi-weights and uncertain couplings under asynchronous switching, Applied Mathematics and Computation 370 (2020) 124929.

[156] L. Zhang, J. Zhong, J. Lu, Intermittent control for finite-time synchronization of fractional-order complex networks, Neural Networks 144 (2021) 11–20.

Theoretical foundations of fractional-order systems

2.1 Theory of fractional calculus

Fractional calculus, which represents a direct and significant extension of traditional calculus, serves as the foundational theoretical underpinning for contemporary system theory and advanced control methods. The systematic development of fractional calculus can be traced back to the late 19th century, primarily owing to the seminal contributions of several prominent mathematicians: the French mathematician Joseph Liouville, the Czech mathematician Antonin Grünwald, the Russian mathematician Aleksey Letnikov, and the German mathematician Bernhard Riemann. Through their collective efforts, the theoretical framework of fractional calculus was gradually established, laying the groundwork for its subsequent advancements and applications. In this chapter, we will delve into several meaningful and widely recognized definitions of fractional calculus that have emerged from this foundational work. Each definition will be introduced and examined in detail, providing a comprehensive understanding of the core concepts and their implications. This systematic introduction aims to elucidate the versatile nature of fractional calculus and its pivotal role in the analysis and control of complex dynamic systems.

2.1.1 Typical functions

(1) The Gamma function.

Definition 2.1 ([1]). The Gamma function $\Gamma(z)$ is defined by the integral

$$\Gamma(z) = \int_0^{+\infty} t^{z-1} e^{-t} \mathrm{d}t, \quad \mathrm{Re}\, z > 0. \tag{2.1}$$

Property 2.1. *One has $\Gamma(z+1) = z\Gamma(z)$ for $\mathrm{Re}\, z > 0$.*

Sliding Mode Control of Fractional-order Systems
https://doi.org/10.1016/B978-0-44-331696-8.00006-9

(2) Mittag-Leffler function.

Definition 2.2 ([1]). Let $\alpha > 0$ and $\beta > 0$. The function $E_{\alpha,\beta}(z)$ is called the Mittag-Leffler function with two-parameters if

$$E_{\alpha,\beta}(z) = \sum_{k=0}^{+\infty} \frac{z^k}{\Gamma(k\alpha + \beta)}, \quad z \in \mathbb{C}. \tag{2.2}$$

From Definition 2.2, one can obtain the following results:

$$E_{1,1}(z) = \sum_{k=0}^{+\infty} \frac{z^k}{\Gamma(k+1)} = \sum_{k=0}^{+\infty} \frac{z^k}{k!} = e^z, \tag{2.3}$$

$$E_{1,2}(z) = \sum_{k=0}^{+\infty} \frac{z^k}{\Gamma(k+2)} = \sum_{k=0}^{+\infty} \frac{z^k}{(k+1)!} = \frac{1}{z} \sum_{k=0}^{+\infty} \frac{z^{k+1}}{(k+1)!} = \frac{e^z - 1}{z}, \tag{2.4}$$

$$E_{1,3}(z) = \sum_{k=0}^{+\infty} \frac{z^k}{\Gamma(k+3)} = \sum_{k=0}^{+\infty} \frac{z^k}{(k+2)!} = \frac{1}{z^2} \sum_{k=0}^{+\infty} \frac{z^{k+2}}{(k+2)!} = \frac{e^z - 1 - z}{z^2}, \tag{2.5}$$

and in general,

$$E_{1,\beta}(z) = \sum_{k=0}^{+\infty} \frac{z^k}{\Gamma(k+\beta)} = \frac{1}{z^{\beta-1}} \left\{ e^z - \sum_{k=0}^{\beta-2} \frac{z^k}{k!} \right\}. \tag{2.6}$$

It is worth noting that the hyperbolic sine and cosine can be also seen as particular cases of the Mittag-Leffler function:

$$E_{2,1}(z) = \sum_{k=0}^{+\infty} \frac{z^{2k}}{\Gamma(2k+1)} = \sum_{k=0}^{+\infty} \frac{z^{2k}}{(2k)!} = \cosh(z) \tag{2.7}$$

and

$$E_{2,2}(z) = \sum_{k=0}^{+\infty} \frac{z^{2k}}{\Gamma(2k+12)} = \frac{1}{z} \sum_{k=0}^{+\infty} \frac{z^{2k+1}}{(2k+1)!} = \frac{\sinh(z)}{z}. \tag{2.8}$$

2.1.2 Typical definitions

From the formal point of view, the order of the fractional order differential or integral operator can be any real or imaginary number but, limited to the current research status of the theory of fractional order calculus, the order used in this chapter is only in the real category, and its operator can be denoted as ${}_{t_0}D_t^\alpha$, where t_0 is its lower limit, t is its upper limit, and $\alpha \in \mathbb{R}$

is its order. According to the value of the order α, the fractional-order differential and integral operators can be expressed in the following form:

$$
{}_{t_0}D_t^{\alpha} f(t) = \begin{cases} \frac{\mathrm{d}^{\alpha}}{\mathrm{d}t^{\alpha}} f(t), & \alpha > 0; \\ f(t), & \alpha = 0; \\ \int_{t_0}^{t} f(\tau)\mathrm{d}\tau^{-\alpha}, & \alpha < 0. \end{cases} \tag{2.9}
$$

As mentioned above, in the development of the theory of fractional order calculus, there exist several different definitions, all of them with their justifications and applications, among which the Riemann–Liouville, Caputo, and Grünwald–Letnikov definitions are the most representative [1–5].

(1) Riemann–Liouville Fractional Integral and Derivative.

Definition 2.3. The α-order Riemann–Liouville integral of $f(t)$ is defined by

$$
{}_{t_0}^{\mathcal{RL}}D_t^{-\alpha} f(t) = \frac{1}{\Gamma(\alpha)} \int_{t_0}^{t} \frac{f(\tau)}{(t-\tau)^{1-\alpha}} \mathrm{d}\tau, \tag{2.10}
$$

where $\alpha > 0$ and the $\mathcal{RL}$ superscript explicitly specifies the Riemann–Liouville definition.

Definition 2.4. The α-order Riemann–Liouville derivative of $f(t)$ is defined by

$$
{}_{t_0}^{\mathcal{RL}}D_t^{\alpha} f(t) = \frac{1}{\Gamma(n-\alpha)} \frac{\mathrm{d}^n}{\mathrm{d}t^n} \int_{t_0}^{t} \frac{f(\tau)}{(t-\tau)^{1+\alpha-n}} \mathrm{d}\tau, \tag{2.11}
$$

where $n - 1 < \alpha \le n, n \in \mathbb{N}^+$.

(2) Caputo Fractional Integral and Derivative.

Definition 2.5. The α-order Caputo integral of $f(t)$ is defined by

$$
{}_{t_0}^{\mathcal{C}}D_t^{-\alpha} f(t) = \frac{1}{\Gamma(\alpha)} \int_{t_0}^{t} \frac{f(\tau)}{(t-\tau)^{1-\alpha}} \mathrm{d}\tau, \tag{2.12}
$$

where $\alpha > 0$ and the superscript $\mathcal{C}$ marks the Caputo definition.

Definition 2.6. The α-order Caputo derivative of $f(t)$ is defined by

$$
{}_{t_0}^{\mathcal{C}}D_t^{\alpha} f(t) = \frac{1}{\Gamma(n-\alpha)} \int_{t_0}^{t} \frac{f^{(n)}(\tau)}{(t-\tau)^{1+\alpha-n}} \mathrm{d}\tau, \tag{2.13}
$$

where $n - 1 < \alpha \le n, n \in \mathbb{N}^+$.

(3) Grünwald–Letnikov Fractional Derivative.

Definition 2.7. The α-order Grünwald–Letnikov derivative of $f(t)$ is defined by

$$
{}^{\mathcal{GL}}_{t_0} D_t^{\alpha} f(t) = \lim_{h \to 0} \frac{1}{h^{\alpha}} \sum_{r=0}^{[(t-t_0)/h]} (-1)^r \binom{\alpha}{r} f(t - rh), \tag{2.14}
$$

where $\alpha > 0$, $\binom{\alpha}{r} = \frac{\Gamma(1+\alpha)}{\Gamma(1+r)\Gamma(1+\alpha-r)}$, and the superscript $\mathcal{GL}$ marks the Grünwald–Letnikov definition.

Particularly, for the Grünwald–Letnikov definition, (2.14) can also be used to define the integral and original function forms for $\alpha < 0$ and $\alpha = 0$, respectively.

Although fractional operators are widely used to describe engineering and physical phenomena because of their unique long memory characteristics, the existing research results show that there are some complex physical phenomena in nature, with their memory and nonlocal characteristics depending on the changes of time, space, or other conditions. Therefore, it may be more accurate to use the theory of variable-order fractional calculus to describe these complex processes and phenomena.

(4) Variable-Order Fractional Integral

Definition 2.8. The $\alpha(t)$-order integral of $f(t)$ is defined by

$$
{}_{t_0} D_t^{-\alpha(t)} f(t) = \frac{1}{\Gamma(\alpha(t))} \int_{t_0}^{t} \frac{f(\tau)}{(t-\tau)^{1-\alpha(t)}} \, d\tau, \tag{2.15}
$$

where $\alpha(t) > 0$.

(5) Variable-Order Riemann–Liouville Fractional Derivative

Definition 2.9. The $\alpha(t)$-order Riemann–Liouville derivative of $f(t)$ is defined by

$$
{}^{\mathcal{RL}}_{t_0} D_t^{\alpha(t)} f(t) = \frac{1}{\Gamma(n-\alpha(t))} \frac{d^n}{dt^n} \int_{t_0}^{t} \frac{f(\tau)}{(t-\tau)^{1+\alpha(t)-n}} \, d\tau, \tag{2.16}
$$

where $n - 1 < \alpha(t) \leq n$, $n \in \mathbb{N}^{+}$.

(6) Variable-Order Caputo Fractional Derivative

Definition 2.10. The $\alpha(t)$-order Caputo derivative of $f(t)$ is defined by

$$
{}^{\mathcal{C}}_{t_0} D_t^{\alpha(t)} f(t) = \frac{1}{\Gamma(n-\alpha(t))} \int_{t_0}^{t} \frac{f^{(n)}(\tau)}{(t-\tau)^{1+\alpha(t)-n}} \, d\tau, \tag{2.17}
$$

where $n - 1 < \alpha(t) \leq n$, $n \in \mathbb{N}^{+}$.

2.1.3 Basic properties of fractional derivatives

(1) Linearity.

Similarly to integer-order differentiation, the fractional differentiation is a linear operation:

$$_{t_0}D_t^{\alpha(t)}(af(t)+bg(t)) = a\,_{t_0}D_t^{\alpha(t)}f(t) + b\,_{t_0}D_t^{\alpha(t)}g(t), \qquad (2.18)$$

for any constants a, b and appropriate functions f and g. In the absence of controversy, $_{t_0}D_t^{\alpha(t)}$ can be seen as the Riemann–Liouville, Grünwald–Letnikov or Caputo fractional differential operator. Particularly, (2.18) still holds when $\alpha(t) = \alpha$.

(2) Relationship between the definitions of each fractional order calculus.

Lemma 2.1 ([6]). *For a given order $\alpha \in (m-1, m) \subset [0, n]$, the Riemann–Liouville definition is equivalent to the Grünwald–Letnikov definition if the function $f(t)$ is continuous and $(n-1)$-order differentiable, $f^{(n)}(t)$ is integrable on (t_0, T).*

Lemma 2.2 ([6]). *For a given order $\alpha \in (n-1, n]$, $n \in \mathbb{N}^+$, if the function $f(t)$ is continuous and n-order differentiable, then one has that*

$$_{t_0}^{\mathcal{RL}}D_t^\alpha f(t) = {}_{t_0}^{\mathcal{C}}D_t^\alpha f(t) + \sum_{k=0}^{n-1} \frac{f^{(k)}(t_0)}{\Gamma(k+1-\alpha)}(t-t_0)^{k-\alpha}. \qquad (2.19)$$

(3) Fractional Laplace transformation

On has the following results:

$$\mathcal{L}[_0D_t^{-\alpha}f(t)] = s^{-\alpha}F(s), \qquad (2.20)$$

$$\mathcal{L}[_0^{\mathcal{RL}}D_t^\alpha f(t)] = s^\alpha F(s) - \sum_{k=0}^{n-1} s^k\,_0^{\mathcal{RL}}D_t^{\alpha-k-1}f(t)\Big|_{t=0}, \qquad (2.21)$$

$$\mathcal{L}[_0^{\mathcal{C}}D_t^\alpha f(t)] = s^\alpha F(s) - \sum_{k=0}^{n-1} s^{\alpha-k-1}f^{(k)}(0), \qquad (2.22)$$

$$\mathcal{L}[_0^{\mathcal{GL}}D_t^\alpha f(t)] = s^\alpha F(s), \qquad (2.23)$$

where $\mathcal{L}[f(t)] = F(s)$, $n-1 < \alpha \leq n$, $n \in \mathbb{N}^+$.

Remark 2.1. It is not difficult to see from the above formulas that:

(1) For fractional integral operators, due to Definitions 2.3, 2.5 and Lemma 2.1, the Laplace transformations are the same for time-varying fractional calculus operators, as shown in (2.20).

(2) For fractional differential operators, it can be seen from Eq. (2.21) that the Riemann–Liouville fractional derivative transformation requires the fractional derivative of a function at the initial time $t = 0$, which is difficult to obtain in practical applications. Therefore, compared with the Laplace transform of a Caputo fractional derivative in (2.22), which only needs the information of the function and its integer-order derivative at the initial moment, the Caputo definition is more suitable for fractional differential equations or systems with nonzero initial values.

(3) Note from (2.21)–(2.23) that if the function $f(t)$ satisfies $f(0) = 0$ and $f^{(n)}(0) = 0$, then the Laplace transformations of the three fractional derivatives are equivalent.

2.2 Stability theory of fractional-order systems

2.2.1 Description of fractional-order control systems

Fractional-order control systems have attracted the attention of scholars at home and abroad because they can solve some problems that cannot be solved by traditional control methods or are difficult to solve. At present, the description methods of fractional-order systems are mainly divided into three categories, namely input–output description, transfer function description, and pseudo-state space model description:

(1) Input–output description.

Consider the following fractional-order system, described by a single-input single-output fractional calculus equation:

$$f({}_{t_0}D_t^{\alpha_1}y(t), \ldots, {}_{t_0}D_t^{\alpha_n}y(t), y(t), t) = g({}_{t_0}D_t^{\beta_1}u(t), \ldots, {}_{t_0}D_t^{\beta_m}u(t), u(t), t),$$

$$(2.24)$$

where $f(\cdot)$ and $g(\cdot)$ are general linear or nonlinear functions, $u(t)$ and $y(t)$ denote the input and output signals, respectively, fractional-orders α_i, $i = 1, 2, \ldots, n$ and β_j, $j = 1, 2, \ldots, m$ are given nonnegative constants.

In particular, if the functions $f(\cdot)$ and $g(\cdot)$ are chosen as general linear functions, then the fractional-order differential equation can be written as

$$\sum_{i=0}^{n} a_i {}_{t_0}D_t^{\alpha_i}y(t) = \sum_{j=0}^{n} b_j {}_{t_0}D_t^{\beta_j}u(t), \tag{2.25}$$

where a_i and b_i are given constants.

Furthermore, for given constants a_i and b_i, if there exists a factor γ such that $\alpha_i = i\gamma$ and $\beta_j = j\gamma$, then (2.25) is called the commensurate fractional-order mode, and its model can be described as

$$\sum_{i=0}^{n} a_i {}_{t_0}\mathrm{D}_t^{i\gamma}y(t) = \sum_{j=0}^{n} b_j {}_{t_0}\mathrm{D}_t^{j\gamma}u(t). \tag{2.26}$$

(2) Transfer function description.

Considering the linear fractional order system (2.25), if it satisfies the zero initial condition, then the Laplace transform of the fractional order derivative is adopted. Then the transfer function model of the system (2.25) can be obtained as follows:

$$G(s) = \frac{Y(s)}{U(s)} = \frac{b_1 s^{\beta_1} + b_2 s^{\beta_2} + \cdots + b_m s^{\beta_m}}{a_1 s^{\alpha_1} + a_2 s^{\alpha_2} + \cdots + a_n s^{\alpha_n}},$$

where $Y(s) = \mathcal{L}[y(t)]$ and $U(s) = \mathcal{L}[u(t)]$.

(3) Pseudo-state space model description.

For a given linear time-invariant commensurate fractional order system (2.26), the pseudo-state space model can be written as follows:

$$\begin{cases} {}_{t_0}D_t^{\gamma} x(t) = Ax(t) + Bu(t), \\ y(t) = Cx(t) + Du(t), \end{cases} \tag{2.27}$$

where $\gamma \in (0, 2)$, and $x(t)$, $u(t)$, $y(t)$ denote the pseudo-state single, input single, and output single variables, respectively. In particular, if $\gamma = 1$, (2.27) degenerates into a traditional control system.

Remark 2.2. Different from the traditional system model, the state variable $x(t)$ in Eq. (2.27) is not the state defined in the traditional system theory, that is, the state of the system at a certain moment cannot be uniquely determined by the current input and output signals, so the state $x(t)$ in system (2.27) can only be called a pseudo-state. This is caused by the long memory characteristic inherent in fractional calculus theory. In particular, the objects studied in this book are fractional-order systems, so they are collectively referred to as system states without causing controversy.

2.2.2 Basic definitions of stability

In this section, the following definitions are given based on the traditional concept of system stability, combined with fractional calculus theory.

Definition 2.11 ([7]). Consider the Caputo or Riemann–Liouville fractional nonautonomous system

$$_{t_0}D_t^{\alpha} x(t) = f(x(t), t), \tag{2.28}$$

where $\alpha \in (0, 1)$, $f : \Omega \times [t_0, +\infty] \to \mathbb{R}^n$ is a piecewise continuous function of t and locally Lipschitz in x on $\Omega \times [t_0, +\infty]$, and $\Omega \in \mathbb{R}^n$ is a domain that contains the origin $x = 0$. If there exists a constant $x(s)$ satisfying $f(x(s), s) = {}_{t_0}D_t^{\alpha}x(t)$, then the constant $x(s)$ is an equilibrium point of fractional dynamic system (2.28). Without loss of generality, let the equilibrium point be $x(s) = 0$.

Definition 2.12 ([8]). For a given fractional-order system (2.28), let $x(s)$ denote the equilibrium point, and the function $y(t, x(t_0), t_0)$ be the solution of system (2.28). For any given real number $\varepsilon > 0$, if there exists a positive real number $\delta(\varepsilon, t_0)$ such that the following inequality is satisfied:

$$\|x(s) - x(t_0)\| < \delta(\varepsilon, t_0), \tag{2.29}$$

and the solution path starting from any initial time state $x(t_0)$ satisfies

$$\|y(t, x(t_0), t_0) - x(s)\| < \varepsilon \quad \forall\, t \geq t_0, \tag{2.30}$$

then the equilibrium point $x(s)$ is said to be stable in the sense of Lyapunov.

Definition 2.13 ([8]). For a given fractional-order system (2.28), let $x(s)$ denote the equilibrium point, and the function $y(t, x(t_0), t_0)$ be the solution of system (2.28). If the following conditions are met:

(1) The equilibrium point $x(s)$ is stable in the sense of Lyapunov;
(2) For any given real number η and $\delta(\varepsilon, t_0)$, if there exists a positive constant which satisfies the following inequality:

$$\|y(t, x(t_0), t_0) - x(s)\| < \eta \quad \forall\, t \geq t_0 + T(\eta, \delta, t_0), \tag{2.31}$$

then the equilibrium point $x(s)$ is said to be asymptotically stable in the sense of Lyapunov.

Definition 2.14 ([7]). The solution of (2.28) is said to be Mittag-Leffler stable if

$$\|x(t)\| < \left(L(x(t_0)) E_\alpha(-\beta(t - t_0)^\alpha) \right)^q, \tag{2.32}$$

where t_0 is the initial time, $\alpha \in (0, 1)$, $\beta > 0$, $q > 0$, $L(0) = 0$, $L(x(t)) \geq 0$, and $L(x(t))$ is locally Lipschitz whenever $x(t) \in \Omega \in \mathbb{R}^n$ with Lipschitz constant $L(0)$.

2.2.3 Stability criteria

Theorem 2.1 ([9]). *Consider the following fractional-order system:*

$$\begin{cases} {}^{C}_{t_0} D^\alpha_t x(t) = A x(t), \\ x(t_0) = x_0, \end{cases} \tag{2.33}$$

where $x(t) \in \mathbb{R}^n$ denotes the pseudo-state vector, matrix $A = [a_{ij}] \in \mathbb{R}^{n \times n}$, and $\alpha \in (0, 2)$. Then the equilibrium point of system (2.33) is asymptotically stable if and on if $\left|\arg(\lambda_i(A))\right| > \frac{\alpha\pi}{2}$, $i = 1, 2, \ldots, n$, where $\arg(\cdot)$ is the argument function.

Remark 2.3. Based on Theorem 2.1, the stability region of the fractional order system (2.33) is shown in Fig. 2.1. It is obvious that the system state is always divergent when the order $\alpha > 2$, so it is meaningless to consider such a system in practical engineering.

Definition 2.15 ([22]). A continuous function $\alpha : [0, t) \to [0, \infty)$ is said to belong to class-κ if it is strictly increasing and $\alpha(0) = 0$.

Theorem 2.2 ([7]). *Let $x = 0$ be an equilibrium point for the nonautonomous fractional-order system* (2.28). *Assume that there exists a Lyapunov function $V(t, x(t))$ and class-κ functions κ_i, $i = 1, 2, 3$, satisfying*

$$\kappa_1(\|x(t)\|) \leq V(t, x(t)) \leq \kappa_2(\|x(t)\|), \ _{t_0}D_t^\alpha V(t, x(t)) \leq -\kappa_3(\|x(t)\|),$$

where $\alpha \in (0, 1)$. Then the system (2.28) *is asymptotically stable.*

Theorem 2.3 ([8]). *Mittag-Leffler stability implies asymptotic stability.*

2.3 Some important lemmas

As mentioned above, compared with the Riemann–Liouville and Grünwald–Letnikov definitions, Caputo definition has two obvious properties:

(1) Its Laplace transform requires only the integer-order derivative information of the function at the origin;
(2) The Caputo fractional derivative of any constant is zero.

Therefore, in this chapter, we mainly use the Caputo fractional calculus definition to study the sliding mode control of fractional-order systems. In addition, in this section, some important lemmas that will be used in subsequent studies will be introduced, and more details can be referred to references [1,10–17].

Lemma 2.3 ([1]). *For a given continuous and differentiable function $x(t) \in \mathbb{R}$, the following inequality holds:*

$$_{t_0}^{\mathcal{C}}D_t^\alpha x^2(t) \leq 2x(t)_{t_0}^{\mathcal{C}}D_t^\alpha x(t),$$

where order $\alpha \in (0, 1)$, $t \geq t_0$.

Lemma 2.4 ([16]). *For a given continuous and differentiable function $x(t) \in \mathbb{R}^n$, the following inequality holds:*

$$_{t_0}^{\mathcal{C}}D_t^\alpha\left(x^{\mathrm{T}}(t)\boldsymbol{\Omega}x(t)\right) \leq 2x^{\mathrm{T}}(t)\boldsymbol{\Omega}\left(_{t_0}^{\mathcal{C}}D_t^\alpha x(t)\right),$$

where order $\alpha \in (0, 1)$, $t \geq t_0$, and $\boldsymbol{\Omega}$ is a symmetric positive definite matrix.

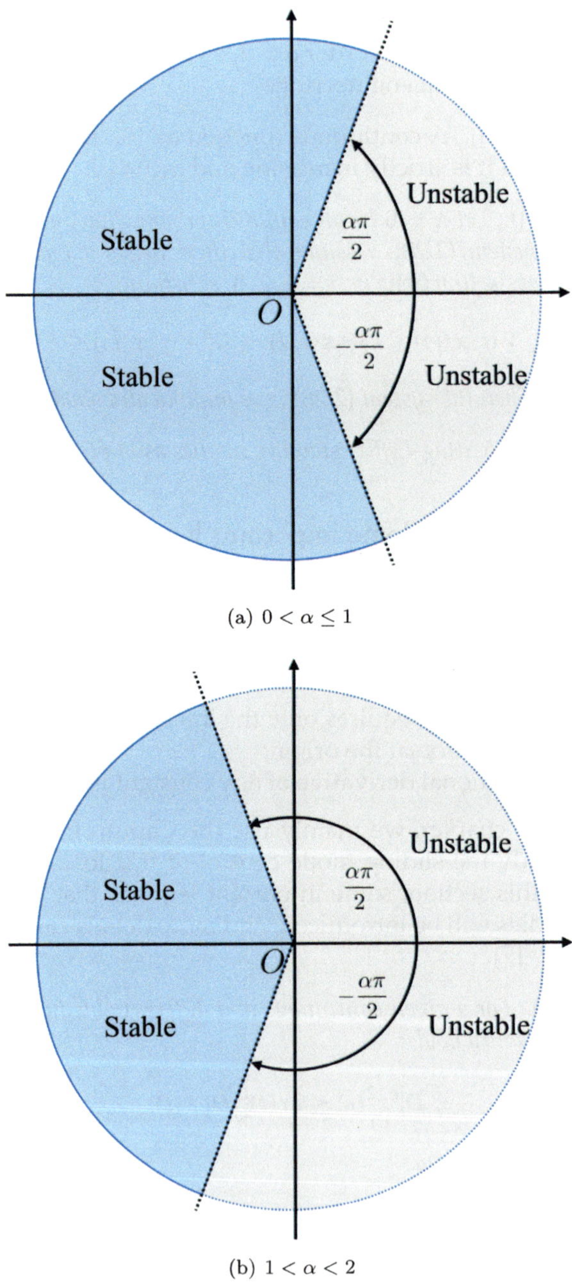

(a) $0 < \alpha \leq 1$

(b) $1 < \alpha < 2$

FIGURE 2.1   Stability regions of fractional-order systems (2.33) for different orders.

Lemma 2.5 ([3]). *For Caputo-type variable-order fractional derivative of order $\alpha(t) \in (0, 1)$, one has*

$$\,^{\mathcal{C}}_{t_0} D_t^{\alpha(t)} x^2(t) \le 2x(t) \,^{\mathcal{C}}_{t_0} D_t^{\alpha(t)} x(t).$$

Lemma 2.6 ([16]). *For a given continuous and differentiable function $x(t) \in \mathbb{R}^n$, the following equality holds:*

$$\,_{t_0} D_t^{-\alpha} \left(\,^{\mathcal{C}}_{t_0} D_t^{\alpha} x(t) \right) = x(t) - x(t_0),$$

where the order $\alpha \in (0, 1)$.

Remark 2.4. It should be pointed out in Lemma 2.6 that if the initial time t_0 of fractional-order integration operator differs from that in the fractional-order derivative operator, i.e., there exists a different initial time t_0' with $t_0' \ne t_0$, then Lemma 2.6 does not hold, that is to say,

$$\,_{t_0'} D_t^{-\alpha} \left(\,^{\mathcal{C}}_{t_0} D_t^{\alpha} x(t) \right) \ne x(t) - x(t_0').$$

However, the results presented in [18,11] did not consider this problem, which may lead to some possibly wrong conclusions.

Lemma 2.7 ([11]). *Let $y(t)$ be a continuous function on $t \in [t_0, +\infty)$, $t_0 > 0$. If there exist $\alpha \in (0, 1)$, $p \in \mathbb{R}$, and $q > 0$ satisfying*

$$\,^{\mathcal{C}}_{t_0} D_t^{\alpha} y(t) \le py(t) + q,$$

then

$$y(t) \le y(t_0) E_{\alpha,1}(p(t - t_0)^{\alpha}) + q(t - t_0)^{\alpha} E_{\alpha,\alpha+1}(p(t - t_0)^{\alpha}),$$

where $E_{\alpha,\beta}(z) = \sum_{k=0}^{\infty} \frac{z^k}{\Gamma(\alpha k + \beta)}$ denotes the Mittag-Leffler function.

Lemma 2.8 ([13]). *For a given continuous function $\phi(t)$, assume that $f(\phi(t))$ is C-regular. Set $V(t) = f(\phi(t))$, $\alpha, \beta \in (0, 1)$ and $\eta > 0$. If*

$$\,^{\mathcal{C}}_{t_0} D_t^{\alpha} V(t) \le -\eta V^{\beta}(t)$$

holds for all $t \ge t_0$, then

$$\lim_{t \to T_s} V(t) = 0, \ V(t) = 0, \ t \ge T_s \quad \text{and} \quad \lim_{t \to T_s} \phi(t) = 0, \ \phi(t) = 0, \ t \ge T_s,$$

which imply that $V(t)$ and $\phi(s)$ converge to 0 in finite time T_s, where the settling time is

$$T_s = \left(V^{1-\beta}(t_0) \frac{\Gamma(\frac{3-\beta}{2} - \alpha)\Gamma(1 + \alpha)}{2\eta\Gamma(\frac{3-\beta}{2})} \right)^{\frac{1}{\alpha}}.$$

Lemma 2.9 ([19]). *For an integrable function $g(t)$, if there exists at least one $\tau \in (0, t)$ such that $g(\tau) \neq 0$, then there is a constant $M > 0$ such that $_0D_t^{-\alpha}|g(t)| \geq M$ holds.*

Lemma 2.10 ([20]). *Assume that $V(t) \in \mathbb{C}([t_0, +\infty], \mathbb{R})$ satisfies $_{t_0}^{C}D_t^{\alpha}V(t) \leq -\xi V(t)$, with $\alpha \in (0, 1)$, $\xi > 0$. Then*

$$V(t) \leq V(t_0)E_\alpha(-\xi(t - t_0)^\alpha),$$

where $E_\alpha(z) = \sum_{k=0}^{\infty} \frac{z^k}{\Gamma(\alpha k + 1)}$ denotes the Mittag-Leffler function.

Lemma 2.11 ([17]). *Assume that $f(t)$ is a nondecreasing continuous function, $h(t)$ is a nonnegative function, and $x(t)$ is a real-valued continuous function on $[t_0, +\infty)$. If the following inequality holds:*

$$x(t) \leq f(t) + \int_{t_0}^{t} h(s)x(s)\mathrm{d}s \quad \forall t \in [t_0, +\infty),$$

then one has

$$x(t) \leq f(t)\exp\left(\int_{t_0}^{t} h(s)\mathrm{d}s\right) \quad \forall t \in [t_0, +\infty).$$

Lemma 2.12 ([14]). *For a given constant $\alpha \in (0, 1)$ and positive sequence $\{b_k\}_{k=1}^{n}$, one has*

$$\sum_{k=1}^{n} b_k^{\alpha} \leq n^{1-\alpha}\left(\sum_{k=1}^{n} b_k\right)^{\alpha}.$$

Lemma 2.13 ([12]). *For a given order $\alpha \in (0, 1)$, if there exists a positive constant γ satisfying $\frac{\alpha\pi}{2} < \gamma < \min\{\pi, \alpha\pi\}$, then one has*

$$\left|E_{\alpha,\beta}(\theta)\right| \leq \frac{\lambda}{1 + |\theta|},$$

where λ is a positive scalar, $\gamma \leq |\arg(\theta)| \leq \pi$, $|\theta| \geq 0$, and $\beta \in \mathbb{R}$.

Lemma 2.14 ([10]). *For a given order $\alpha \in (0, 1)$, there exists a scalar $\eta > 0$ satisfying*

$$\left|t^{\alpha} E_{\alpha,\alpha+1}(\phi t^{\alpha})\right| \leq \eta t^{\alpha} e^{\phi t},$$

where $\phi \in \mathbb{R}$ and $t \in [0, +\infty)$.

Lemma 2.15 ([21]). *For given matrices E, F, and H, with $F^{\mathrm{T}}F \leq I$, there exists a scalar $\lambda > 0$ satisfying*

$$E^{\mathrm{T}}FH + H^{\mathrm{T}}F^{\mathrm{T}}E \leq \lambda E^{\mathrm{T}}E + \lambda^{-1}H^{\mathrm{T}}H.$$

2.4 Conclusion

In this chapter, we presented a thorough review of fundamental concepts in fractional calculus and stability theory, with particular emphasis on the common definitions and their intrinsic properties. This foundational overview included detailed discussions of various fractional derivatives and integrals, such as Riemann–Liouville, Grünwald–Letnikov, and Caputo definitions, highlighting their respective characteristics and applications. Concurrently, we presented several critical theorems and relevant lemmas specific to Caputo-type fractional-order systems. These theoretical results were meticulously examined to provide essential analytical tools for the subsequent research discussed in this book. The inclusion of these theorems and lemmas not only strengthens the theoretical framework but also facilitates a deeper understanding of the dynamic behaviors and stability properties of fractional-order systems. Through this comprehensive exploration, we aim to equip readers with a robust theoretical foundation that is crucial for advancing research in fractional calculus and its applications. The methodologies and analytical techniques presented herein serve as indispensable resources for scholars and practitioners engaged in the study and application of fractional-order dynamic systems.

References

[1] I. Podlubny, Fractional Differential Equations. An Introduction to Fractional Derivatives, Fractional Differential Equations, to Methods of Their Solution and Some of Their Applications, Mathematics in Science and Engineering, vol. 198, Academic Press, 1998.

[2] S.G. Samko, Fractional integration and differentiation of variable order, Analysis Mathematica 21 (3) (1995) 213–236.

[3] J. Jiang, D. Cao, H. Chen, Sliding mode control for a class of variable-order fractional chaotic systems, Journal of the Franklin Institute 357 (15) (2020) 10127–10158.

[4] I. Podlubny, Geometric and physical interpretation of fractional integration and fractional differentiation, arXiv preprint, arXiv:math/0110241, 2001.

[5] C.A. Monje, Y.Q. Chen, B.M. Vinagre, D. Xue, V. Feliu-Batlle, Fractional-Order Systems and Controls: Fundamentals and Applications, Springer Science & Business Media, 2010.

[6] M.D. Ortigueira, J.A. Tenreiro Machado, What is a fractional derivative?, Journal of Computational Physics 293 (2015) 4–13.

[7] Y. Li, Y.Q. Chen, I. Podlubny, Mittag-Leffler stability of fractional order nonlinear dynamic systems, Automatica 45 (8) (2009) 1965–1969.

[8] Y. Li, Y.Q. Chen, I. Podlubny, Stability of fractional-order nonlinear dynamic systems: Lyapunov direct method and generalized Mittag-Leffler stability, Computers & Mathematics with Applications 59 (5) (2010) 1810–1821.

[9] I. Podlubny, Fractional-order systems and $PI^{\lambda}D^{\mu}$-controllers, IEEE Transactions on Automatic Control 44 (1) (1999) 208–214.

[10] Q. Li, S. Liu, Y. Chen, Combination event-triggered adaptive networked synchronization communication for nonlinear uncertain fractional-order chaotic systems, Applied Mathematics and Computation 333 (2018) 521–535.

[11] Q. Chang, A. Hu, Y. Yang, L. Li, Pinning exponential boundedness of fractional-order multi-agent systems with intermittent combination event-triggered protocol, International Journal of Systems Science 52 (4) (2021) 874–888.

[12] L. Wang, J. Dong, Adaptive fuzzy consensus tracking control for uncertain fractional-order multi-agent systems with event-triggered input, IEEE Transactions on Fuzzy Systems (2020).

[13] Z. Gao, H. Zhang, Y. Cai, Y. Mu, Finite-time event-triggered output consensus of heterogeneous fractional-order multiagent systems with intermittent communication, IEEE Transactions on Cybernetics 53 (4) (2021) 2164–2176.

[14] L. Liu, X. Cao, Z. Fu, S. Song, H. Xing, Finite-time control of uncertain fractional-order positive impulsive switched systems with mode-dependent average dwell time, Circuits, Systems, and Signal Processing 37 (9) (2018) 3739–3755.

[15] W. Yang, W. Yu, Y. Lv, L. Zhu, T. Hayat, Adaptive fuzzy tracking control design for a class of uncertain nonstrict-feedback fractional-order nonlinear SISO systems, IEEE Transactions on Cybernetics 51 (6) (2019) 3039–3053.

[16] L. Chen, T. Li, Y.Q. Chen, R. Wu, S. Ge, Robust passivity and feedback passification of a class of uncertain fractional-order linear systems, International Journal of Systems Science 50 (6) (2019) 1149–1162.

[17] H. Ye, J. Gao, Y. Ding, A generalized Gronwall inequality and its application to a fractional differential equation, Journal of Mathematical Analysis and Applications 328 (2) (2007) 1075–1081.

[18] T. Feng, Y.-E. Wang, L. Liu, B. Wu, Observer-based event-triggered control for uncertain fractional-order systems, Journal of the Franklin Institute 357 (14) (2020) 9423–9441.

[19] D. Deepika, S. Kaur, S. Narayan, Uncertainty and disturbance estimator based robust synchronization for a class of uncertain fractional chaotic system via fractional order sliding mode control, Chaos, Solitons and Fractals 115 (2018) 196–203.

[20] K. Mathiyalagan, G. Sangeetha, Finite-time stabilization of nonlinear time delay systems using LQR based sliding mode control, Journal of the Franklin Institute 356 (7) (2019) 3948–3964.

[21] X. Meng, C. Gao, Z. Liu, B. Jiang, Robust H_∞ control for a class of uncertain neutral-type systems with time-varying delays, Asian Journal of Control 23 (3) (2021) 1454–1465.

[22] H.K. Khalil, Nonlinear Systems, third edition, Prentice Hall, 2002.

Finite-time projective synchronization control of variable-order fractional chaotic systems via sliding mode approach

3.1 Introduction

During the past few decades, the fractional calculus theory has been widely addressed in the literature [1,2]. It has often been used as an effective tool to describe the natural behaviors in engineering, physics, finance, etc. However, more research results indicate that the memory and non-locality properties of some complex physical phenomena in nature may depend on the change of time, space, or other conditions. Therefore, the theory of variable-order fractional (VOF) calculus can be more accurate than constant-order (CO) theory to describe these processes and phenomena. In 1995, the definition and related properties of Riemann–Liouville VOF calculus were firstly given by Samko [3]. Afterwards, in [4], Lorenzo and Hartley made a profound study and presented different concepts of the VOF calculus, providing calculation methods for these definitions. Since then, the VOF calculus is widely applied to many fields of science and engineering [5,6].

Put forward by Pecora and Carroll in [7], the issue of chaotic synchronization has received great attention. The research on synchronization control of chaotic systems has been widely applied to communication security, image processing, and other fields [8–15]. In [9], for fractional-order chaotic systems with uncertain term and time-delay, synchronization criteria were obtained with the aid of the adaptive fuzzy SMC strategy. Meanwhile, Tolba et al. considered the synchronization problem of fractional-order chaotic systems which were applied to Izhikevich neuron model in [10]. In addition, the SMC approach [16–21], due to its advantages, has also been widely applied to deal with complex nonlinear systems, such as the application in the stream water quality standards' regulation [22], the projective synchronization problem [23], the controller design issue for Chua's circuit with time-delay [24], etc.

For the VOF chaotic systems, several Mittag-Leffler stability criteria were presented by using an SMC approach in [6]. However, it should be pointed out that the method proposed in [6] is only appropriate for the problem of stabilization for specific VOF chaotic systems.

Based on the above analysis, compared with the previous investigations, the contributions are:

(1) The projective synchronization problem of the VOF chaotic systems is turned into the asymptotical stability of the error systems;
(2) By designing a novel VOF integral- and derivative-type sliding surface and corresponding control law, several asymptotical stability criteria are presented to achieve projective synchronization;
(3) For the issue of finite-time stability, a novel criterion is provided using the theory of VOF calculus.

3.2 Problem statement

In this part, consider the following n-dimensional VOF chaotic system as the master system:

$$\,_0^{\mathcal{C}}D_t^{\alpha_i(t)}y_i(t) = f_i(\boldsymbol{Y}(t), t), \tag{3.1}$$

where $\boldsymbol{Y} = [y_1, y_2, \ldots, y_n]^{\mathrm{T}}$, $\alpha_i(t) \in [\alpha_0, \alpha_1] \subset (0, 1)$, and functions $f_i(\cdot, \cdot) \in \mathbb{C}(\mathbb{R}^n \times \mathbb{R}, \mathbb{R})$. In addition, consider the following VOF system as the slave system:

$$\,_0^{\mathcal{C}}D_t^{\beta_i(t)}x_i(t) = g_i(\boldsymbol{X}(t), t) + u_i(t), \tag{3.2}$$

where $\boldsymbol{X} = [x_1, x_2, \ldots, x_n]^{\mathrm{T}}$, $\beta_i(t) \in [\beta_0, \beta_1] \subset (0, 1)$, functions $g_i(\cdot, \cdot) \in \mathbb{C}(\mathbb{R}^n \times \mathbb{R}, \mathbb{R})$, and $u_i(t)$ is the control input.

In this chapter, the main aim is to realize finite-time projective synchronization for different VOF chaotic systems using the SMC approach. Hence we denote by $e_i(t) = y_i(t) - \rho x_i(t)$ the projective synchronization error, with $\rho \neq 0$ being the projection coefficient.

Therefore, the projective error system for the VOF chaotic systems (3.1) and (3.2) is obtained as follows:

$$
\begin{aligned}
{}^{C}_{0}D^{\beta_i(t)}_{t}e_i(t) &= {}^{C}_{0}D^{\beta_i(t)}_{t}(y_i(t) - \rho x_i(t)) \\
&= {}^{C}_{0}D^{\beta_i(t)}_{t}y_i(t) - \rho(g_i(X,t) + u_i(t)).
\end{aligned}
\tag{3.3}
$$

3.3 Main results

3.3.1 Sliding mode surface design and stability analysis

In this section, based on the VOF calculus theory, the following two sliding mode surface design methods and their corresponding sliding mode control strategies are given:

(1) Consider the following novel VOF integral-type sliding surface function:

$$
s_{1,i}(t) = e_i(t) + k_i D^{-\beta_i(t)}_{t}e_i(t),
\tag{3.4}
$$

where $k_i > 0$, $i \in \mathbb{L}$.

Then, due to $s_{1,i}(t) = 0$ and ${}^{C}_{0}D^{\beta_i(t)}_{t}s_{1,i}(t) = 0$, one has an equivalent control as

$$
u_{1i}(t) = -\frac{1}{\rho}{}^{C}_{0}D^{\beta_i(t)}_{t}y_i(t) - g_i(X,t) + \frac{k_i}{\rho}e_i(t).
\tag{3.5}
$$

Theorem 3.1. *Given a projection coefficient $\rho \neq 0$, the projective error system (3.3) with the sliding surface function (3.4), and the controller (3.5), the VOF error system (3.3) is asymptotically stable.*

Proof. For each subsystem of the projective error system (3.3), construct the Lyapunov function

$$
V_{1,i}(t) = \frac{1}{2}e_i^2(t).
\tag{3.6}
$$

Taking the VOF $\beta_i(t)$-order derivative with respect to time of (3.6), and combining with Lemma 2.3, one has

$$
\begin{aligned}
{}^{C}_{0}D^{\beta_i(t)}_{t}V_{1,i}(t) &\leq e_i(t){}^{C}_{0}D^{\beta_i(t)}_{t}e_i(t) \\
&= e_i(t)\left({}^{C}_{0}D^{\beta_i(t)}_{t}y_i(t) - \rho(g_i(X,t) + u_{1i}(t))\right) \\
&= -\frac{k_i}{\rho}e_i^2(t) \leq 0.
\end{aligned}
\tag{3.7}
$$

Hence, based on Lemma 2.6 in [6], it is concluded that the projective error system (3.3) is asymptotically stable. $\square$

Remark 3.1. From the sliding surface function (3.4), $^{C}_{0}D_t^{\beta_i(t)}e_i(t) = -k_i e_i(t)$ can be derived from $^{C}_{0}D_t^{\beta_i(t)}s_{1,i}(t) = 0$. Therefore, the obtained projective error system is the simplest linear structure.

Theorem 3.2. *For the given projective error system (3.3) and sliding surface function (3.4), the error trajectories maintain sliding motion after reaching the sliding surface by utilizing the control rule below:*

$$\begin{cases} u_i(t) = u_{1i}(t) + u_{2i}(t), \\ u_{2i}(t) = \frac{\lambda_{1i}}{\rho}s_{1,i}(t) + \frac{\lambda_{2i}}{\rho}\mathrm{sgn}(s_{1,i}(t)), \end{cases} \tag{3.8}$$

where λ_{1i}, $\lambda_{2i} > 0$, $i \in \mathbb{L}$.

Proof. For each subsystem of the projective error system (3.3), choose the Lyapunov function as follows:

$$V_{2,i}(t) = \frac{1}{2}s_{1,i}^2(t). \tag{3.9}$$

Taking the VOF $\beta_i(t)$-order derivative with respect to time of (3.9), we have

$$\begin{aligned} ^{C}_{0}D_t^{\beta_i(t)}V_{2,i}(t) &\le s_{1,i}(t)^{C}_{0}D_t^{\beta_i(t)}s_{1,i}(t) \\ &= s_{1,i}(t)\left(^{C}_{0}D_t^{\beta_i(t)}y_i(t) - \rho(g_i(X,t) + u_i(t)) + k_i e_i(t)\right) \quad (3.10) \\ &\le -\lambda_{1i}s_{1,i}^2(t) - \lambda_{2i}|s_{1,i}(t)| \le 0. \end{aligned}$$

Thus, we know that the projective error trajectories converge to the sliding surface asymptotically by utilizing the control strategy (3.8), that is to say, two VOF systems (3.1) and (3.2) can realize projective synchronization. $\square$

(2) Another novel VOF derivative-type sliding surface function is proposed as follows:

$$s_{2,i}(t) = {}^{C}_{0}D_t^{\beta_i(t)}e_i(t) + k_i e_i(t), \tag{3.11}$$

where $k_i > 0$, $i \in \mathbb{L}$.

Theorem 3.3. *Consider the VOF derivative-type sliding surface function (3.11) and equivalent control (3.5), then the error trajectories will converge to $s_{2,i}(t) = 0$ asymptotically by applying the following control law:*

$$\begin{cases} u_i(t) = u_{1i}(t) + u_{2i}(t), \\ ^{C}_{0}D_t^{\beta_i(t)}u_{2i}(t) = \frac{\lambda_{1i}}{\rho}s_{2,i}(t) + \frac{\lambda_{2i}}{\rho}\mathrm{sgn}(s_{2,i}(t)), \end{cases} \tag{3.12}$$

where λ_{1i}, $\lambda_{2i} > 0$, $i \in \mathbb{L}$.

Proof. Consider the following Lyapunov function:

$$V_{3,i}(t) = \frac{1}{2}s_{2,i}^2(t). \tag{3.13}$$

Then, from Lemma 2.3, one has

$$\begin{aligned}
{}_0^C D_t^{\beta_i(t)} V_{3,i}(t) &\leq s_{2,i}(t) {}_{t_0}^C D_t^{\beta_i(t)} s_{2,i}(t) \\
&= s_{2,i}(t) {}_0^C D_t^{\beta_i(t)} \left[{}_0^C D_t^{\beta_i(t)} e_i(t) + k_i e_i(t) \right] \\
&\leq -\lambda_{1i} s_{2,i}^2(t) - \lambda_{2i} |s_{2,i}(t)| \leq 0.
\end{aligned} \tag{3.14}$$

Thus, based on the fractional-order Barbalat lemma in [6], the projective error trajectories converge to zero asymptotically in a finite-time. $\qquad\square$

Remark 3.2. Noticing (3.4) and (3.11), for selecting the parameters k_i, it can be easily obtained that ${}_0^C D_t^{\beta_i(t)} s_{1,i}(t) = s_{2,i}(t)$ by taking the VOF $\beta_i(t)$-order derivative of (3.4). Thus, under the same conditions, assuming that the sliding surface converges to zero under the proposed control strategy, i.e., ${}_0^C D_t^{\beta_i(t)} s_{1,i}(t) = s_{2,i}(t) \to 0$, from Definition 2.6 we know that the convergence rate of the derivative-type sliding surface (3.11) is larger than that for the integral-type sliding surface in (3.4). As a consequence, the integral-type sliding surface requires less computation and has a slower convergence rate, while the derivative-type sliding surface has a larger computation burden and a faster convergence rate.

3.4 Finite-time stability analysis

Theorem 3.4. *Assume there exists a continuous-time positive-definite function $V(t)$ satisfying*

$$\,_0^C D_t^{\varphi(t)} V(t) \leq -\lambda_1 V(t) - \lambda_2 |V^{\frac{1}{2}}(t)|, \tag{3.15}$$

where $\varphi(t) \in [\varphi_0, \varphi_1] \subset (0, 1)$ and $\lambda_1,\ \lambda_2 > 0$. Then $V(t)$ is stable in finite-time with

$$t_r \leq \left\{ \delta \left[\frac{\Gamma(1-\varphi_1)}{\Gamma(1-\varphi_0)} \left(\lambda_1 M_1 + \lambda_2 M_2 \right) - V(0) \right]^{-1} \right\}^{\frac{1}{1-\varphi_0}}, \tag{3.16}$$

with $\delta \triangleq \dfrac{V^{\varphi_0-1}(0)-V(0)}{\Gamma(\varphi_0)}$ and $M_1,\ M_2 > 0$.

Proof. According to [6], since $\frac{\Gamma(1-\varphi_0)}{\Gamma(1-\varphi_1)} {}^{C}_{0}D^{\varphi_0}_t V(t) \leq {}^{C}_{0}D^{\varphi(t)}_t V(t)$, one has

$$
\begin{aligned}
{}^{C}_{0}D^{\varphi_0}_t V(t) &\leq \frac{\Gamma(1-\varphi_1)}{\Gamma(1-\varphi_0)} {}^{C}_{0}D^{\varphi(t)}_t V(t) \\
&\leq \frac{\Gamma(1-\varphi_1)}{\Gamma(1-\varphi_0)}\left(-\lambda_1 V(t) - \lambda_2 |V^{\frac{1}{2}}(t)|\right) \leq 0.
\end{aligned}
\tag{3.17}
$$

In addition, from Lemma 2.2, we have

$$
{}^{R\mathcal{L}}_{0}D^{\varphi_0}_t V(t) = {}^{C}_{0}D^{\varphi_0}_t V(t) + \frac{V(0)}{\Gamma(1-\varphi_0)} t^{-\varphi_0}.
\tag{3.18}
$$

Thus,

$$
{}^{R\mathcal{L}}_{0}D^{\varphi_0}_t V(t) \leq \frac{V(0)}{\Gamma(1-\varphi_0)} t^{-\varphi_0} + \frac{\Gamma(1-\varphi_1)}{\Gamma(1-\varphi_0)}\left(-\lambda_1 V(t) - \lambda_2 |V^{\frac{1}{2}}(t)|\right).
\tag{3.19}
$$

Taking the CO φ_0-order integral with respect to time in (3.19), we obtain that

$$
\begin{aligned}
D^{-\varphi_0}\left({}^{R\mathcal{L}}_{0}D^{\varphi_0}_t V(t)\right) &\leq \frac{V(0)}{\Gamma(1-\varphi_0)} D^{-\varphi_0} t^{-\varphi_0} \\
&\quad + \frac{\Gamma(1-\varphi_1)}{\Gamma(1-\varphi_0)}\left(-\lambda_1 D^{-\varphi_0} V(t) - \lambda_2 D^{-\varphi_0} |V^{\frac{1}{2}}(t)|\right).
\end{aligned}
\tag{3.20}
$$

Considering $D^{-\varphi_0} t^{-\varphi_0}$ and denoting $f(t) = t^{\mu}$, one gets

$$
{}^{R\mathcal{L}}_{0}D^{\varphi_0}_t f(t) = \frac{\Gamma(\mu+1)}{\Gamma(\mu+1-\varphi_0)} t^{\mu-\varphi_0},
\tag{3.21}
$$

where $\mu > -1, 0 < \varphi_0 < 1$.

Meanwhile, taking the CO φ_0-order integral with respect to time from 0 to t_r in (3.21), we obtain

$$
\begin{aligned}
D^{-\varphi_0}\left({}^{R\mathcal{L}}_{0}D^{\varphi_0}_t f(t)\right) &= f(t_r) - f^{\varphi_0}(0)\frac{t_r^{\varphi_0-1}}{\Gamma(\varphi_0)} \\
&= \frac{\Gamma(\mu+1)}{\Gamma(\mu+1-\varphi_0)} D^{-\varphi_0} t^{\mu-\varphi_0}.
\end{aligned}
\tag{3.22}
$$

Thus, letting $\mu = 0$, we obtain

$$
D^{-\varphi_0} t^{-\varphi_0} = \Gamma(1-\varphi_0)\left(1 - \frac{t_r^{\varphi_0-1}}{\Gamma(\varphi_0)}\right).
\tag{3.23}
$$

From (3.21) and (3.23), we have

$$V(t_{\mathrm{r}}) - V^{\varphi_0-1}(0)\frac{t_{\mathrm{r}}^{\varphi_0-1}}{\Gamma(\varphi_0)} \leq V(0)\left(1 - \frac{t_{\mathrm{r}}^{\varphi_0-1}}{\Gamma(\varphi_0)}\right)$$
$$+ \frac{\Gamma(1-\varphi_1)}{\Gamma(1-\varphi_0)}\left(-\lambda_1 D^{-\varphi_0}V(t) - \lambda_2 D^{-\varphi_0}|V^{\frac{1}{2}}(t)|\right).$$

$$(3.24)$$

In addition, using Lemma 1 in [8], there exist positive constants M_1 and M_2 such that $D^{-\varphi_0}V(t) \geq M_1$ and $D^{-\varphi_0}|V^{\frac{1}{2}}(t)| \geq M_2$, thus,

$$\delta\, t_{\mathrm{r}}^{\varphi_0-1} \geq \frac{\Gamma(1-\varphi_1)}{\Gamma(1-\varphi_0)}\left(\lambda_1 M_1 + \lambda_2 M_2\right) - V(0), \tag{3.25}$$

i.e.,

$$t_{\mathrm{r}} \leq \left\{\delta\left[\frac{\Gamma(1-\varphi_1)}{\Gamma(1-\varphi_0)}\left(\lambda_1 M_1 + \lambda_2 M_2\right) - V(0)\right]^{-1}\right\}^{\frac{1}{1-\varphi_0}}. \tag{3.26}$$

$\square$

Remark 3.3. Notice that the core of finite-time stability criterion in this chapter composes two transformations. The first is to turn the VOF derivative into the COF derivative as shown in (3.18). The second is to turn the C-type fractional derivative into RL-type as shown in (3.19). It should be pointed out that both transformations depend on the same time domain. Therefore, the final results will not be changed by those transformations.

Remark 3.4. Jiang et al. [6] discussed the stabilization problem of FVO chaotic systems via SMC approach. However, the finite-time stability problem of VOF system has not been completely discussed in the literature. Therefore, compared with [6], the superiority of the proposed control strategy is that a novel finite-time stability criterion is developed for the VOF control systems. Theorem 3.4 implies that the reaching time t_{r} has an upper bound. Thus, how to minimize the bound is a challenging problem that should be intensively studied in the future.

3.5 Numerical example

Example 3.1. To illustrate the validity of this work, consider the following VOF-Lorenz chaotic system [6] as the master system:

$$\begin{cases} {}_0^{\mathcal{C}}D_t^{\alpha_1(t)}y_1(t) = -10(y_1 - y_2), \\ {}_0^{\mathcal{C}}D_t^{\alpha_2(t)}y_2(t) = 28y_1 - y_2 - y_1 y_3, \\ {}_0^{\mathcal{C}}D_t^{\alpha_3(t)}y_3(t) = -\frac{8}{3}y_3 + y_1 y_2, \end{cases} \tag{3.27}$$

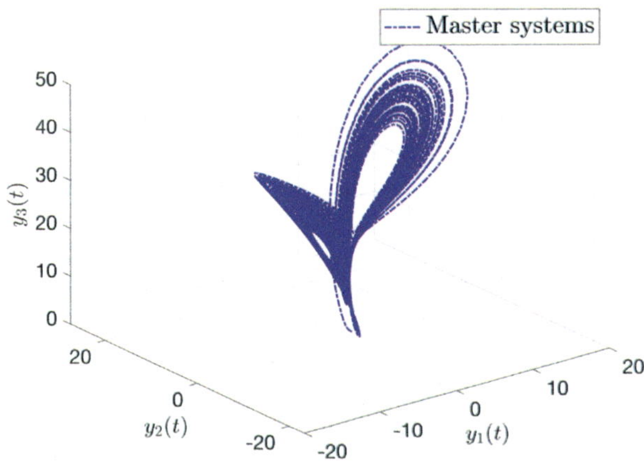

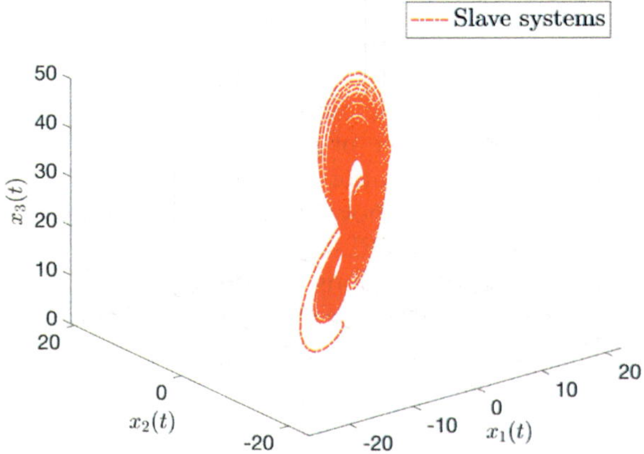

FIGURE 3.1   Chaotic behaviors of VOF-Lorenz (3.27) and VOF-Chen (3.28) systems without controllers.

with

$$\alpha_i(t) = \begin{cases} 0.990 + 0.005\sin(\frac{\pi t}{8}), & i = 1, \\ 0.991 - 0.002\cos(\frac{\pi t}{16}), & i = 2, \\ 0.988 + 0.004\frac{t}{100}, & i = 3. \end{cases}$$

Meanwhile, choose the following VOF-Chen chaotic system as the slave system:

$$\begin{cases} {}_0^C D_t^{\beta_1(t)} x_1(t) = -7x_2 - x_2 x_3 + 28x_1 + u_1, \\ {}_0^C D_t^{\beta_2(t)} x_2(t) = 35(x_1 - x_2) + u_2, \\ {}_0^C D_t^{\beta_3(t)} x_2(t) = x_1 x_2 - 3x_3 + u_3, \end{cases} \tag{3.28}$$

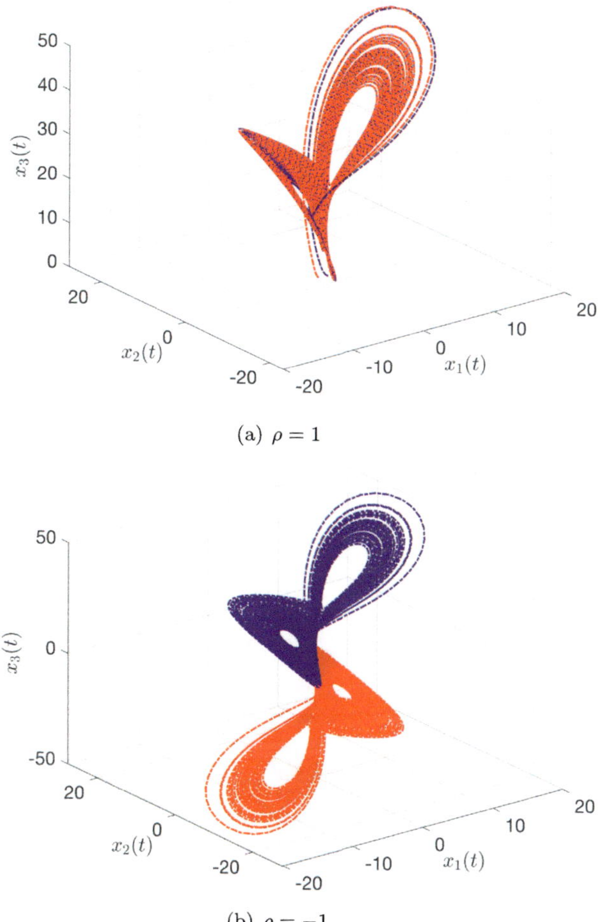

(a) $\rho = 1$

(b) $\rho = -1$

FIGURE 3.2 Chaotic behaviors of (3.27) and (3.28) with controllers as in (3.8) under different projection coefficients.

with

$$\beta_i(t) = \begin{cases} 0.976 + 0.012\frac{t}{100}, & i = 1, \\ 0.985 - 0.005\sin(\frac{\pi t}{16}), & i = 2, \\ 0.978 - 0.002\cos(\frac{\pi t}{8}), & i = 3. \end{cases}$$

The initial conditions are given by $y(0) = [1\ 1\ 1]^{\mathrm{T}}$ and $x(0) = [-1\ 0\ 2]^{\mathrm{T}}$. And, according to Theorem 3.2, some parameters are chosen as follows: $k_1 = 2.61$, $k_2 = 1.88$, $k_3 = 0.74$, $\lambda_{11} = 7.97$, $\lambda_{21} = 2.05$, $\lambda_{12} = 2.45$, $\lambda_{22} = 7.03$, $\lambda_{13} = 1.09$, and $\lambda_{23} = 5.79$. In order to reduce the chattering effect from the

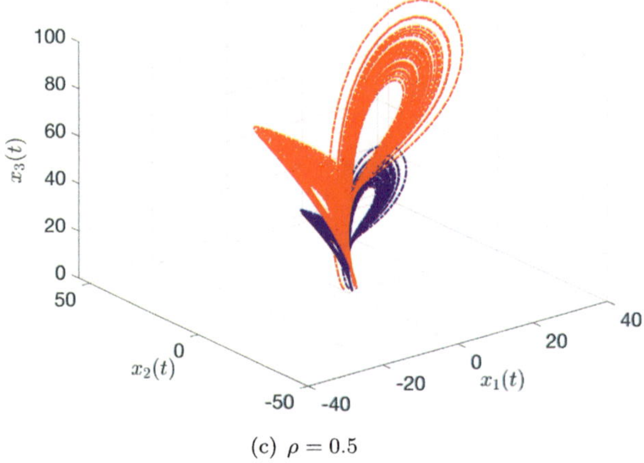

(c) $\rho = 0.5$

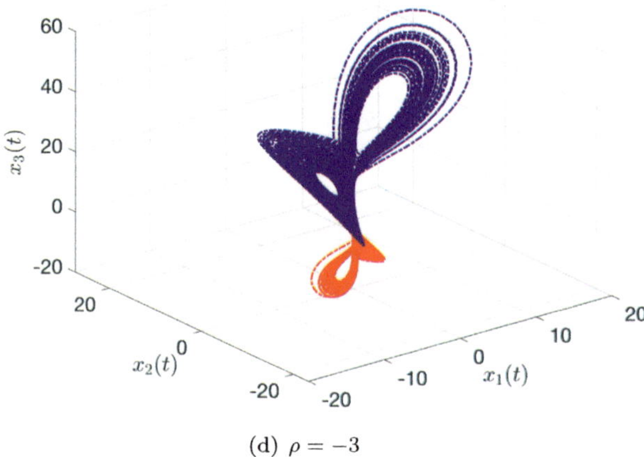

(d) $\rho = -3$

FIGURE 3.2   (*continued*)

switching signal, we replace sgn($s(t)$) with $\frac{s(t)}{\|s(t)\|+0.01}$. The simulation re-
sults are depicted in Figs. 3.1–3.5. Figs. 3.1 and 3.2 show the chaotic trajectories of the open- and closed-loop VOF-Lorenz and VOF-Chen systems,
respectively; Figs. 3.3 and 3.4 show the VOF integral-type sliding surface,
controllers, and the projective synchronization error curves, respectively;
Finally, the response of VOF derivative-type sliding surface function (3.11)
is presented in Fig. 3.5.

Based on the above simulation results, the slave system state $x_i(t)$ approaches the master system state $y_i(t)$ and achieves good projective synchronization, including complete synchronization and asynchronization,

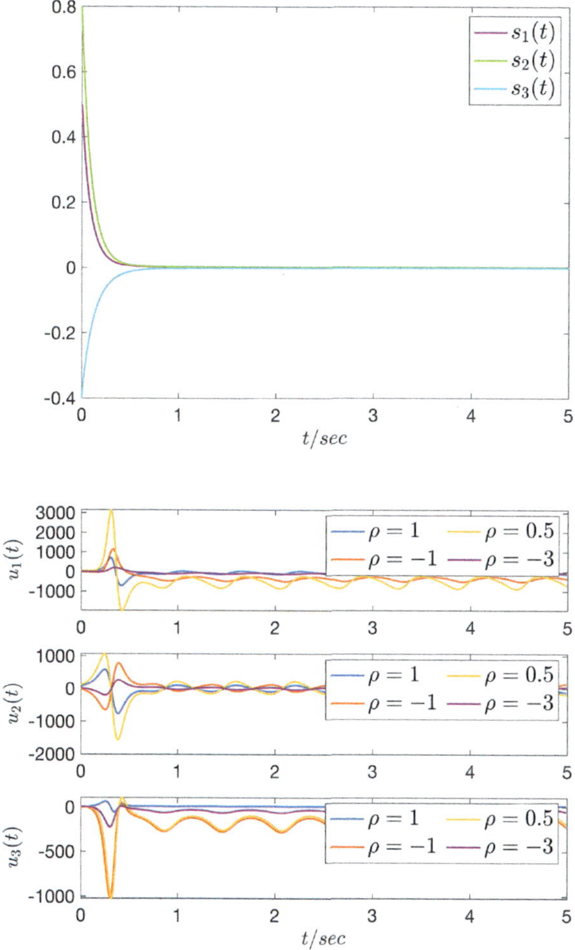

FIGURE 3.3 VOF integral-type sliding surfaces (3.4) and controllers (3.8).

i.e., $\rho = 1$ and $\rho = -1$, respectively. Besides, from Figs. 3.3 and 3.5, we can see that with the same initial condition in this example, the VOF derivative-type sliding surface needs less time to reach the sliding surface than the VOF integral-type sliding surface.

Now, choose the slave system as follows:

$$\begin{cases} {}_{0}^{C}D_t^{\alpha_1(t)}x_1(t) = -10(x_1 - x_2), \\ {}_{0}^{C}D_t^{\alpha_2(t)}x_2(t) = 28x_1 - x_2 - x_1x_3, \\ {}_{0}^{C}D_t^{\alpha_3(t)}x_3(t) = -\frac{8}{3}x_3 + x_1x_2. \end{cases} \tag{3.29}$$

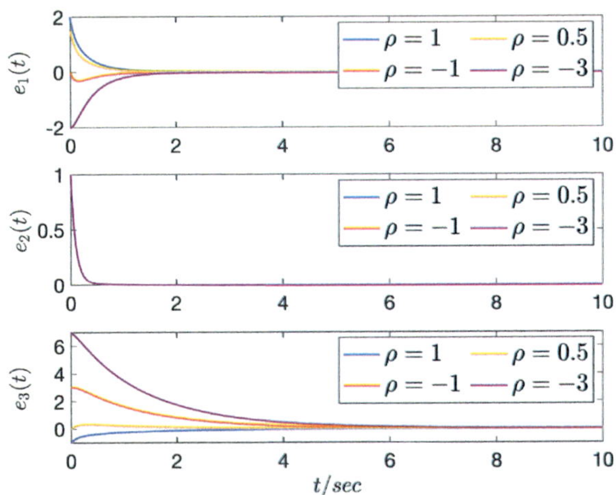

FIGURE 3.4 Projective synchronization error curves for different projection coefficients.

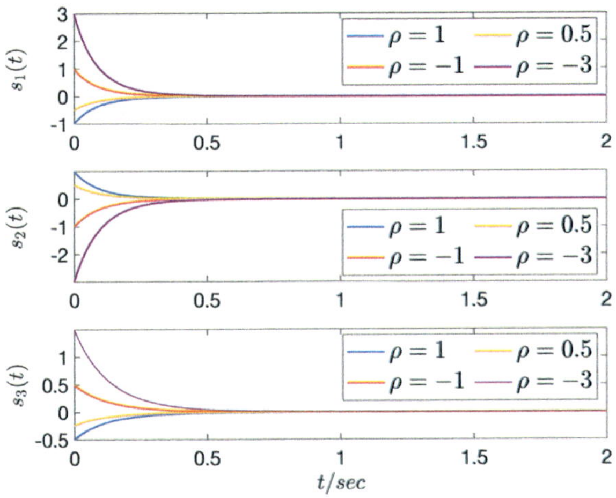

FIGURE 3.5 VOF derivative-type sliding surface (3.11).

A comparison with [10] is given to illustrate the advantages of the proposed control strategy in this work. Fig. 3.6 shows the synchronization behaviors between the proposed SMC strategy and the method proposed in [10], respectively; Fig. 3.6 presents the curves of synchronization errors and compares the sums of each individual error in mean-square with these in [10] under the same initial condition within 5 s, which shows that better synchronization performance in this chapter is achieved than that in [10] as the time increases.

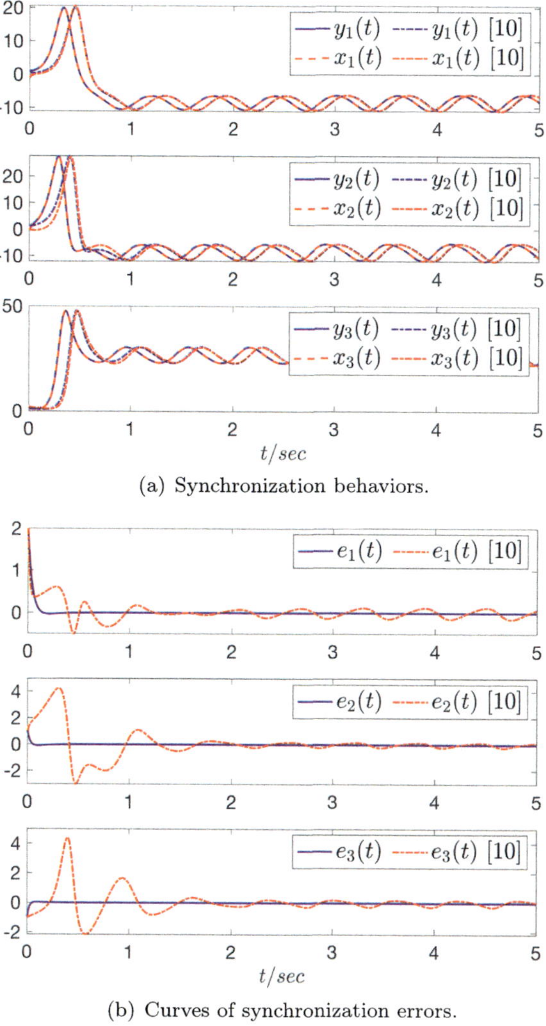

(a) Synchronization behaviors.

(b) Curves of synchronization errors.

FIGURE 3.6 Comparison with the methods from [10].

3.6 Conclusion

This chapter has dealt with the issue of finite-time projective synchronization of VOF chaotic systems via the SMC approach. First, according to the VOF calculus theory, novel VOF integral- and derivative-type sliding surfaces have been proposed, respectively. Second, the VOF controllers have been synthesized to realize the projective synchronization for the VOF unperturbed chaotic systems. Third, an upper bound of the reaching time has been given by presenting a novel finite-time criterion. Finally, a numerical simulation has been added to show the effectiveness of the proposed SMC method.

References

[1] M.D. Ortigueira, J.A. Tenreiro Machado, What is a fractional derivative?, Journal of Computational Physics 293 (2015) 4–13.

[2] M.D. Ortigueira, D. Valério, J.A. Tenreiro Machado, Variable order fractional systems, Communications in Nonlinear Science and Numerical Simulation 71 (2019) 231–243.

[3] S.G. Samko, Fractional integration and differentiation of variable order, Analysis Mathematica 21 (3) (1995) 213–236.

[4] C.F. Lorenzo, T.T. Hartley, Variable order and distributed order fractional operators, Nonlinear Dynamics 29 (2002) 57–98.

[5] J. Jiang, H. Chen, J.L.G. Guirao, D. Cao, Existence of the solution and stability for a class of variable fractional order differential systems, Chaos, Solitons and Fractals 128 (2019) 269–274.

[6] J. Jiang, D. Cao, H. Chen, Sliding mode control for a class of variable-order fractional chaotic systems, Journal of the Franklin Institute 357 (15) (2020) 10127–10158.

[7] L.M. Pecora, T.L. Carroll, Synchronization in chaotic systems, Physical Review Letters 64 (8) (1990) 821.

[8] D. Deepika, S. Kaur, S. Narayan, Uncertainty and disturbance estimator based robust synchronization for a class of uncertain fractional chaotic system via fractional order sliding mode control, Chaos, Solitons and Fractals 115 (2018) 196–203.

[9] T.-C. Lin, T.-Y. Lee, Chaos synchronization of uncertain fractional-order chaotic systems with time delay based on adaptive fuzzy sliding mode control, IEEE Transactions on Fuzzy Systems 19 (4) (2011) 623–635.

[10] M.F. Tolba, A.H. Elsafty, M. Armanyos, L.A. Said, A.H. Madian, A.G. Radwan, Synchronization and FPGA realization of fractional-order Izhikevich neuron model, Microelectronics Journal 89 (2019) 56–69.

[11] H. Ren, H.R. Karimi, R. Lu, Y. Wu, Synchronization of network systems via aperiodic sampled-data control with constant delay and application to unmanned ground vehicles, IEEE Transactions on Industrial Electronics 67 (6) (2019) 4980–4990.

[12] Y. Wang, H.R. Karimi, H. Yan, et al., An adaptive event-triggered synchronization approach for chaotic Lur'e systems subject to aperiodic sampled data, IEEE Transactions on Circuits and Systems. II, Express Briefs 66 (3) (2019) 442–446.

[13] W. Zou, P. Shi, Z. Xiang, Y. Shi, Finite-time consensus of second-order switched nonlinear multi-agent systems, IEEE Transactions on Neural Networks and Learning Systems 31 (5) (2019) 1757–1762.

[14] X. Zhang, L. Tang, J. Lü, Synchronization analysis on two-layer networks of fractional-order systems: intralayer and interlayer synchronization, IEEE Transactions on Circuits and Systems. I, Regular Papers 67 (7) (2020) 2397–2408.

[15] M. Chen, S.-Y. Shao, P. Shi, Y. Shi, Disturbance-observer-based robust synchronization control for a class of fractional-order chaotic systems, IEEE Transactions on Circuits and Systems. II, Express Briefs 64 (4) (2016) 417–421.

[16] X. Meng, C. Gao, Z. Liu, B. Jiang, Robust H_∞ control for a class of uncertain neutral-type systems with time-varying delays, Asian Journal of Control 23 (3) (2021) 1454–1465.

[17] P. Zhu, Y. Chen, M. Li, P. Zhang, Z. Wan, Fractional-order sliding mode position tracking control for servo system with disturbance, ISA Transactions 105 (2020) 269–277.

[18] G. Sun, Z. Ma, J. Yu, Discrete-time fractional order terminal sliding mode tracking control for linear motor, IEEE Transactions on Industrial Electronics 65 (4) (2017) 3386–3394.

[19] A. Modiri, S. Mobayen, Adaptive terminal sliding mode control scheme for synchronization of fractional-order uncertain chaotic systems, ISA Transactions 105 (2020) 33–50.

[20] Y. Han, C.-Y. Su, Y. Kao, C. Gao, Non-fragile sliding mode control of discrete switched singular systems with time-varying delays, IET Control Theory & Applications 14 (5) (2020) 726–737.

[21] W. Jie, Z. Yudong, B. Yulong, H.H. Kim, M.C. Lee, Trajectory tracking control using fractional-order terminal sliding mode control with sliding perturbation observer for a 7-DOF robot manipulator, IEEE/ASME Transactions on Mechatronics 25 (4) (2020) 1886–1893.

[22] K. Khandani, V.J. Majd, M. Tahmasebi, Robust stabilization of uncertain time-delay systems with fractional stochastic noise using the novel fractional stochastic sliding approach and its application to stream water quality regulation, IEEE Transactions on Automatic Control 62 (4) (2016) 1742–1751.

[23] X. Meng, Y. Kao, C. Gao, B. Jiang, Projective synchronisation of variable-order systems via fractional sliding mode control approach, IET Control Theory & Applications 14 (1) (2020) 12–18.

[24] R. Nie, S. He, F. Liu, X. Luan, Sliding mode controller design for conic-type nonlinear semi-Markovian jumping systems of time-delayed Chua's circuit, IEEE Transactions on Systems, Man, and Cybernetics: Systems 51 (4) (2019) 2467–2475.

Finite-time synchronization of variable-order fractional uncertain coupled systems via adaptive sliding mode control

4.1 Introduction

Fractional calculus has been widely applied to describing natural behaviors and complex phenomena in practical problems such as engineering, mathematics, physics, and finance in the past decades; see [1–4]. But it is worth pointing out that more and more research results indicated that the memory and nonlocality properties of some complex physical systems in nature may depend on the change of time, space, or other conditions. Therefore, utilizing the theory of variable-order fractional (VOF) calculus could be more accurate than constant-order (CO) calculi to describe these processes and phenomena. In 1995, the definition and related properties of Riemann–Liouville VOF calculus were firstly given by Samko [5]. Since then, Lorenzo and Hartley [6] made a profound study and presented several different concepts of the VOF calculus, and provided calculation methods for these definitions. Whereafter, the VOF calculus has been widely applied to many fields of science and engineering [7–9]. In [7], the control design issue of nonlinear viscoelastic oscillators was considered via VOF operators. And in [8], the existence and stability of the solution for VOF differential equation was investigated via the Arzela–Ascoli theory.

Recently, the stability and stabilizability problem of CO coupled systems have been studied in [10–12]. In [11], Xu et al. considered the synchronization problem for a class of nonlinear fractional-order coupled systems with time-delay by utilizing a periodically intermittent control

Sliding Mode Control of Fractional-order Systems
https://doi.org/10.1016/B978-0-44-331696-8.00008-2

57

method. Chen et al. have also investigated the synchronization problem of nonlinear fractional-order systems through sliding mode control (SMC) approach in [12]. It should be noted that the theory of VOF calculus has not been applied to the controller design of coupled systems so far since the Lyapunov functional method, which is widely used in CO coupled systems, cannot be extended to VOF coupled systems directly. Further, in [10], the authors considered the finite-time projective synchronization control problem for a class of VOF chaotic systems by means of the SMC approach. However, for an uncertain complex physical model, an integer- or fractional-order system is not enough to describe the dynamics effectively or globally. Thus, applying VOF derivative in coupled systems is more realistic and reasonable when considering the issues of controller design and stability. On the other hand, in most existing literature, the considered coupled systems are all CO derivative and unperturbed terms dependent. Hence, the unperturbed terms should be taken into account to study the synchronization of VOF coupled systems.

SMC has good robustness and transient properties, which have been widely applied to the analysis and synthesis of complex nonlinear systems, such as the application in the stream water quality standards regulation [13], the electrical motor model [14], the steering of unmanned underwater vehicles [15], and other synchronization issues [16,17]; for more details, please refer to [18,16,13–15,17,19,20] and the references therein. In [21], the finite-time robust synchronization analysis for fractional-order chaotic systems with uncertain terms was obtained by the fractional SMC method. In [22], the stabilization and synchronization for fractional-order chaotic systems were obtained with the aid of the SMC approach. In [23], for two different fractional-order chaotic systems with the uncertainty and time-delay, synchronization criteria were obtained via an adaptive fuzzy SMC strategy. Since the fuzzy algorithm is also model-free and robust against disturbances, the authors [24] presented an adaptive fuzzy method to investigate the synchronization control problem, and applied it in the formation control of unmanned surface vehicles with actuator saturation and unknown nonlinear items.

Synthesizing the above analysis, this chapter deals with the synchronization control for a class of network-based VOF uncertain coupled systems. The primary novelties and contributions of this chapter are:

(1) The synchronization issue of the VOF uncertain coupled systems is turned into the asymptotical stability of the synchronization error systems;

(2) By designing a novel VOF integral-type sliding surface and adaptive SMC law, several asymptotical stability criteria are presented to achieve projective synchronization;

(3) Synchronization criteria are obtained using VOF calculus and a strongly connected network.

4.2 Problem statement

In this section, several synchronization SMC strategies for a class of VOF uncertain coupled systems will be proposed, i.e., the VOF integral-type sliding surface design, asymptotically stability analysis, and adaptive control law design.

Consider the following VOF coupled systems on a digraph $\mathcal{G}$ composed of n nodes, which is described by

$$
{}_{0}^{C}D_{t}^{\alpha(t)}x_i(t) = f(x_i(t)) + \sum_{j=1}^{n} \beta_{ij}g_{ij}(x_i(t) - x_j(t)) \tag{4.1}
$$
$$
+ \Delta h_i(X,t) + \delta_i(t) + u_i(t), \quad i \in \mathbb{L},
$$

where $0 < \alpha_0 \leq \alpha(t) \leq \alpha_1 < 1$ is the system variable-order continuous function, $x_i(t) = [x_{i1}(t), x_{i2}(t), \ldots, x_{im}(t)]^{\mathrm{T}} \in \mathbb{R}^m$ denotes the state vector of the ith node, $\Delta h_i(X,t) \in \mathbb{R}^m$ and $\delta_i(t) \in \mathbb{R}^m$ demonstrate the unknown parametric uncertainties and exterior disturbances, respectively, and $u_i(t) \in \mathbb{R}^m$ is the control input. In addition, $f(\cdot) \in \mathbb{R}^m$ is a continuous vector-valued function, and $g_{ij}(\cdot) : \mathbb{R}^m \to \mathbb{R}^m$ denotes the coupling term reflecting the influence of the jth node on the ith node, the nonnegative constant β_{ij} represents the coupling strength.

In addition, suppose that $y(t) \in \mathbb{R}^m$ is a solution of an isolated node satisfying

$$
{}_{0}^{C}D_{t}^{\alpha(t)}y(t) = f(y(t)). \tag{4.2}
$$

It should be pointing that the state trajectory $y(t)$ could be chosen as the equilibrium point, the periodic solutions, or even the chaotic orbit. In addition, define the synchronization error vector as $e_i(t) = x_i(t) - y(t)$, $i \in \mathbb{L}$. Then the synchronization error system can be described as

$$
{}_{0}^{C}D_{t}^{\alpha(t)}e_i(t) = \hat{f}(e_i(t)) + \sum_{j=1}^{n} \beta_{ij}g_{ij}(e_i(t) - e_j(t)) \tag{4.3}
$$
$$
+ \Delta h_i(X,t) + \delta_i(t) + u_i(t), \quad i \in \mathbb{L},
$$

where $\hat{f}(e_i(t)) \triangleq f(x_i(t)) - f(y(t))$.

Assumption 4.1. For any $i, j \in \mathbb{L}$, there exist positive constants η and η_{ij} such that

$$
|\hat{f}(\cdot)| \leq \eta| \cdot |, \quad |g_{ij}(\cdot)| \leq \eta_{ij}| \cdot |. \tag{4.4}
$$

Assumption 4.2. The uncertainties $\Delta h_i(X, t)$ and the exterior disturbance term $\delta_i(t)$ are bounded:

$$|\Delta h_i(X, t)| \leq \upsilon_i, \quad |\delta_i(t)| \leq \nu_i, \quad i \in \mathbb{L}. \tag{4.5}$$

Remark 4.1. In [10,12], the authors also considered the nonlinear terms in coupled systems. However, the main technique in dealing with the considered nonlinear terms is different from ours in this chapter. Generally speaking, the state trajectory on each node of the network could be chosen as an m-dimensional vector. Therefore, in this chapter, we select the nonlinear terms $f(y)$ and $[\partial f/\partial y](y)$ which are continuous on W, thus f is locally Lipschitz in y on W. Further, the Lipschitz constant δ can be taken as $\eta = \max\limits_{y \in W}\{\|J\|_\infty\}$, where $J = \left[\frac{\partial f}{\partial y}\right]$ denotes the Jacobian matrix of function f, and the convex set is $W = \{y \in \mathbb{R}^m \mid |y_j| \leq w_j, j = 1, 2, \ldots, m\}$.

<h3 style="text-align:center">4.3 Main results</h3>

4.3.1 VOF sliding mode controller design

In this section, a novel VOF integral-type sliding surface function is proposed as follows:

$$s_i(t) = e_i(t) + D_t^{-\alpha(t)}\left[-\hat{f}(e_i(t)) \right.$$
$$\left. - \sum_{j=1}^{n} \beta_{ij} g_{ij}(e_i(t) - e_j(t)) + k_i e_i(t) \right], \quad i \in \mathbb{L}. \tag{4.6}$$

Using the equivalent control theory of SMC, i.e., $s_i(t) = 0$ and ${}_0^C D_t^{\alpha(t)} s_i(t) = 0$, the sliding dynamics can be obtained as follows:

$${}_0^C D_t^{\alpha(t)} e_i(t) = -k_i e_i(t) + \hat{f}(e_i(t)) + \sum_{j=1}^{n} \beta_{ij} g_{ij}. \tag{4.7}$$

Theorem 4.1. *Suppose that Assumption 4.1 holds, and digraph $(\mathcal{G}, \mathbf{B})$ with $\mathbf{B} = (\beta_{ij}\eta_{ij})_{n \times n}$ is a strongly connected network. Take the VOF integral-type sliding surface function proposed in (4.6). Then the systems (4.1) and (4.2) can achieve global synchronization in finite-time if there exist constants k_i such that*

$$\varepsilon \triangleq \min_{i \in \mathbb{L}} \left\{ k_i - \eta - 2 \sum_{j=1}^{n} \beta_{ij} \eta_{ij} \right\} > 0. \tag{4.8}$$

Proof. For each node of the synchronization error system, choose the following Lyapunov function:

$$V_1(\boldsymbol{E}(t), t) = \sum_{i=1}^{n} c_i V_{1i}(e_i(t), t) = \sum_{i=1}^{n} \frac{c_i}{2} e_i^2(t), \tag{4.9}$$

where $c_i > 0$ is the cofactor of the ith diagonal element in the Laplacian matrix of the strongly connected network $(\mathcal{G}, \boldsymbol{B})$. Then, taking the VOF $\alpha(t)$-order derivative with respect to time of (4.9), and combining with Lemma 2.3, we have

$$\begin{aligned}
{}_{0}^{C}D_t^{\alpha(t)} V_1 &\leq \sum_{i=1}^{n} c_i e_i(t) {}_{0}^{C}D_t^{\alpha(t)} e_i(t) \\
&= \sum_{i=1}^{n} c_i e_i(t) \left[-k_i e_i(t) + \hat{f}(e_i(t)) + \sum_{j=1}^{n} \beta_{ij} g_{ij} \right] \\
&\leq \sum_{i=1}^{n} c_i \left\{ -k_i e_i^2(t) + \eta e_i^2(t) + 2 \sum_{j=1}^{n} \beta_{ij} \eta_{ij} e_i^2(t) \right. \\
&\quad \left. + \frac{1}{2} \sum_{j=1}^{n} \beta_{ij} \eta_{ij} (e_j^2(t) - e_i^2(t)) \right\}.
\end{aligned} \tag{4.10}$$

In the meantime, let $F_{ij}(e_i(t), e_j(t)) \triangleq \frac{1}{2}(e_j^2(t) - e_i^2(t))$. Then, based on Lemma 2 in [11] and Condition A2 of Theorem 4.1 in [25], we have that $\sum_{i=1}^{n}\sum_{j=1}^{n} c_i \beta_{ij} \eta_{ij} F_{ij}(e_i(t), e_j(t)) \leq 0$, hence

$$\begin{aligned}
{}_{0}^{C}D_t^{\alpha(t)} V_1 &\leq \sum_{i=1}^{n} c_i \left(-k_i + \eta + 2 \sum_{j=1}^{n} \beta_{ij} \eta_{ij} \right) e_i^2(t) \\
&\leq -\varepsilon V_1(\boldsymbol{E}(t), t) \leq 0.
\end{aligned} \tag{4.11}$$

Hence, based on Lemma 2.6 in [4], the latter implies that the synchronization error trajectories of system (4.7) converge to zero asymptotically.

On the other hand, since

$$\begin{aligned}
{}_{0}^{C}D_t^{\alpha_0} V_1(t) \leq \frac{\Gamma(1-\alpha_1)}{\Gamma(1-\alpha_0)} {}_{0}^{C}D_t^{\alpha(t)} V_1(t), \tag{4.12}
\end{aligned}$$

combining with Lemma 2.2, one has

$$\begin{aligned}
{}_{0}^{\mathcal{RL}}D_t^{\alpha_0} V_1(t) \leq \frac{V_1(0)}{\Gamma(1-\alpha_0)} t^{-\alpha_0} - \varepsilon \frac{\Gamma(1-\alpha_1)}{\Gamma(1-\alpha_0)} V_1(t). \tag{4.13}
\end{aligned}$$

By taking the VOF integral $D_t^{-\alpha_0}$ from 0 to t_e on both sides with $E(t_e) = 0$, we have

$$D_t^{-\alpha_0}\left({}_0^{\mathcal{RL}}D_t^{\alpha_0}V_1(t)\right) \leq \frac{V_1(0)}{\Gamma(1-\alpha_0)}D_t^{-\alpha_0}t^{-\alpha_0} - \varepsilon\frac{\Gamma(1-\alpha_1)}{\Gamma(1-\alpha_0)}I_t^{\alpha_0}V_1(t). \quad (4.14)$$

Since $D_t^{-\alpha_0}t^{-\alpha_0} = \Gamma(1-\alpha_0)\left[1 - \frac{t_e^{\alpha_0-1}}{\Gamma(\alpha_0)}\right]$, one gets

$$V_1(t_e) - V_1^{\alpha_0-1}(0)\frac{t_e^{\alpha_0-1}}{\Gamma(\alpha_0)} \leq V_1(0)\left[1 - \frac{t_e^{\alpha_0-1}}{\Gamma(\alpha_0)}\right] - \varepsilon\frac{\Gamma(1-\alpha_1)}{\Gamma(1-\alpha_0)}I_t^{\alpha_0}V_1(t). \quad (4.15)$$

In addition, recalling Lemma 2.7, we know that there exists a positive constant M_0 such that $D_t^{-\alpha_0}V_1(t) \geq M_0$, thus, the finite-time t_e (time in which the synchronization errors reach the origin) can be bounded as

$$t_e \leq \left[\zeta \cdot \left[\varepsilon\frac{\Gamma(1-\alpha_1)}{\Gamma(1-\alpha_0)}M_0 - V_1(0)\right]^{-1}\right]^{\frac{1}{1-\alpha_0}}, \quad (4.16)$$

where $\zeta \triangleq \frac{V_1(0)^{\alpha_0-1}-V_1(0)}{\Gamma(\alpha_0)}$. $\qquad\qquad\square$

Remark 4.2. Noticing that the core of finite-time stability criterion above is composed of two transformations. The first is to turn the VOF derivative into the COF derivative as in (4.12). The second is to turn the Caputo fractional derivative into that of RL type shown in (4.13). It should be pointed out that both transformations depend on the same time domain. Therefore, the final result will be not changed by those transformations.

Theorem 4.2. *Suppose that Assumption 4.2 holds, and digraph $(\mathcal{G}, \mathbf{B})$ is a strongly connected network. Take the VOF sliding surface function and synchronization error dynamics proposed in (4.6) and (4.7), respectively. Then the synchronization error states converge to the VOF sliding surface $s_i(t) = 0$ in finite-time when using the following VOF SMC law:*

$$u_i(t) = -k_ie_i(t) - \lambda_is_i(t) - (\upsilon_i + v_i)\mathrm{sgn}(s_i(t)). \quad (4.17)$$

Proof. Choose the following Lyapunov function:

$$V_2(t) = \sum_{i=1}^{n}\frac{1}{2}s_i^2(t). \quad (4.18)$$

Then, taking the VOF $\alpha(t)$-order derivative with respect to time of (4.18), one has

$$
{}_{0}^{\mathcal{C}}D_t^{\alpha(t)}V_2(t) \leq \sum_{i=1}^{n} s_i(t){}_{0}^{\mathcal{C}}D_t^{\alpha(t)}s_i(t)
$$

$$
= \sum_{i=1}^{n} s_i(t)[\Delta h_i(X,t) + \delta_i(t) + k_i e_i(t) + u_i(t)]
$$

$$
\leq \sum_{i=1}^{n} \left\{ |s_i(t)|(|\Delta h_i(X,t)| + |\delta_i(t)|) - \lambda_i s_i^2(t) - (\upsilon_i + \nu_i)|s_i(t)| \right\}
$$

$$
\leq -\frac{\hat{\lambda}}{2} \sum_{i=1}^{n} s_i^2(t) \leq 0,
$$

where $\hat{\lambda} \triangleq \min_{i \in \mathbb{L}}\{\lambda_i\} \geq 0$. Thus, we see that the synchronization error states converge to the VOF sliding surface $s_i(t) = 0$ in finite-time. Meanwhile, similar to the proof of Theorem 4.1, we can also bound the reaching time t_s as

$$
t_s \leq \left[\hat{\zeta} \cdot \left[\frac{\hat{\lambda}}{2} \frac{\Gamma(1-\alpha_1)}{\Gamma(1-\alpha_0)} M_1 - V_2(0) \right]^{-1} \right]^{\frac{1}{1-\alpha_0}}, \tag{4.19}
$$

where $\hat{\zeta} \triangleq \frac{V_2(0)^{\alpha_0-1} - V_2(0)}{\Gamma(\alpha_0)}$. $\qquad\qquad\square$

Remark 4.3. In the above analysis, the asymptotical stability criteria of the sliding mode and error dynamics are given based on the condition that upper bounds of the uncertain term are known. However, in real-world systems, such information is often not available as desired due to various circumstances. Therefore, further investigation on this issue should be undertaken. As we know, the adaptive control approach [26,27] is an effective tool to study uncertain systems. Thus, combining the adaptive SMC theory and the VOF calculus, a novel VOF adaptive SMC strategy is presented in the following. By utilizing an adaptive SMC law, the synchronization error trajectories can be forced to the origin in finite-time.

4.3.2 Adaptive VOF controller design

Assumption 4.3. The uncertainties $\Delta h_i(X,t)$ and the exterior disturbance term $\delta_i(t)$ are bounded:

$$
|\Delta h_i(X,t) + \delta_i(t)| \leq \gamma_i, \quad i \in \mathbb{L}. \tag{4.20}
$$

Theorem 4.3. *Suppose that Assumption 4.3 holds. Consider the synchronization error dynamics (4.3) and the VOF sliding surface (4.6). Design the adaptive VOF controller as follows:*

$$u_i(t) = -k_i e_i(t) - \lambda_{1i} s_i(t) - \lambda_{2i} \operatorname{sgn}(s_i(t))|s_i(t)|^z - \hat{\gamma}_i \operatorname{sgn}(s_i(t)), \qquad (4.21)$$

with the adaptive law

$$\,^{C}_0 D_t^{\alpha(t)} \hat{\gamma}_i = \mu_i |s_i(t)|, \quad i \in \mathbb{L}, \qquad (4.22)$$

in which $\hat{\gamma}_i$ denotes the estimate of γ_i with initial value $\gamma_i(0)$. Then the synchronization error trajectories of the system (4.3) reach the sliding surface in finite time.

Proof. For each node system, choose the following Lyapunov function:

$$V_3(t) = \sum_{i=1}^{n} \left\{ \frac{1}{2} s_i^2(t) + \frac{1}{2\mu_i} (\hat{\gamma}_i - \gamma_i)^2 \right\}. \qquad (4.23)$$

Then, taking the VOF $\alpha(t)$-order derivative with respect to time of (26), we have

$$
\begin{aligned}
\,^{C}_0 D_t^{\alpha(t)} V_3(t) &\leq \sum_{i=1}^{n} \left\{ s_i(t) \,^{C}_0 D_t^{\alpha(t)} s_i(t) + \frac{1}{\mu_i} (\hat{\gamma}_i - \gamma_i) \,^{C}_0 D_t^{\alpha(t)} \hat{\gamma}_i \right\} \\
&= \sum_{i=1}^{n} \left\{ s_i(t) \left[\,^{C}_0 D_t^{\alpha(t)} e_i(t) - \hat{f}(e_i(t)) - \sum_{j=1}^{n} \beta_{ij} g_{ij} + k_i e_i(t) \right] \right. \\
&\qquad \left. + (\hat{\gamma}_i - \gamma_i)|s_i(t)| \right\} \\
&\leq \sum_{i=1}^{n} \left\{ |s_i(t)||\Delta h_i(X,t) + \delta_i(t)| - \lambda_{1i} s_i^2(t) \right. \\
&\qquad \left. - \lambda_{2i}|s_i(t)|^{1+z} - \hat{\gamma}_i |s_i(t)| + (\hat{\gamma}_i - \gamma_i)|s_i(t)| \right\} \\
&\leq \sum_{i=1}^{n} \left\{ -\lambda_{1i} s_i^2(t) - \lambda_{2i}|s_i(t)|^{1+z} \right\} \leq 0.
\end{aligned}
\qquad (4.24)
$$

Therefore, we easily obtain that the synchronization error trajectories converge to the sliding surface asymptotically. Therefore, the VOF uncertain coupled systems and the isolated system can realize the synchronization by utilizing the proposed control strategy. □

Remark 4.4. Compared with [9], the purpose of Assumption 4.3 is to reduce the computational burden of numerical simulation. In the following numerical example, we select 15 nodes to constitute a strongly connected network. And in each node, the state trajectory $x_i(t) \in \mathbb{R}^3$, thus, if we choose Assumption 4.2 to consider the adaptive control issue, we must give two adaptive laws for each uncertainty $\Delta h_i(X, t)$ and exterior disturbance term $\delta_i(t)$, respectively. In theory, it makes no influence, but it will add more computational burden of numerical simulation. Hence, we propose Assumption 4.3 for those uncertainties and exterior disturbance terms to reduce the running time.

4.4 Numerical examples

In this section, for the purpose of illustrating the effectiveness of the theoretical results and showing the feasibility of an adaptive VOF controller, a numerical simulation is given as follows.

Example 4.1. Consider the following VOF Chen coupled system with unknown parameter uncertainties and exterior disturbances modeled on a digraph $(\mathcal{G}, \boldsymbol{B})$ with 15 nodes (see Fig. 4.1):

$$
{}_{0}^{\mathcal{C}}D_t^{\alpha(t)} x_i(t) = f(x_i(t)) + \sum_{j=1}^{n} \beta_{ij} g_{ij}(x_i(t) - x_j(t))
$$
$$
+ \Delta h_i(X, t) + \delta_i(t) + u_i(t), \quad i = 1, 2, \ldots, 15,
$$

(4.25)

where

$$
f(x_i(t)) = \begin{pmatrix} -7x_{i2}(t) - x_{i2}(t)x_{i3}(t) + 28x_{i1}(t) \\ 35(x_{i1}(t) - x_{i2}(t)) \\ x_{i1}(t)x_{i2}(t) - 3x_{i3}(t) \end{pmatrix},
$$

$$
\alpha(t) = 0.980 + 0.005 \sin(\frac{\pi t}{8}), \quad g_{ij}(\cdot) = \sin(\cdot),
$$

and

$$
\begin{aligned}
&\beta_{18} = 0.0914, \quad &&\beta_{28} = 0.0166, \quad &&\beta_{3,10} = 0.0258, \quad &&\beta_{4,10} = 0.0402, \\
&\beta_{5,12} = 0.0154, \quad &&\beta_{6,12} = 0.0871, \quad &&\beta_{71} = 0.0531, \quad &&\beta_{76} = 0.0832, \\
&\beta_{8,13} = 0.0453, \quad &&\beta_{92} = 0.0743, \quad &&\beta_{93} = 0.0424, \quad &&\beta_{10,14} = 0.0290, \\
&\beta_{11,4} = 0.0936, \quad &&\beta_{11,5} = 0.0458, \quad &&\beta_{12,15} = 0.0741, \quad &&\beta_{13,14} = 0.0463, \\
&\beta_{13,7} = 0.0212, \quad &&\beta_{14,15} = 0.0164, \quad &&\beta_{14,9} = 0.0666, \quad &&\beta_{15,11} = 0.0154, \\
&\beta_{15,13} = 0.0953.
\end{aligned}
$$

The uncertainties and exterior disturbances terms are described as

$$
\Delta h_1(X, t) + \delta_1(t) = -0.25 \sin(\pi x_1(t)) - 0.12 \sin(t),
$$

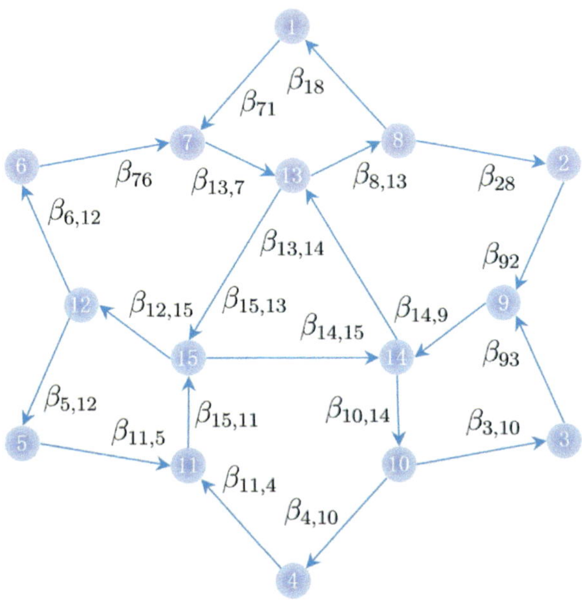

FIGURE 4.1 Digraph $(\mathcal{G}, \boldsymbol{B})$ with 15 nodes.

$$\Delta h_2(X, t) + \delta_2(t) = 0.25 \cos(2\pi x_2(t)) + 0.12 \sin(3t),$$
$$\Delta h_3(X, t) + \delta_3(t) = 0.1 \sin(3\pi x_3(t)) - 0.1 \cos(2t),$$
$$\Delta h_4(X, t) + \delta_4(t) = -0.1 \cos(2\pi x_4(t)) + 0.1 \cos(t),$$
$$\Delta h_5(X, t) + \delta_5(t) = 0.28 \sin(2\pi x_5(t)) + 0.02 \sin(2t),$$
$$\Delta h_6(X, t) + \delta_6(t) = -0.03 \cos(\pi x_6(t)) - 0.02 \sin(2t),$$
$$\Delta h_7(X, t) + \delta_7(t) = 0.13 \sin(\pi x_7(t)) - 0.1 \cos(2t),$$
$$\Delta h_8(X, t) + \delta_8(t) = 0.15 \cos(3\pi x_8(t)) + 0.15 \cos(t),$$
$$\Delta h_9(X, t) + \delta_9(t) = -0.21 \sin(3\pi x_9(t)) + 0.06 \sin(t),$$
$$\Delta h_{10}(X, t) + \delta_{10}(t) = 0.08 \cos(2\pi x_{10}(t)) + 0.1 \sin(3t),$$
$$\Delta h_{11}(X, t) + \delta_{11}(t) = -0.3 \sin(\pi x_{11}(t)) - 0.24 \cos(2t),$$
$$\Delta h_{12}(X, t) + \delta_{12}(t) = -0.15 \cos(\pi x_{12}(t)) + 0.13 \cos(2t),$$
$$\Delta h_{13}(X, t) + \delta_{13}(t) = 0.12 \sin(\pi x_{13}(t)) - 0.08 \sin(2t),$$
$$\Delta h_{14}(X, t) + \delta_{14}(t) = -0.14 \cos(\pi x_{14}(t)) + 0.1 \sin(t),$$
$$\Delta h_{15}(X, t) + \delta_{15}(t) = 0.32 \sin(2\pi x_{15}(t)) + 0.14 \cos(t).$$

Obviously, digraph $(\mathcal{G}, \boldsymbol{B})$ is a strongly connected network. Suppose that $y(t)$ is a solution of an isolated node which satisfies

$$\substack{C\\0} D_t^{\alpha(t)} y(t) = f(y(t)). \tag{4.26}$$

Fig. 4.2 shows the state trajectories of the system (4.25) and (4.26) with and without the proposed control strategy, respectively. From Fig. 4.2, we can find that $|y_1(t)| \leq 33.61$, $|y_2(t)| \leq 28.41$, and $|y_3(t)| \leq 50.83$. Thus, based on Remark 4.1, we can get the Lipschitz constant as $\eta = 114.2317$. In addition, choosing $\eta_{ij} = 1$, by a simple calculation, we can derive the values of parameters k_i, λ_{1i}, and λ_{2i} which are listed in Table 4.1.

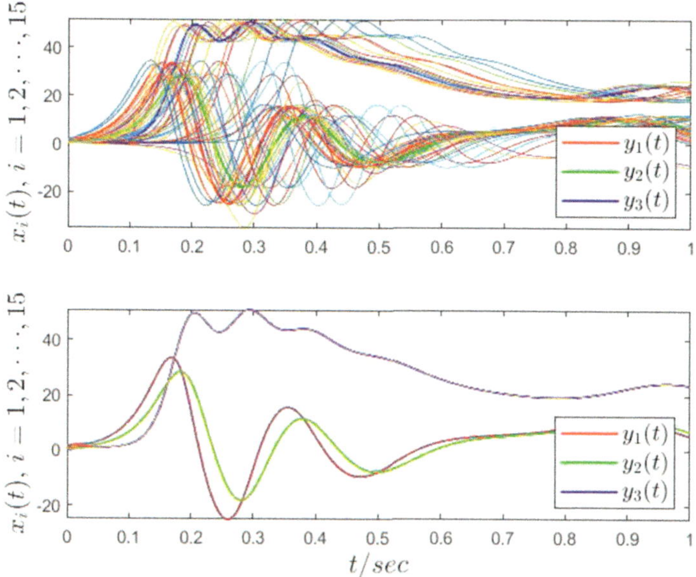

FIGURE 4.2   Response curves of VOF-Chen system (4.25).

TABLE 4.1   Parameters of VOF coupled systems.

node	k_i	λ_{1i}	λ_{2i}	node	k_i	λ_{1i}	λ_{2i}
$i=1$	115.030	39.045	21.940	$i=2$	114.871	69.155	11.930
$i=3$	114.654	5.180	47.887	$i=4$	115.255	28.153	43.447
$i=5$	114.629	11.280	53.068	$i=6$	114.670	47.988	20.590
$i=7$	114.966	23.472	20.946	$i=8$	114.700	25.172	39.082
$i=9$	114.778	8.741	1.710	$i=10$	114.578	45.758	66.986
$i=11$	114.612	53.473	53.153	$i=12$	115.049	7.414	47.709
$i=13$	114.398	57.650	12.251	$i=14$	115.209	36.159	49.189
$i=15$	114.883	47.581	2.559				

Fig. 4.3 shows the sliding surface function and controller, respectively. Fig. 4.4 depicts the synchronization errors between system (4.25) and (4.26) and the adaptive estimates, respectively.

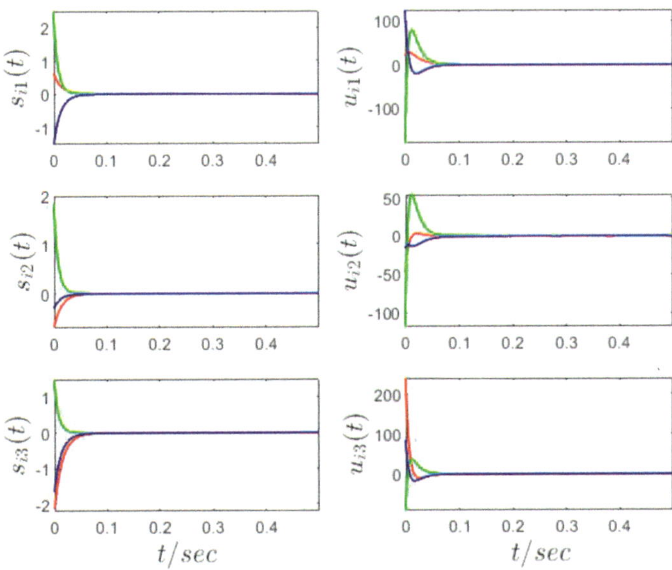

FIGURE 4.3   Response curves of the sliding surface (4.6) and controller (4.17).

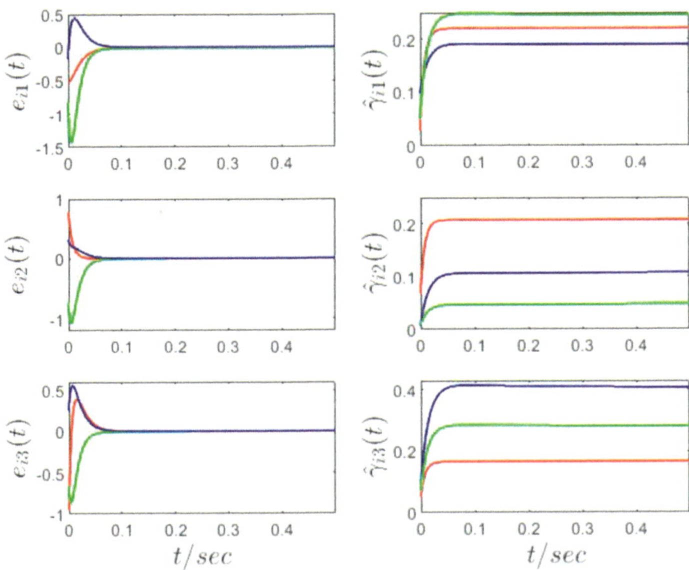

FIGURE 4.4   Response curves of synchronization errors and adaptive estimates.

Example 4.2. Consider the following hyperchaotic VOF system as the master system:

$$\begin{cases} {}^{\mathcal{C}}_{0}D_t^{\alpha(t)}x_1(t) = x_2(t) - x_1(t) + 1.5x_4(t), \\ {}^{\mathcal{C}}_{0}D_t^{\alpha(t)}x_2(t) = x_1(t)(26 - x_3(t)) - x_2(t), \\ {}^{\mathcal{C}}_{0}D_t^{\alpha(t)}x_3(t) = x_1(t)x_2(t) - 0.7x_3(t), \\ {}^{\mathcal{C}}_{0}D_t^{\alpha(t)}x_4(t) = -x_1(t) - x_4(t), \end{cases} \tag{4.27}$$

with $\alpha(t) = 0.97 + 0.02\sin(t)$, and the slave system

$$\begin{cases} {}^{\mathcal{C}}_{0}D_t^{\alpha(t)}y_1(t) = y_2(t) - y_1(t) + 1.5y_4(t) + u_1(t) + \delta_1(t), \\ {}^{\mathcal{C}}_{0}D_t^{\alpha(t)}y_2(t) = y_1(t)(26 - y_3(t)) - y_2(t) + u_2(t) + \delta_2(t), \\ {}^{\mathcal{C}}_{0}D_t^{\alpha(t)}y_3(t) = y_1(t)y_2(t) - 0.7y_3(t) + u_3(t) + \delta_3(t), \\ {}^{\mathcal{C}}_{0}D_t^{\alpha(t)}y_4(t) = -y_1(t) - y_4(t) + u_4(t) + \delta_4(t), \end{cases} \tag{4.28}$$

where $\delta_1(t) = \sin^2(3t) + 0.5\cos^2(2t)$, $\delta_2(t) = 2\cos(0.1t^2)$, $\delta_3(t) = 1 + \sin(5t)$, and $\delta_4(t) = 0.5 - \cos(5t)$. The initial conditions are $x(0) = [5, 5, 5, 5]^{\mathrm{T}}$ and $y(0) = [0, 0, 0, 0]^{\mathrm{T}}$, respectively.

Meanwhile, the parameters of the controller in this chapter and those in [28] are listed in Table 4.2.

TABLE 4.2 Parameters in [28] and Chapter 4.

δ	κ_r	κ_l	ξ_r	ξ_l	k	λ_1	λ_2
10	1	1	1	1	59.45	28.12	12.54

Fig. 4.5 shows the response curves of systems (4.27) and (4.28) by utilizing the method in this chapter and that in [28], respectively. The red line denotes state trajectories $x(t)$ of system (4.27), while the blue and green lines depict the response curves of system (4.28) by means of the method in this chapter and that in [28], respectively. Meanwhile, the corresponding synchronization error trajectories are shown in Fig. 4.6. From Figs. 4.5 and 4.6, we see that a better synchronization performance in this chapter can be achieved compared to the method in [28].

4.5 Conclusion

In this chapter, the synchronization problem for a class of VOF uncertain coupled systems on a network with strong connectedness has been discussed via the adaptive SMC approach. First, based on the theory of VOF calculus, a novel VOF integral-type sliding surface has been designed. Second, a VOF controller has been synthesized based on proposed sliding surface to ensure the asymptotical stability of the closed-loop error

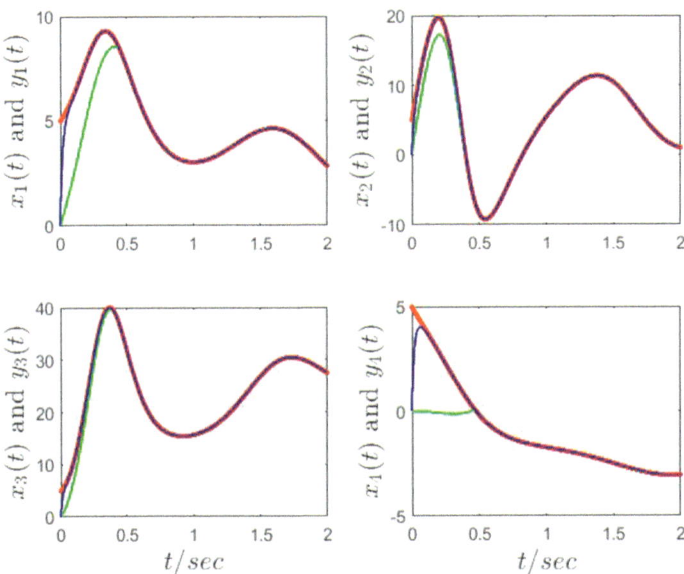

FIGURE 4.5   State trajectories of systems (4.27) and (4.28).

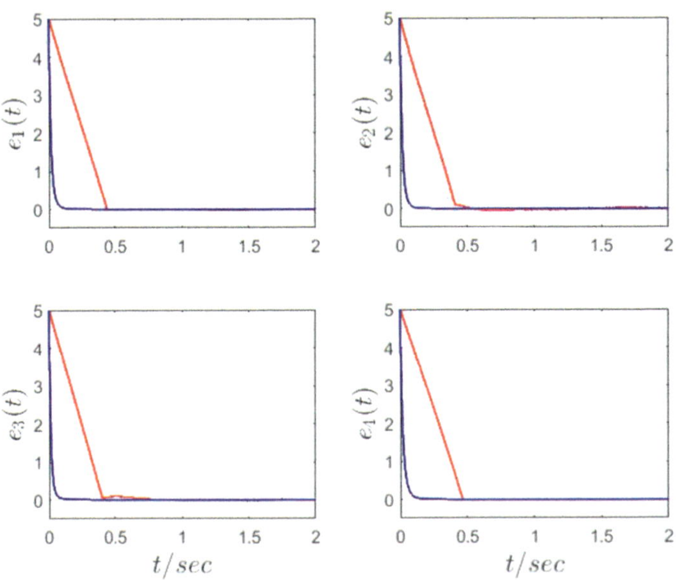

FIGURE 4.6   State trajectories of systems (4.27) and (4.28) with controllers (4.21).

systems in finite-time. Furthermore, an adaptive control strategy has been proposed to realize the synchronization for the VOF coupled systems with uncertain terms. Finally, two numerical examples have been presented to verify the effectiveness of the proposed method. In the future, we will extend the current result to VOF coupled systems on a network without strong connectedness.

References

[1] S.G. Samko, B. Ross, Integration and differentiation to a variable fractional order, Integral Transforms and Special Functions 1 (4) (1993) 277–300.

[2] J. Cheng, J.H. Park, X. Zhao, H.R. Karimi, J. Cao, Quantized nonstationary filtering of networked Markov switching RSNSs: a multiple hierarchical structure strategy, IEEE Transactions on Automatic Control 65 (11) (2019) 4816–4823.

[3] J. Cheng, W. Huang, H.-K. Lam, J. Cao, Y. Zhang, Fuzzy-model-based control for singularly perturbed systems with nonhomogeneous Markov switching: a dropout compensation strategy, IEEE Transactions on Fuzzy Systems 30 (2) (2020) 530–541.

[4] J. Jiang, D. Cao, H. Chen, K. Zhao, The vibration transmissibility of a single degree of freedom oscillator with nonlinear fractional order damping, International Journal of Systems Science 48 (11) (2017) 2379–2393.

[5] S.G. Samko, Fractional integration and differentiation of variable order, Analysis Mathematica 21 (3) (1995) 213–236.

[6] C.F. Lorenzo, T.T. Hartley, Variable order and distributed order fractional operators, Nonlinear Dynamics 29 (2002) 57–98.

[7] G. Diaz, C.F.M. Coimbra, Nonlinear dynamics and control of a variable order oscillator with application to the van der Pol equation, Nonlinear Dynamics 56 (2009) 145–157.

[8] J. Jiang, H. Chen, J.L.G. Guirao, D. Cao, Existence of the solution and stability for a class of variable fractional order differential systems, Chaos, Solitons and Fractals 128 (2019) 269–274.

[9] J. Jiang, D. Cao, H. Chen, Sliding mode control for a class of variable-order fractional chaotic systems, Journal of the Franklin Institute 357 (15) (2020) 10127–10158.

[10] X. Meng, Z. Wu, C. Gao, B. Jiang, H.R. Karimi, Finite-time projective synchronization control of variable-order fractional chaotic systems via sliding mode approach, IEEE Transactions on Circuits and Systems. II, Express Briefs 68 (7) (2021) 2503–2507.

[11] Y. Xu, Y. Li, W. Li, Graph-theoretic approach to synchronization of fractional-order coupled systems with time-varying delays via periodically intermittent control, Chaos, Solitons and Fractals 121 (2019) 108–118.

[12] J. Chen, C. Li, X. Yang, Global Mittag-Leffler projective synchronization of nonidentical fractional-order neural networks with delay via sliding mode control, Neurocomputing 313 (2018) 324–332.

[13] K. Khandani, V.J. Majd, M. Tahmasebi, Robust stabilization of uncertain time-delay systems with fractional stochastic noise using the novel fractional stochastic sliding approach and its application to stream water quality regulation, IEEE Transactions on Automatic Control 62 (4) (2016) 1742–1751.

[14] Z. Cao, Y. Niu, H.R. Karimi, Dynamic output feedback sliding mode control for Markovian jump systems under stochastic communication protocol and its application, International Journal of Robust and Nonlinear Control 30 (17) (2020) 7307–7325.

[15] A.J. Healey, D. Lienard, Multivariable sliding mode control for autonomous diving and steering of unmanned underwater vehicles, IEEE Journal of Oceanic Engineering 18 (3) (1993) 327–339.

[16] H. Ren, H.R. Karimi, R. Lu, Y. Wu, Synchronization of network systems via aperiodic sampled-data control with constant delay and application to unmanned ground vehicles, IEEE Transactions on Industrial Electronics 67 (6) (2019) 4980–4990.

[17] H.R. Karimi, A sliding mode approach to H_∞ synchronization of master-slave time-delay systems with Markovian jumping parameters and nonlinear uncertainties, Journal of the Franklin Institute 349 (4) (2012) 1480–1496.

[18] X. Meng, C. Gao, Z. Liu, B. Jiang, Robust H_∞ control for a class of uncertain neutral-type systems with time-varying delays, Asian Journal of Control 23 (3) (2021) 1454–1465.

[19] B. Jiang, H.R. Karimi, S. Yang, C. Gao, Y. Kao, Observer-based adaptive sliding mode control for nonlinear stochastic Markov jump systems via T–S fuzzy modeling: applications to robot arm model, IEEE Transactions on Industrial Electronics 68 (1) (2020) 466–477.

[20] Z. Liu, H.R. Karimi, J. Yu, Passivity-based robust sliding mode synthesis for uncertain delayed stochastic systems via state observer, Automatica 111 (2020) 108596.

[21] D. Deepika, S. Kaur, S. Narayan, Uncertainty and disturbance estimator based robust synchronization for a class of uncertain fractional chaotic system via fractional order sliding mode control, Chaos, Solitons and Fractals 115 (2018) 196–203.

[22] M.P. Aghababa, Robust stabilization and synchronization of a class of fractional-order chaotic systems via a novel fractional sliding mode controller, Communications in Nonlinear Science and Numerical Simulation 17 (6) (2012) 2670–2681.

[23] T.-C. Lin, T.-Y. Lee, Chaos synchronization of uncertain fractional-order chaotic systems with time delay based on adaptive fuzzy sliding mode control, IEEE Transactions on Fuzzy Systems 19 (4) (2011) 623–635.

[24] W. Zhou, Y. Wang, C.K. Ahn, J. Cheng, C. Chen, Adaptive fuzzy backstepping-based formation control of unmanned surface vehicles with unknown model nonlinearity and actuator saturation, IEEE Transactions on Vehicular Technology 69 (12) (2020) 14749–14764.

[25] Y. Xu, Q. Li, W. Li, Periodically intermittent discrete observation control for synchronization of fractional-order coupled systems, Communications in Nonlinear Science and Numerical Simulation 74 (2019) 219–235.

[26] Y. Wang, H.R. Karimi, H. Yan, et al., An adaptive event-triggered synchronization approach for chaotic Lur'e systems subject to aperiodic sampled data, IEEE Transactions on Circuits and Systems. II, Express Briefs 66 (3) (2019) 442–446.

[27] A. Modiri, S. Mobayen, Adaptive terminal sliding mode control scheme for synchronization of fractional-order uncertain chaotic systems, ISA Transactions 105 (2020) 33–50.

[28] J.-F. Li, H. Jahanshahi, S. Kacar, Y.-M. Chu, J.F. Gómez-Aguilar, N.D. Alotaibi, K.H. Alharbi, On the variable-order fractional memristor oscillator: data security applications and synchronization using a type-2 fuzzy disturbance observer-based robust control, Chaos, Solitons and Fractals 145 (2021) 110681.

Sliding mode projective synchronization for fractional-order coupled systems based on network without strong connectedness

5.1 Introduction

Since being proposed by Emelyanov in the 1960s [1,2], SMC has been used in neutral-type systems [3], singular systems [4], Markovian jump systems [5], uncertain stochastic systems [6], and other practical engineering systems [7] due to its fast response, good transient performance, and other features. For example, Liu et al. investigated the problem of fault estimation sliding-mode observer design for linear systems over digital communication channels in [8]; Han et al. studied the robust SMC issue for a class of discrete singular systems with time-varying delays, parameter uncertainties in [9].

On the other hand, fractional-order coupled systems (FOCS) have gotten more attention in the last few decades due to large applications in various fields. Sun et al. systematically discussed the development and application of the fractional calculus theory in physics, dynamical systems, and other branches of engineering and science in [10]. For example, in the research of biomedical and underwater sediment, the model with a fractional derivative could better describe the power law attenuation phenomenon than other integral-order models. As we all know, Lyapunov method has always been an important and useful tool to analyze the stability of dynamic systems. Meanwhile, a new method was introduced by Li and Shuai [11] to investigate the stability of integer-order coupled systems that are built on a strongly connected network, namely by designing a novel vertex Lyapunov function combined with the graph theory. Based on their work, some scholars have begun to investigate other dynamical behaviors of FOCS, for instance, the global Mittag-Leffler stability (MLS) problem for a class of FOCS was studied in [12]. Meanwhile, the MLS problem of FOCS with impulses on networks via the graph theory was also investigated in [13].

In 1996, Pecora and Carroll proposed the chaotic synchronization problem for the first time [14], and related results have been reported in many fields. It is worth pointing out that there are different kinds of synchronization problems, such as the issues of complete synchronization and α-synchronization, the projective synchronization problem [15,16]. Meng et al. considered the projective synchronization issue for a class of fractional variable-order couple system via SMC in [15], through which the fractional master–slave system achieved projective synchronization, both complete synchronization and antisynchronization. However, only a few scholars considered the control issues of coupled systems on a network with strong connectedness [17,8,18], such as through impulsive control, periodically intermittent control, feedback control, etc. So far, for the FOCS on a network without strong connectedness, the problem of synchronization still needs to be studied. On the other hand, most scholars chose the Caputo fractional derivative operator or the Riemann–Liouville derivative operator to study the stability of fractional-order systems, while Munoz et al. presented an extended Caputo fractional derivative definition in [17]. In their work, for a Lipschitz continuous function $f(x)$, the authors obtained that the Caputo operator was the same as the extended Caputo operator. From the work in [11], it can be seen that the solution of the coupled system will achieve synchronization at the single point when the network is strongly connected. Nevertheless, networks are not all strongly connected. Therefore, Du and Li [19] introduced a partial order relationship to prove that all solutions of a coupled system on a network which is not strongly connected will converge to a unique point. Meanwhile, they proved that the solutions on each strongly connected component could converge to a unique equilibrium. Based on the above work, Wang et al. [20] proposed a novel hierarchical approach by using the theory of asymptotically autonomous systems to study the stability of coupled systems on a network without strong connectivity.

Hence, motivated by the above analysis, we aim to investigate the Mittag-Leffler projective synchronization issue of the considered systems by mean of SMC. Some novel synchronization criteria are derived by utilizing a vertex Lyapunov function combined with graph theory. The main contributions of our work can be established as follows:

(1) Based on the graph theory, a new fractional SMC scheme is applied to achieve projective synchronization for the FOCS on a network without strong connectedness;

(2) Projection coefficient, nonlinear coupling, and external disturbance are also considered in our work, which is more comprehensive in theoretical analysis;

(3) By utilizing the hierarchical method, projective synchronization conditions are obtained for the considered coupled systems.

5.2 Problem statement

In this section, we consider the following FOCS on a weighted network $(\mathcal{G}, A)$ as the master system:

$$\begin{cases} {}^{C}_{0}D^{\alpha}_{t} x_k(t) = -\alpha_k x_k + f_k(x_k, x_k(t-\tau)) + \sum_{h=1}^{n} b_{kh} \tilde{f}_{kh}(x_k, x_h) + l_k, \\ x_k(s) = \phi_k(s), \ s \in [-\tau, 0], \ k = 1, 2, \ldots, n, \end{cases} \tag{5.1}$$

where ${}^{C}_{0}D^{\alpha}_{t}$ denotes the Caputo derivative operator, $\alpha \in (0, 1)$, and $x_k(t) \in \mathbb{R}$ is the pseudo-state vector, $l_k \in \mathbb{R}$ is the disturbance vector, $\phi_k(t) \in \mathbb{C}([-\tau, 0], \mathbb{R})$ is the initial function of the FOCS (5.1), and α_k, b_{kh} are constants.

Meanwhile, the corresponding slave system of system (5.1) is

$$\begin{cases} {}^{C}_{0}D^{\alpha}_{t} y_k(t) = -\beta_k y_k + g_k(y_k, y_k(t-\tau)) + \sum_{h=1}^{n} s_{kh} \tilde{g}_{kh}(y_k, y_h) + w_k + u_k, \\ y_k(s) = \psi_k(s), \ s \in [-\tau, 0], \ k = 1, 2, \ldots, n, \end{cases}$$
$$\tag{5.2}$$

where $y_k(t) \in \mathbb{R}$, $u_k(t) \in \mathbb{R}$ are the pseudo-state and control vectors, respectively, and β_k, s_{kh} are constants.

Now, some necessary definitions are presented as follows:

Definition 5.1 ([11]). A digraph $\mathcal{G}$ is strongly connected if, for any pair of distinct vertices, there exists a directed path from one to the other.

Definition 5.2 ([21]). The MSS (5.1) and (5.2) achieve projective synchronization if there exists a constant $\rho \in \mathbb{R} \setminus \{0\}$ such that

$$\lim_{t \to +\infty} \|y_k(t) - \rho x_k(t)\| = 0 \tag{5.3}$$

holds, where ρ is the projection coefficient.

From the MSS (5.1) and (5.2), let $e_k(t) = y_k(t) - \rho x_k(t)$ denote the projective error (PE), i.e., $E(t) = [e_1(t), e_2(t), \ldots, e_n(t)]^{\mathrm{T}}$. Then, the projective error system can be obtained as

$$\,_0^C D_t^\alpha e_k(t) = -\beta_k e_k + \Phi_k(e_k, e_k(t - \tau)) + \sum_{h=1}^{n} s_{kh}\tilde{\Phi}(e_k, e_h) + (w_k - \rho l_k) + u_k$$

$$+ \rho(\alpha_k - \beta_k)x_k + g_k(\rho x_k, \rho x_k(t - \tau)) - \rho f_k(x_k, x_k(t - \tau))$$

$$+ \sum_{h=1}^{n} [s_{kh}\tilde{g}_{kh}(\rho x_k, \rho x_h) - \rho b_{kh}\tilde{f}_{kh}(x_k, x_h)],$$

$$(5.4)$$

where $\Phi_k(e_k, e_k(t - \tau)) \triangleq g_k(y_k, y_k(t - \tau)) - g_k(\rho x_k, \rho x_k(t - \tau))$ and $\tilde{\Phi}(e_k, e_h) \triangleq \tilde{g}_{kh}(y_k, y_h) - \tilde{g}_{kh}(\rho x_k, \rho x_h)$. By means of Lemmas 2.1 and 2.4, one has

$$e_k(t) = e_k(0) + \,_0^C D_t^{-\alpha}\Big\{ -\beta_k e_k + \Phi_k(e_k, e_k(t - \tau)) + \sum_{h=1}^{n} s_{kh}\tilde{\Phi}(e_k, e_h)$$

$$+ (w_k - \rho l_k) + u_k + \rho(\alpha_k - \beta_k)x_k + g_k(\rho x_k, \rho x_k(t - \tau))$$

$$- \rho f_k(x_k, x_k(t - \tau)) + \sum_{h=1}^{n} [s_{kh}\tilde{g}_{kh}(\rho x_k, \rho x_h) - \rho b_{kh}\tilde{f}_{kh}(x_k, x_h)]\Big\},$$

$$(5.5)$$

Definition 5.3 ([12]). For any solution $e_k(t)$ of (5.5) with the initial value, denote by e^* the equilibrium point of (5.5), and assume that there exist constants η, β, and $\gamma > 0$ such that

$$\|e(t) - e^*\| \leq \{\eta\|e(t_0)\|^\beta E_\alpha[-\gamma(t - t_0)^\alpha]\}^{\frac{1}{\beta}}, \quad t \geq t_0,$$

holds, then we say that the equilibrium point of (5.5) is globally MLS.

Assumption 5.1. There exist nonnegative constants L_{1k}, L_{2k}, L_{3k}, L_{4k} such that, for any $u, v \in \mathbb{R}$ and $k, h \in \mathbb{L}$, the following inequalities always hold:

(1) $|u\varphi_k(u, v)| \leq L_{1k}u^2 + L_{2k}v^2$;
(2) $|u\tilde{\Phi}_{kh}(u, v)| \leq L_{3k}u^2 + L_{4k}v^2.$

5.3 Main results

5.3.1 Sliding surface design

In this section, the sliding mode surface function is designed as follows:

$$\varphi_k(t) = e_k + \,_0 D_t^{-\alpha}\Big[\pi_k e_k - \Phi_k(e_k, e_k(t - \tau)) - \sum_{h=1}^{n} s_{kh}\tilde{\Phi}_{kh}(e_k, e_h)\Big],$$

$$(5.6)$$

$$k = 1, 2, \ldots, n,$$

where $\pi_k \in \mathbb{R}$ is a gain coefficient. Then, based on (5.5) and (5.6), we have

$$
\begin{aligned}
{}_0^C D_t^\alpha \varphi_k(t) &= {}_0^C D_t^\alpha e_k + \left[\pi_k e_k - \Phi_k(e_k, e_k(t-\tau)) - \sum_{h=1}^n s_{kh} \tilde{\Phi}_{kh}(e_k, e_h) \right] \\
&= (\pi_k - \beta_k) e_k + (w_k - \rho l_k) + u_k + \rho(\alpha_k - \beta_k) x_k \\
&\quad + g_k(\rho x_k, \rho x_k(t-\tau)) - \rho f_k(x_k, x_k(t-\tau)) \\
&\quad + \sum_{h=1}^n \left[s_{kh} \tilde{g}_{kh}(\rho x_k, \rho x_h) - \rho b_{kh} \tilde{f}_{kh}(x_k, x_h) \right].
\end{aligned}
$$

Since ${}_0^C D_t^\alpha \varphi_k(t) = 0$, one obtains

$$
\begin{aligned}
\bar{u}_k = {}&-(\pi_k - \beta_k) e_k - (w_k - \rho l_k) - \rho(\alpha_k - \beta_k) x_k - g_k(\rho x_k, \rho x_k(t-\tau)) \\
&+ \rho f_k(x_k, x_k(t-\tau)) - \sum_{h=1}^n \left[s_{kh} \tilde{g}_{kh}(\rho x_k, \rho x_h) - \rho b_{kh} \tilde{f}_{kh}(x_k, x_h) \right].
\end{aligned} \tag{5.7}
$$

Therefore, combining the FOCS (5.4) with the controller (5.7), the error system can be described as

$$
{}_0^C D_t^\alpha e_k(t) = -\pi_k e_k + \Phi_k(e_k, e_k(t-\tau)) + \sum_{h=1}^n s_{kh} \tilde{\Phi}_{kh}(e_k, e_h). \tag{5.8}
$$

Remark 5.1. On the jth layer, let $\mathcal{R}_{ji}$ denote the ith strongly connected components (SCC), and assume $V(\mathcal{R}_{ji})$ denotes the vertex set of the SCC $\mathcal{R}_{ji}$ while N_{ji} denotes the number of vertices of the SCC $V(\mathcal{R}_{ji})$. Obviously, $\sum_{j,i} N_{ji} = n$ holds. Then the sliding dynamic (5.8) can be written as follows:

(1) On the first layer, we have the system (5.8) as

$$
{}_0^C D_t^\alpha e_k(t) = -\pi_k e_k + \Phi_k(e_k, e_k(t-\tau)) + \sum_{h \in V(\mathcal{R}_{1i})} s_{kh} \tilde{\Phi}_{kh}(e_k, e_h), \tag{5.9}
$$
$$
k \in V(\mathcal{R}_{1i});
$$

(2) On the jth layer, we have the error system (5.8) as

$$
\begin{aligned}
{}_0^C D_t^\alpha e_k(t) = {}&-\pi_k e_k + \Phi_k(e_k, e_k(t-\tau)) + \sum_{h \in V(\mathcal{R}_{ji})} s_{kh} \tilde{\Phi}_{kh}(e_k, e_h) \\
&+ \sum_{h \in \hat{V}} s_{kh} \tilde{\Phi}_{kh}(e_k, e_h), \quad k \in V(\mathcal{R}_{ji}),
\end{aligned} \tag{5.10}
$$

where $\hat{V} = \bigcup_{1 \le m < j} \bigcup_n V(\mathcal{R}_{mn})$.

Remark 5.2. In [22], Liu et al. considered the stochastic stabilization problem of the complex network without strong connectedness via the hierarchical method. Therefore, combining Assumption 5.2 and the theory of asymptotic autonomous systems, the system (5.8) is equivalent to (5.10) on the jth layer, $j \geq 2$.

Assumption 5.2. There exists $e_h(t) = 0$ such that $\tilde{\Phi}_{kh}(e_k, e_h) = 0$ holds for all $k \in V(\mathcal{R}_{ji})$.

Assumption 5.3. There exist p_{ji}, $q_{ji} > 0$ and $\eta > 1$ such that $p_{ji}e_k^2(t) \leq V_{ji}(t) \leq q_{ji}e_k^2(t)$ and $\sup\limits_{t-\tau \leq s \leq t} V_{1i}(s) < \eta V_{1i}(t)$ hold.

5.3.2 Mittag-Leffler stability analysis

In this section, it will be shown that the solution of the sliding modes (5.8) is the adequacy criterion for Mittag-Leffler stability (MLS).

Theorem 5.1. *The equilibrium point of the error system (5.8) is globally MLS if*

(1) *$(\mathcal{G}, A)$ is not a strongly connected network;*
(2) *$(\mathcal{R}_{ji}, H_{ji})$ is an SCC, where $H_{ji} = (\theta_{kh})_{k,h \in V(\mathcal{R}_{ji})}$ is such that $\theta_{kk} = 0$, $\theta_{kh} = \theta_{hk}$ if $k \neq h$;*
(3) *$\theta_{kh} = |s_{kh}|L_{4k}$ holds for all $k, h \in V(\mathcal{R}_{ji})$;*
(4) *$-\lambda_1^{ji} + \lambda_2^{ji}\eta < 0$ holds, where*

$$\lambda_1^{ji} = \min_{k \in V(\mathcal{R}_{ji})} \left\{ 2\left[\pi_k - L_{1k} - \sum_{h \in V(\mathcal{R}_{ji})} |s_{kh}|(L_{3k} + L_{4k}) \right] \right\} > 0$$

and $\lambda_2^{ji} = \max\limits_{k \in V(\mathcal{R}_{ji})} \left\{ 2L_{2k} \right\} > 0$, with $\eta > 1$.

Proof. On the first layer, for the error system (5.8) that is restricted to $(\mathcal{R}_{1i}, H_{1i})$, we construct the following vertex Lyapunov function:

$$V_{1i}(t) = \sum_{k \in V(\mathcal{R}_{1i})} c_k^{1i} V_k(t) = \frac{1}{2} \sum_{k \in V(\mathcal{R}_{1i})} c_k^{1i} e_k^2(t), \tag{5.11}$$

where c_k^{1i} denotes the algebraic cofactor of the kth diagonal element of L_{1i}. Due to the strong connectedness of $(\mathcal{R}_{1i}, H_{1i})$, we get $c_k^{1i} > 0$ for every

$k \in V(\mathcal{R}_{1i})$. Then, by utilizing (5.10), we have

$$\begin{aligned}
{}_0^{\mathcal{C}} D_t^{\alpha} V_{1i}(t) &= \frac{1}{2} \sum_{k \in V(\mathcal{R}_{1i})} c_k^{1i} {}_0^{\mathcal{C}} D_t^{\alpha} e_k^2(t) \leq \sum_{k \in V(\mathcal{R}_{1i})} c_k^{1i} e_k(t) {}_0^{\mathcal{C}} D_t^{\alpha} e_k(t) \\
&= \sum_{k \in V(\mathcal{R}_{1i})} c_k^{1i} e_k \Bigg\{ -\pi_k e_k + \Phi_k(e_k, e_k(t-\tau)) \\
&\qquad\qquad + \sum_{h \in V(\mathcal{R}_{1i})}^{n} s_{kh} \tilde{\Phi}_{kh}(e_k, e_h) \Bigg\} \\
&\leq \sum_{k \in V(\mathcal{R}_{1i})} c_k^{1i} \Bigg\{ -\pi_k e_k^2 + L_{1k} e_k^2 + L_{2k} e_k^2(t-\tau) \\
&\qquad\qquad + \sum_{h \in V(\mathcal{R}_{1i})} |s_{kh}|(L_{3k} e_k^2 + L_{4k} e_k^2(t-\tau)) \Bigg\} \\
&= \sum_{k \in V(\mathcal{R}_{1i})} c_k^{1i} \Bigg\{ \Big[-\pi_k + L_{1k} + \sum_{h \in V(\mathcal{R}_{1i})} |s_{kh}|(L_{3k} + L_{4k}) \Big] e_k^2 \\
&\qquad\qquad + L_{2k} e_k^2(t-\tau) \Bigg\} + \sum_{k \in V(\mathcal{R}_{1i})} c_k^{1i} |s_{kh}| L_{4k}(e_h^2 - e_k^2) \\
&\leq -\lambda_1^{1i} V_{1i}(t) + \lambda_2^{1i} V_{1i}(t-\tau) + \sum_{k \in V(\mathcal{R}_{1i})} c_k^{1i} \theta_{kh}(P_h(e_h) - P_k(e_k)),
\end{aligned}$$

$$(5.12)$$

where $\lambda_1^{1i} = \min\limits_{k \in V(\mathcal{R}_{1i})} \Big\{ 2\Big[\pi_k - L_{1k} - \sum_{h \in V(\mathcal{R}_{1i})} |s_{kh}|(L_{3k} + L_{4k}) \Big] \Big\} > 0$, $\lambda_2^{1i} = \max\limits_{k \in V(\mathcal{R}_{1i})} \big\{ 2L_{2k} \big\} > 0$, and $P_k(e_k(t)) = e_k^2(t)$.

Here, let $\theta_{kh} = |s_{ku}| L_{4h}$. Because $(\mathcal{R}_{1i}, \boldsymbol{H}_{1i})$ is balanced and strongly connected, from Lemmas 2.1 and 2.4 and the lemmas in [11,23], we obtain

$$\sum_{k \in V(\mathcal{R}_{1i})} c_k^{1i} \theta_{kh}(P_h(e_h) - P_k(e_k))$$
$$= \sum_{Q \in \mathcal{Q}_{1i}} W(Q) \sum_{(s,r) \in E(\mathcal{C}_Q)} (P_r(e_r) - P_s(e_s)) = 0.$$

Then, it holds that

$$\begin{aligned}
{}_0^{\mathcal{C}} D_t^{\alpha} V_{1i}(t) &\leq -\lambda_1^{1i} V_{1i} + \lambda_2^{1i} V_{1i}(t-\tau) \\
&\leq -\lambda_1^{1i} V_{1i}(t) + \lambda_2^{1i} \sup_{t-\tau \leq s \leq t} V_{1i}(s).
\end{aligned} \qquad (5.13)$$

Meanwhile, based on the Razumikhin technique and Assumption 5.3, there exists a constant $\eta > 1$ such that

$$\sup_{t-\tau \le s \le t} V_{1i}(s) < \eta V_{1i}(t). \tag{5.14}$$

Combining (5.13) and (5.14), one obtains

$$_{0}^{C}D_{t}^{\alpha} V_{1i}(t) < (-\lambda_{1}^{1i} + \lambda_{2}^{1i}\eta)V_{1i}(t). \tag{5.15}$$

Then, based on Lemma 2.8 and (5.15), we easily get

$$V_{1i}(t) < V_{1i}(0)E_{\alpha}((-\lambda_{1}^{1i} + \lambda_{2}^{1i}\eta)t^{\alpha}). \tag{5.16}$$

Denote $\boldsymbol{E}_{1i}(t) = [\ldots, e_k(t), \ldots]^{\mathrm{T}}$, $\xi_1^{1i} \triangleq \min_{k}\{c_k^{1i}\}$, $\xi_2^{1i} \triangleq \max_{k}\{c_k^{1i}\}$, where $k \in V(\mathcal{R}_{1i})$. Since

$$\sum_{k \in V(\mathcal{R}_{1i})} e_k^2(t) < \|\boldsymbol{E}_{1i}(t)\|^2 < 2N_{1i} \sum_{k \in V(\mathcal{R}_{1i})} e_k^2(t), \tag{5.17}$$

from (5.15)–(5.17), one obtains

$$\|\boldsymbol{E}_{1i}(t)\|^2 < \frac{2N_{1i}\xi_2^{1i}}{\xi_1^{1i}} \|\boldsymbol{E}_{1i}(0)\|^2 E_{\alpha}((-\lambda_{1}^{1i} + \lambda_{2}^{1i}\eta)t^{\alpha}), \tag{5.18}$$

and then

$$\|\boldsymbol{E}_{1i}(t)\| < \left\{ M_{1i}\|\boldsymbol{E}_{1i}(0)\|^2 E_{\alpha}((-\lambda_{1}^{1i} + \lambda_{2}^{1i}\eta)t^{\alpha}) \right\}^{\frac{1}{2}}, \tag{5.19}$$

where $M_{1i} = \frac{2N_{1i}\xi_2^{1i}}{\xi_1^{1i}}$. Hence, the equilibrium point of the error system (5.8) is MLS on the first layer.

On the second layer, for the error system (5.8) restricted on the ith SCC $(\mathcal{R}_{2i}, \boldsymbol{H}_{2i})$, through the theory of asymptotically autonomous systems, Assumption 5.2 and Eqs. (5.9), (5.10), we have

$$_{0}^{C}D_{t}^{\alpha} e_k(t) = -\pi_k e_k + \Phi_k(e_k, e_k(t-\tau)) + \sum_{h \in V(\mathcal{R}_{2i})} s_{kh}\tilde{\Phi}_{kh}(e_k, e_h), \ k \in V(\mathcal{R}_{2i}). \tag{5.20}$$

Now we construct a Lyapunov function as

$$V_{2i}(t) = \sum_{k \in V(\mathcal{R}_{2i})} c_k^{2i} V_k(t) = \frac{1}{2} \sum_{k \in V(\mathcal{R}_{2i})} c_k^{2i} e_k^2(t). \tag{5.21}$$

Similar to the analysis of the first step, we obtain

$$\|E_{2i}(t)\| < \left\{ M_{2i}\|E_{2i}(0)\|^2 E_\alpha((-\lambda_1^{2i} + \lambda_2^{2i}\eta)t^\alpha) \right\}^{\frac{1}{2}}, \tag{5.22}$$

where $M_{2i} = \frac{2N_{2i}\xi_2^{2i}}{\xi_1^{2i}}$, $\lambda_1^{2i} = \min_k \left\{ 2\left[\pi_k - L_{1k} - \sum_{h \in V(\mathcal{R}_{2i})} |s_{kh}|(L_{3k} + L_{4k}) \right] \right\}$, $\lambda_2^{2i} = \max_k \left\{ 2L_{2k} \right\}$, $\xi_1^{2i} \triangleq \min_k \{c_k^{2i}\}$, $\xi_2^{2i} \triangleq \max_k \{c_k^{2i}\}$, with $k \in V(\mathcal{R}_{2i})$. So, the equilibrium point of system (5.8) on the digraph $(\mathcal{R}_{ji}, H_{ji})$ is MLS on the second layer.

Repeating all the above process until any SCC $(\mathcal{R}_{ji}, H_{ji})$ can be achieved, we know that the solution of the error system (5.8) is MLS in any SCC.

Finally, consider the global MLS issue of the error system (5.8). Following the analysis of the first and second steps, let $\lambda \triangleq \min_{j,i}\{\lambda_1^{ji} - \lambda_2^{ji}\eta\}$. Then one can easily obtain that

$$^C_0 D_t^\alpha V_{ji}(t) < (-\lambda_1^{ji} + \lambda_2^{ji}\eta)V_{ji}(t) \le -\lambda V_{ji}(t). \tag{5.23}$$

Meanwhile, we construct the following vertex Lyapunov function:

$$V(t) = \sum_{j,i} V_{ji}(t) = \sum_{j,i} \sum_{k \in V(\mathcal{R}_{ji})} \frac{1}{2} c_k^{ji} e_k^2(t). \tag{5.24}$$

Combining (5.23) with (5.24), we can easily get that

$$^C_0 D_t^\alpha V(t) \le -\lambda V(t). \tag{5.25}$$

Denote $E(t) = [e_1(t), e_2(t), \ldots, e_n(t)]^T$, $\xi_1 = \min_{j,i}\{\xi_1^{ji}\}$, and $\xi_2 = \max_{j,i}\{\xi_2^{ji}\}$. Since

$$\sum_{k=1}^n e_k^2(t) < \|E(t)\|^2 < 2n \sum_{k=1}^n e_k^2(t), \tag{5.26}$$

one has

$$\|E(t)\| < \left\{ M\|E(0)\|^2 E_\alpha(-\lambda t^\alpha) \right\}^{\frac{1}{2}}, \tag{5.27}$$

where $M = \frac{2n\xi_2}{\xi_1}$. Therefore, the equilibrium point of the error system (5.8) on the digraph $(\mathcal{R}, H)$ is globally MLS. $\qquad\square$

Remark 5.3. In [16], Chen et al. presented an important criterion that $\sup\limits_{t-\tau \leq s \leq t} V(s, x(s)) \leq \eta V(t, x(t))$ holds for a time-delay system satisfying the Razumikhin condition, through which other stability issues can be also solved for time-delay systems; see [24], for instance.

Remark 5.4. The global Mittag-Leffler synchronization problem of fractional-order systems has also been discussed in [16] while, compared with the works in [12,16], we extend the model to the networks without strong connectedness. Therefore, our work has more applicability when considering the applications of the coupled oscillator in engineering and other fields, which also facilitates the development of the theory in some relevant areas.

5.3.3 Reachability of a sliding surface

In this part, we will study the finite-time reachability problem of a sliding mode surface $\varphi_k(t) = 0$. On the other hand, the sliding mode law is chosen as

$$\tilde{u}_k = -\xi \operatorname{sgn}(\varphi_k(t)), \tag{5.28}$$

where $\xi > 0$. Finally, the controller $u_k(t)$ is

$$
\begin{aligned}
u_k &= \bar{u}_k + \tilde{u}_k \\
&= -(\pi_k - \beta_k)e_k - (w_k - \rho l_k) - \rho(\alpha_k - \beta_k)x_k - g_k(\rho x_k, \rho x_k(t-\tau)) \\
&\quad + \rho f_k(x_k, x_k(t-\tau)) - \sum_{h=1}^{n} \left[s_{kh}\tilde{g}_{kh}(\rho x_k, \rho x_h) - \rho b_{kh}\tilde{f}_{kh}(x_k, x_h) \right] \\
&\quad - \xi \operatorname{sgn}(\varphi_k(t)).
\end{aligned}
\tag{5.29}
$$

Theorem 5.2. *Assume that the sliding mode surface function (5.6) and controller (5.29) are selected. Then the trajectories of the error system (5.5) can asymptotically reach the sliding mode surface $\varphi_k(t) = 0$.*

Proof. Select the following Lyapunov function:

$$V_i(t) = \sum_{k=1}^{n} \frac{1}{2}\varphi_k^2(t). \tag{5.30}$$

Then, taking the Caputo derivative of (5.30) and combining it with the sliding mode surface function (5.6), we have

$$\begin{aligned}
{}_0^C D_t^\alpha V_i(t) &\leq \sum_{k=1}^{n} \varphi_k(t)\, {}_0^C D_t^\alpha \varphi_k(t) \\
&= \sum_{k=1}^{n} \varphi_k(t)(-\xi\,\mathrm{sgn}(\varphi_k(t))) \\
&= -\xi \sum_{k=1}^{n} |\varphi_k(t)| \leq -\xi V_i^{\frac{1}{2}}(t) < 0, \quad \forall \varphi_k(t) \neq 0.
\end{aligned} \tag{5.31}$$

Thus, one sees that the states of the error system (5.4) can be globally stable on the predefined sliding mode surface (5.6). $\qquad\square$

5.4 Numerical example

Example 5.1. Consider the following MSS with time-delay on the network without strong connectedness:

$$\begin{cases}
{}_0^C D_t^\alpha x_k(t) = -\alpha_k x_k + f_k(x_k, x_k(t-\tau)) + \sum_{h=1}^{9} b_{kh}\tilde{f}_{kh}(x_k, x_h) + l_k, \\
{}_0^C D_t^\alpha y_k(t) = -\beta_k y_k + g_k(y_k, y_k(t-\tau)) + \sum_{h=1}^{9} s_{kh}\tilde{g}_{kh}(x_k, x_h) + w_k + u_k, \\
x_k(s) = \phi_k(s), \; y_k(s) = \psi_k(s), \; s \in [-\tau, 0],
\end{cases} \tag{5.32}$$

where $\alpha = 0.9$, $\tau = 0.2$, and

$$\begin{aligned}
f_k(x_k, x_k(t-\tau)) &= x_k - \tanh(x_k(t-\tau)), \\
\tilde{f}_{kh}(x_k, x_h) &= 2\cos(x_k(t)) - 2\sin(x_h(t)), \\
g_k(y_k, y_k(t-\tau)) &= 0.6 y_k(t-\tau), \\
\tilde{g}_{kh}(x_k, x_h) &= 0.2 y_k(t) - 0.4 y_h(t).
\end{aligned}$$

Combining (5.32) with Assumption 5.2, let

$$\hat{\beta}_k = \beta_k - 0.2 \sum_{h=1}^{9} s_{kh}, \quad \hat{g}_{kh}(y_k(t), y_h(t)) = -0.4 y_h(t).$$

Then, the slave system can be written as follows:

$$ {}_0^C D_t^\alpha y_k(t) = -\hat{\beta}_k y_k + g_k(y_k, y_k(t-\tau)) + \sum_{h=1}^{9} s_{kh}\hat{g}_{kh}(x_k, x_h) + w_k + u_k.$$

Obviously, Assumption 5.2 still holds for this coupled system (5.32). The other parameters of the system (5.32) are shown as follows:

$$(\alpha_k)_{9\times 1} = \begin{bmatrix} 1.32 \\ 1.34 \\ 1.57 \\ 1.00 \\ 0.70 \\ -1.52 \\ -1.89 \\ 0.45 \\ 0.93 \end{bmatrix},$$

$$(b_{kh})_{9\times 9} = \begin{bmatrix} 1.46 & 0.69 & -0.85 & 1.62 & -1.76 & 0.11 & 0.03 & -0.05 & 1.40 \\ 1.54 & 1.73 & -1.40 & -1.87 & -1.22 & 0.55 & 1.00 & -0.76 & -0.30 \\ 0.51 & 0.49 & 0.36 & -0.77 & -0.70 & 1.78 & 1.47 & 0.53 & 0.23 \\ -0.52 & 0.12 & 0.78 & 0.94 & 0.57 & 1.08 & -1.10 & -0.27 & -0.88 \\ 1.30 & 0.27 & 1.11 & -0.39 & 0.74 & 1.32 & 1.42 & 1.37 & -0.60 \\ 1.76 & -1.71 & -1.23 & 0.20 & -1.70 & -1.70 & -1.89 & -1.79 & -0.42 \\ 1.89 & -0.57 & 0.12 & 1.02 & -0.56 & -1.14 & -1.94 & 0.69 & -1.27 \\ 0.15 & -0.95 & -0.69 & 1.91 & -1.20 & 1.21 & -0.17 & 0.29 & -1.26 \\ -1.52 & -0.96 & 0.47 & -0.01 & 1.95 & -1.43 & -0.58 & 0.12 & 0.74 \end{bmatrix},$$

$$(l_k)_{1\times 9} = \begin{bmatrix} 0.06 & -1.64 & 1.84 & -0.21 & -0.69 & -1.85 & 0.05 & 1.22 & 1.74 \end{bmatrix},$$

$$(\beta_k)_{9\times 1} = \begin{bmatrix} 1,81 \\ 0.51 \\ -0.12 \\ -0.06 \\ 1.47 \\ 1.27 \\ 0.43 \\ 0.27 \\ 0.03 \end{bmatrix},$$

$$(s_k)_{9\times 9} = \begin{bmatrix} 0 & -0.06 & -0.12 & 0 & 0 & 0 & 1.38 & 0 & 0 \\ 0.06 & 0 & -1.52 & 0 & -1.39 & 0 & 0 & 0 & 0 \\ 0.12 & 1.52 & 0 & 0 & 0 & 0 & 0 & 0 & 0.06 \\ 0 & 0 & 0 & 0 & 1.27 & 0 & 0 & 0 & 0 \\ 0 & 0 & 0 & 1.27 & 0 & 0 & 0 & 0 & 0 \\ 0 & 0 & 0 & 0 & 0 & 0 & 1.25 & 0 & 0 \\ 0 & 0 & 0 & 0 & 0 & -1.25 & 0 & 0 & 0 \\ 0 & 0 & 0 & 0 & 0 & 0 & 0 & 0 & 1.49 \\ 0 & 0 & 0 & 0 & 0 & 0 & 0 & -1.49 & 0 \end{bmatrix},$$

$$(w_k)_{1\times 9} = \begin{bmatrix} 0.68 & -1.71 & -1.33 & 0.09 & -1.17 & 0.66 & 1.81 & -0.35 & -1.81 \end{bmatrix},$$

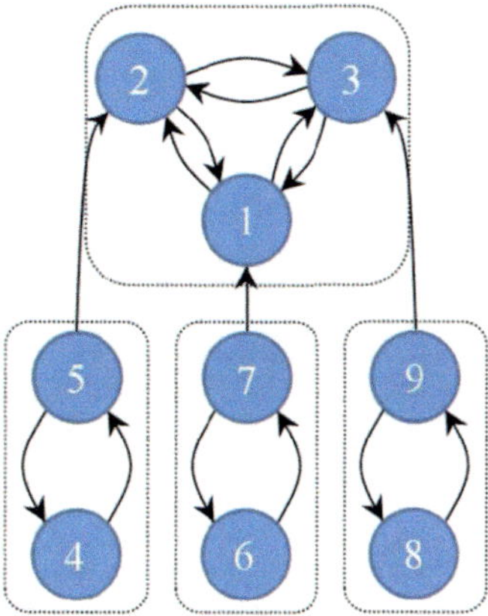

FIGURE 5.1   A network $(\mathcal{G}, \boldsymbol{A})$ and corresponding SCC $(\mathcal{R}_{ji}, \boldsymbol{H}_{ji})$.

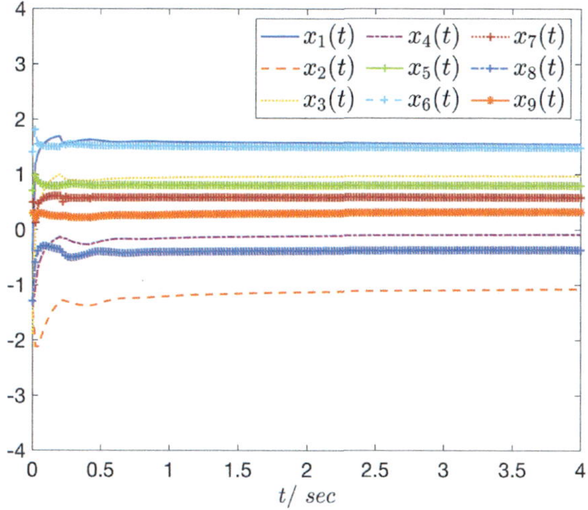

FIGURE 5.2   State trajectories of the master system (5.32) for $\rho = 2$.

$$(\phi_k)_{1\times 9} = \begin{bmatrix} -0.4 & -1.5 & -1.9 & -1.4 & 0.7 & 1.4 & 0.5 & -1.3 & 0.3 \end{bmatrix},$$

$$(\psi_k)_{1\times 9} = \begin{bmatrix} -0.3 & 1.9 & -0.2 & 1.1 & 0.7 & 1.4 & 1.6 & 1.2 & 1.5 \end{bmatrix}.$$

Next, selecting $\pi_k = 10$, $\xi = 10$, by means of the MATLAB$^{\circledR}$ toolbox, we have $c_4^{11} = c_5^{11} = 0.254$, $c_6^{12} = c_7^{12} = 0.25$, $c_8^{13} = c_9^{13} = 0.298$, $c_1^{21} = c_2^{21} = c_3^{21} = 0.0112$, $L_{1k} = 0.3$, $L_{2k} = 0.3$, $L_{3k} = 0.4$, $L_{4k} = 0.2$, with $k \in \mathbb{L}$. Since $\theta_{kh} = |s_{kh}|L_{4k}$ holds for all $k, h \in V(\mathcal{R}_{ji})$, we obtain that

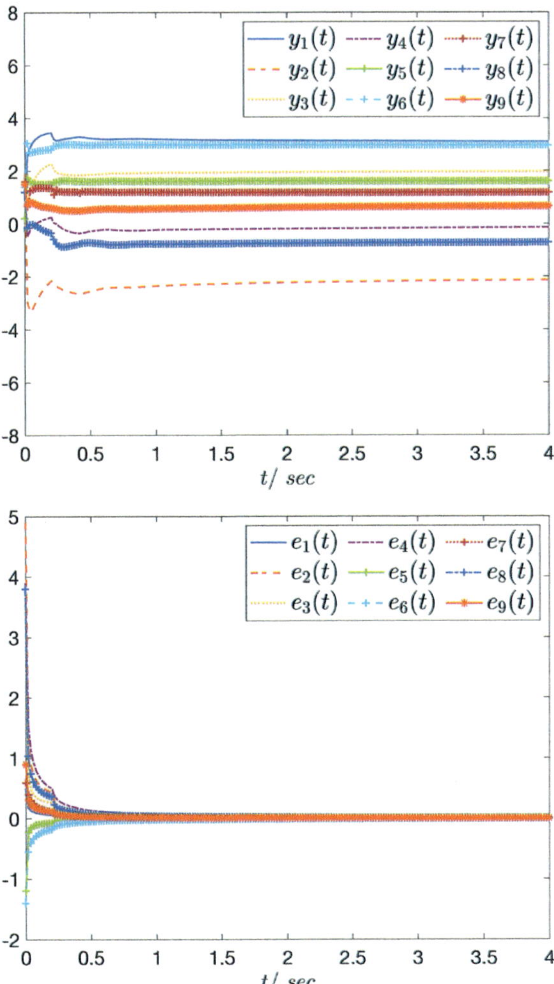

FIGURE 5.3 State trajectories of the slave system (5.32) for $\rho = 2$.

$$A = \begin{bmatrix} 0 & 0.012 & 0.024 & 0 & 0 & 0 & 0.276 & 0 & 0 \\ 0.012 & 0 & 0.304 & 0 & 0.278 & 0 & 0 & 0 & 0 \\ 0.024 & 0.304 & 0 & 0 & 0 & 0 & 0 & 0 & 0.132 \\ 0 & 0 & 0 & 0 & 0.254 & 0 & 0 & 0 & 0 \\ 0 & 0 & 0 & 0.254 & 0 & 0 & 0 & 0 & 0 \\ 0 & 0 & 0 & 0 & 0 & 0 & 0.25 & 0 & 0 \\ 0 & 0 & 0 & 0 & 0 & 0.25 & 0 & 0 & 0 \\ 0 & 0 & 0 & 0 & 0 & 0 & 0 & 0 & 0.298 \\ 0 & 0 & 0 & 0 & 0 & 0 & 0 & 0.298 & 0 \end{bmatrix}.$$

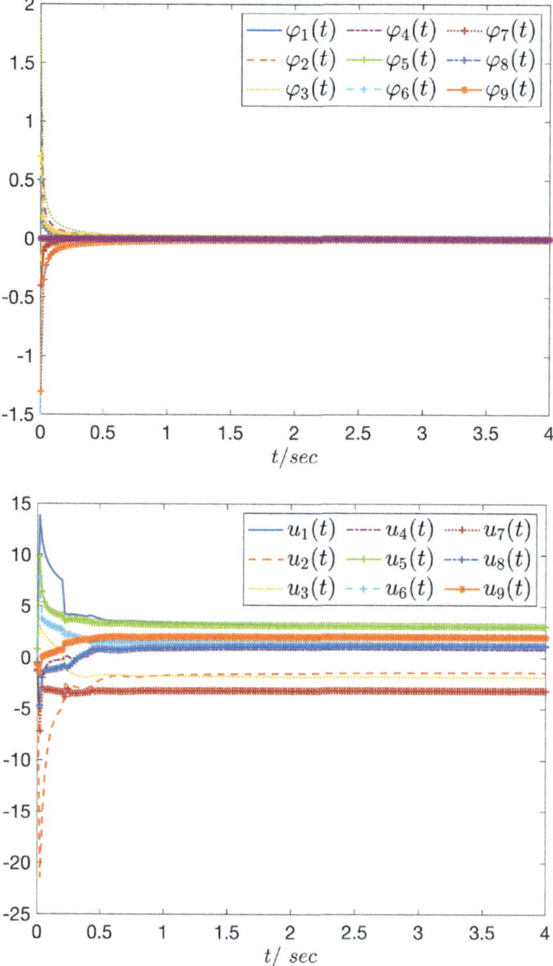

FIGURE 5.4 State trajectories of sliding surface function (5.6) and controllers (5.29) for $\rho = 2$.

Obviously, from Fig. 5.1, we can easily see that $(\mathcal{G}, \boldsymbol{A})$ is not strongly connected; however, it can be divided into four SCCs, that is, $(\mathcal{R}_{11}, \boldsymbol{B}_{11})$, $(\mathcal{R}_{12}, \boldsymbol{B}_{12})$, $(\mathcal{R}_{13}, \boldsymbol{B}_{13})$, and $(\mathcal{R}_{21}, \boldsymbol{B}_{21})$. Select the projection coefficient $\rho = 2$, then Fig. 5.2 shows the state trajectories of the master system (5.32), and Fig. 5.3 illustrates the state trajectories of the slave system and error curves, Fig. 5.4 gives the curves of the sliding surfaces and controllers. Meanwhile, Figs. 5.5 and 5.6 show the state trajectories of (5.32) and the synchronization error curves for $\rho = 1, -1$, respectively.

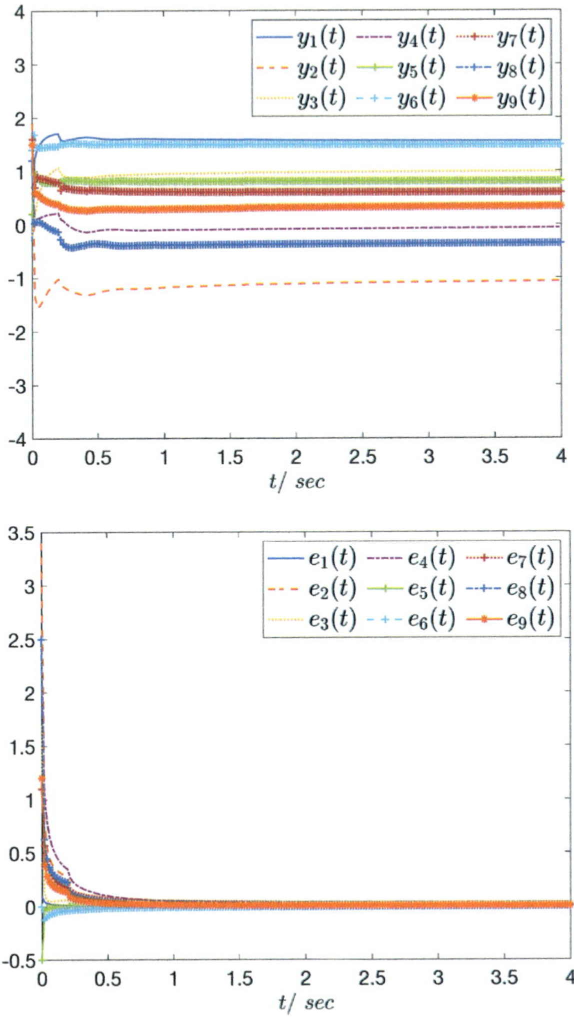

FIGURE 5.5 State trajectories and error curves for $\rho = 1$.

To illustrate the advantages of the presented main results, a comparison with the work in [25] is presented. Fig. 5.7 shows the synchronization behaviors between the proposed SMC strategy and the method proposed in [25], respectively. By comparison, it is found that the synchronization effect of the sliding mode control strategy proposed in this chapter is slightly better than that in [25].

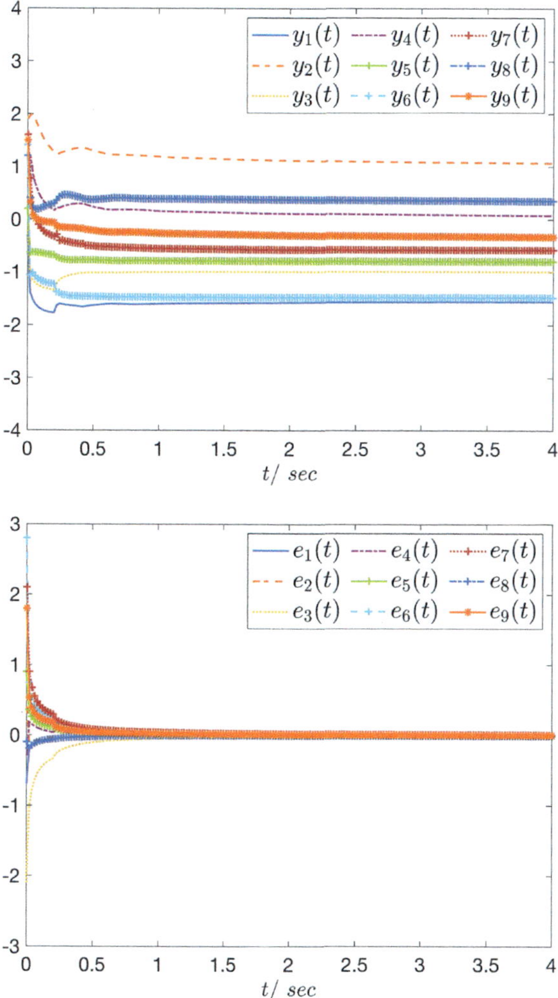

FIGURE 5.6 State trajectories and error curves for $\rho = -1$.

5.5 Conclusion

In this chapter, the projective synchronization issue of the FOCS on the network has been discussed. The sliding mode surface and controller have been designed to realize the projective synchronization between the master and the slave systems. In the meantime, several sufficient conditions have been obtained to guarantee the MLS of the considered FOCS on a network without strong connectedness via the graph theory and hierarchical method.

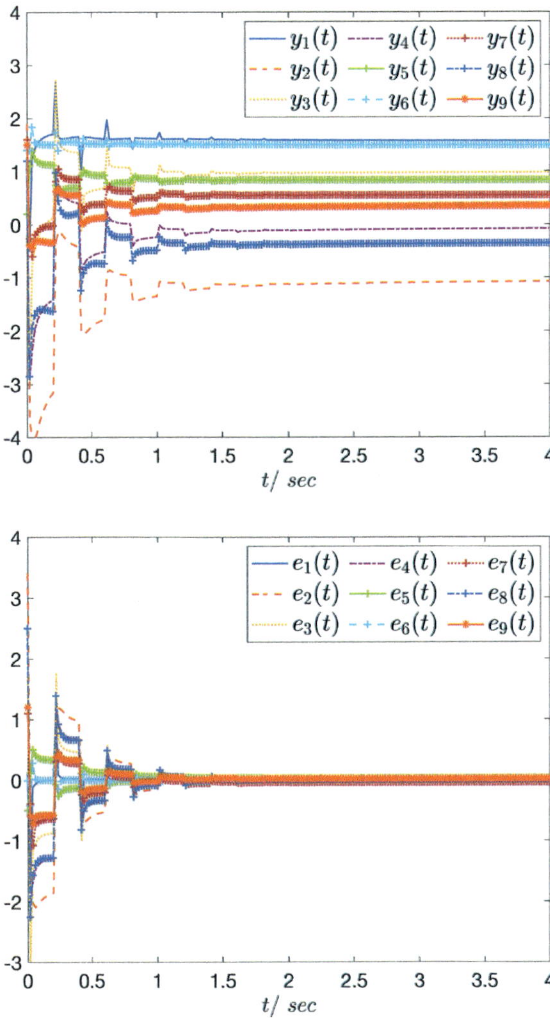

FIGURE 5.7 State trajectories and error curves in [25] for $\rho = 1$.

References

[1] Y. Han, C.-Y. Su, Y. Kao, C. Gao, Non-fragile sliding mode control of discrete switched singular systems with time-varying delays, IET Control Theory & Applications 14 (5) (2020) 726–737.

[2] S.V. Emelyanov, Variable structure control systems, 1967.

[3] X. Meng, C. Gao, Z. Liu, B. Jiang, Robust H_∞ control for a class of uncertain neutral-type systems with time-varying delays, Asian Journal of Control 23 (3) (2021) 1454–1465.

[4] Z. Liu, L. Zhao, C. Gao, Observer-based adaptive H_∞ control of uncertain stochastic singular systems via integral sliding mode technique, IET Control Theory & Applications 11 (5) (2017) 668–676.

[5] H.R. Karimi, A sliding mode approach to H_∞ synchronization of master–slave time-delay systems with Markovian jumping parameters and nonlinear uncertainties, Journal of the Franklin Institute 349 (4) (2012) 1480–1496.

[6] Z. Liu, H.R. Karimi, J. Yu, Passivity-based robust sliding mode synthesis for uncertain delayed stochastic systems via state observer, Automatica 111 (2020) 108596.

[7] C.-Y. Su, T.-P. Leung, Q.-J. Zhou, Force/motion control of constrained robots using sliding mode, IEEE Transactions on Automatic Control 37 (5) (1992) 668–672.

[8] M. Liu, L. Zhang, P. Shi, Y. Zhao, Fault estimation sliding-mode observer with digital communication constraints, IEEE Transactions on Automatic Control 63 (10) (2018) 3434–3441.

[9] Y. Han, Y. Kao, C. Gao, Robust sliding mode control for uncertain discrete singular systems with time-varying delays and external disturbances, Automatica 75 (2017) 210–216.

[10] H.G. Sun, Y. Zhang, D. Baleanu, W. Chen, Y.Q. Chen, A new collection of real world applications of fractional calculus in science and engineering, Communications in Nonlinear Science and Numerical Simulation 64 (2018) 213–231.

[11] M.Y. Li, Z. Shuai, Global-stability problem for coupled systems of differential equations on networks, Journal of Differential Equations 248 (1) (2010) 1–20.

[12] H.-L. Li, Y.-L. Jiang, Z. Wang, L. Zhang, Z. Teng, Global Mittag-Leffler stability of coupled system of fractional-order differential equations on network, Applied Mathematics and Computation 270 (2015) 269–277.

[13] H. Li, Y.G. Kao, Mittag-Leffler stability for a new coupled system of fractional-order differential equations with impulses, Applied Mathematics and Computation 361 (2019) 22–31.

[14] L.M. Pecora, T.L. Carroll, Synchronization in chaotic systems, Physical Review Letters 64 (8) (1990) 821.

[15] X. Meng, Y. Kao, C. Gao, B. Jiang, Projective synchronisation of variable-order systems via fractional sliding mode control approach, IET Control Theory & Applications 14 (1) (2020) 12–18.

[16] J. Chen, C. Li, X. Yang, Global Mittag-Leffler projective synchronization of nonidentical fractional-order neural networks with delay via sliding mode control, Neurocomputing 313 (2018) 324–332.

[17] A.-J. Muñoz-Vázquez, V. Parra-Vega, A. Sánchez-Orta, Non-smooth convex Lyapunov functions for stability analysis of fractional-order systems, Transactions of the Institute of Measurement and Control 41 (6) (2019) 1627–1639.

[18] I. Stamova, G. Stamov, Mittag-Leffler synchronization of fractional neural networks with time-varying delays and reaction–diffusion terms using impulsive and linear controllers, Neural Networks 96 (2017) 22–32.

[19] P. Du, M.Y. Li, Impact of network connectivity on the synchronization and global dynamics of coupled systems of differential equations, Physica D. Nonlinear Phenomena 286 (2014) 32–42.

[20] G. Wang, W. Li, J. Feng, Stability analysis of stochastic coupled systems on networks without strong connectedness via hierarchical approach, Journal of the Franklin Institute 354 (2) (2017) 1138–1159.

[21] J. Yu, C. Hu, H. Jiang, X. Fan, Projective synchronization for fractional neural networks, Neural Networks 49 (2014) 87–95.

[22] Y. Liu, J. Mei, W. Li, Stochastic stabilization problem of complex networks without strong connectedness, Applied Mathematics and Computation 332 (2018) 304–315.

[23] Y. Li, Y.Q. Chen, I. Podlubny, Mittag-Leffler stability of fractional order nonlinear dynamic systems, Automatica 45 (8) (2009) 1965–1969.

[24] C. Zhang, B.-S. Han, Stability analysis of stochastic delayed complex networks with multi-weights based on Razumikhin technique and graph theory, Physica A. Statistical Mechanics and Its Applications 538 (2020) 122827.

[25] C. Wang, Q. Yang, Y. Zhuo, R. Li, Synchronization analysis of a fractional-order non-autonomous neural network with time delay, Physica A. Statistical Mechanics and Its Applications 549 (2020) 124176.

An event-triggered sliding mode control mechanism to exponential consensus of fractional-order descriptor nonlinear multi-agent systems

6.1 Introduction

Multi-agent systems (MASs), as one of the most effective solutions to solve complex tasks, have received great attention from academia and engineering practice due to their high efficiency, low cost, flexibility, and reliability; see, for example, [1–3] and the references therein. MASs have been widely used in computer networks [4], robot collaborative control [5], etc. Consensus, as one of the challenges for coordinated control of MASs, aiming to reach a global agreement on specific features of interest, has been extensively studied [6,7], and used in search and rescue, avoiding collision, etc. Feature-specific MAS consensus approaches include leader-following [8], time-delay [9], etc. Hereafter, some progress on consensus control strategies of nonlinear MASs have been reported in [10,11]. In [10], by designing a state feedback controller, the authors considered a distributed consensus problem of nonlinear MASs with a one-sided Lipschitz condition. However, the consensus control strategies in [10,11] may not be extendable to descriptor systems, which poses further challenges for subsequent studies, including admissible consensus.

Focusing on the descriptor MASs, some results on the consensus control have been reported in [12–14]. In [12], the authors presented necessary and sufficient conditions of admissible consensus for descriptor MASs. Afterward, for homogeneous descriptor MASs with undirected graphs and disturbances, the authors in [14] designed distributed controllers to study the admissible consensus problem. Meanwhile, the admissible output consensus problem of high-order descriptor linear MASs was investigated based on a dynamic output feedback consensus control method in [13].

As we known, fractional-order systems are more suitable to describe complex dynamical phenomena than classical systems. Fractional-order calculus theory has been widely used in physics, automatic control, and biochemistry, due to its memory and genetic properties, and also in engineering during the past few years [15,16]. However, the above consensus control methods were all concentrated on integer-order MAS rather than fractional-order MAS. Paralleled to the above developments, there does exist several results extending MSAs to fractional-order MASs (FO-MASs); see [17–19] and the references therein. In [20], the authors studied the consensus problem of FO-MASs with input delays. Meanwhile, the authors in [17] investigated the robust consensus problem for a class of FO-MASs with external perturbations by designing a pinning control strategy. Furthermore, based on a new observer frame, the leader-following tracking consensus problem of FO-MASs was investigated, in which the given leader system was described by second-order systems [18,21].

Since many practical systems, including power networks with supercoils and supercapacitors, robotic systems operating on dirt roads, etc., can be modeled by fractional-order descriptor linear MASs (FOD-LMASs) [22], the consensus problem of FOD-LMASs has received great attention [23]. For example, the authors in [23] investigated the leader-following admissible consensus problem of FOD-LMASs with $\alpha \in (0, 2)$ based on an undirected graph by designing a distributed consensus control strategy. In [24], the authors presented a sliding mode controller to realize the consensus problem for FOD-LMASs. It should be noted that the above results were developed for the linear case. Therefore they may not function for nonlinear MASs. Compared to the linear case, the study of consensus problems for fractional-order descriptor nonlinear multi-agent systems (FOD-NMASs) presents greater difficulties and challenges such as: (1) complex nonlinear functional forms due to highly complex nonlinear functional forms that are involved in FOD-NMASs, which makes mathematically based analysis and design for such systems more difficult to implement; (2) dynamic coupling, due to the agents in FOD-NMASs being often tightly coupled to each other and this coupling changing dynamically over time. This means that it is difficult to predict when the system will reach a consensus state, and different control strategies need to be employed to achieve consensus; (3) noise and uncertainty as realistic multi-agent systems are subject to various kinds of noise and uncertainty, such as communication delays, distortion, and packet loss. FOD-NMASs have more complex nonlinear characteristics, and this noise and uncertainty can greatly exacerbate the difficulty of the consensus problem. To our knowledge, there are few results on the consensus problem for FOD-NMASs, which motivates us to explore it.

The event-triggered control has attracted more and more scholars' attention in the field of control because of its advantages, such as smaller execution times, less energy consumption, and the ability to ensure a required stability of the sampled-data system; see [25–29] and the references therein. Recently, some consensus control strategies based on event-triggered mechanism, including adaptive fuzzy consensus, finite-time output consensus, leader-following consensus, leader-following exponential consensus, etc., have been developed for both MAS and FO-MASs [30–33]. For example, the authors in [32] considered an adaptive fuzzy consensus problem of FO-MASs based on an event-triggered strategy, and the boundedness of tracking errors was established. As for FOD-LMASs, the authors in [30] investigated the leader-following exponential consensus control problem combined with an event-triggered mechanism. And two different fractional-order event-triggered control methods were proposed to achieve the exponential consensus problem of FO-MASs. It is worth pointing out that each agent was treated as a descriptor linear system. However, it should be even reasonable to treat each agent as a descriptor nonlinear system. If this is the case, the method proposed in [30] needs to be revisited, which provides a further challenge. Besides, research results on the consensus control problem, if an event-trigger mechanism is adopted, are sparse for FOD-NMASs.

Based on the above analysis, this chapter designs an event-triggered sliding mode control strategy to explore the exponential consensus problem for FOD-NMASs. Under the Mittag-Leffler stability, such an exponential consensus is achieved by designing a suitable event-triggered mechanism that contains sampling of state trajectory for each agent, which is the main difference from the linear combination measurement event-triggering mechanism-based adopted in [30]. Due to the memory nature of the fractional-order calculus operator, we propose a new condition to avoid the occurrence of Zeno behavior, which is different from existing event-triggered schemes.

6.2 Problem statement

6.2.1 Graph theory

In this section, let $\hat{\mathcal{G}} = \{\mathcal{V}_0\} \cup \mathcal{G}$ denote the interaction topology, where $\mathcal{V}_0$ is the set of a leader, $\mathcal{G} = \{\mathcal{V}, \mathcal{E}, \mathcal{P}\}$ is the exchange information among the followers, in which $\mathcal{V} = \{1, 2, \ldots, \mathcal{N}\}, \mathcal{E} \subseteq \mathcal{V} \times \mathcal{V}$, and $\mathcal{P} = [p_{ij}] \in \mathbb{R}^{\mathcal{N} \times \mathcal{N}}$ are the vertex set, edge set, and the weighted adjacency matrix, respectively.

The Laplacian matrix of $\mathcal{G}$ is denoted as $\hat{P} = [\hat{p}_{ij}] \in \mathbb{R}^{\mathcal{N} \times \mathcal{N}}$ with

$$\hat{p}_{ij} = \begin{cases} \sum_{j=1,\ j\neq i}^{\mathcal{N}} p_{ij}, & i = j, \\ -p_{ij}, & i \neq j. \end{cases}$$

Denote $\mathcal{L} = \hat{P} + \mathcal{Q}$, where $\mathcal{Q} = \mathrm{diag}\{q_1, q_2, \ldots, q_{\mathcal{N}}\}$ is the communication topology between the leader and followers, and $q_i \geq 0$ is the weight coefficient of $(\mathcal{V}_i, \mathcal{V}_0)$. Obviously, according to [34, Lemma 1], we know that $\mathcal{L}$ is a symmetric positive definite matrix.

Assumption 6.1. In this chapter, the considered topology $\mathcal{G}$ among those followers is undirected and connected.

6.2.2 Model description

Definition 6.1 ([30]). A given fractional-order descriptor system $E_{t_0}^{\mathcal{C}} D_t^{\alpha} x(t) = A x(t)$ is

(1) regular if $\det(s^{\alpha} E - A)$ is not identically zero, where $s \in \mathbb{C}$;
(2) impulse-free if it is regular and $\deg(\det(s E - A)) = \mathrm{rank}(E)$, where $s \in \mathbb{C}$;
(3) admissible if it is regular, impulse-free, and stable.

Definition 6.2 ([30]). If, for any given initial condition, there exist positive constants a, b, T, and $\beta \in (0, 1)$ such that

$$\|x_i(t) - x_0(t)\| \leq a(E_{\alpha}(-b(t - t_0)^{\alpha}))^{\beta},$$

holds for $t > T$, where $E_{\alpha}[\cdot]$ is the Mittag-Leffler function, then MAS is said to achieve exponential consensus.

Consider the fractional-order descriptor linear MASs given in [30] with nonlinear term. Choose the leader system as

$$E_{t_0}^{\mathcal{C}} D_t^{\alpha} x_0(t) = A x_0(t) + f(x_0(t)), \tag{6.1}$$

where order $\alpha \in (0, 1)$, the pseudo-state vector of leader system $x_0(t) \in \mathbb{R}^n$, $f(\cdot)$ denotes a nonlinear function, and $\mathrm{rank}(E) \leq n$.

The follower systems are given as

$$E_{t_0}^{\mathcal{C}} D_t^{\alpha} x_i(t) = A x_i(t) + f(x_i(t)) + B u_i(t), \tag{6.2}$$

where the pseudo-state vectors are $x_i(t) \in \mathbb{R}^n$ and control input signal is $u_i(t) \in \mathbb{R}^m$, $i = 1, 2, \ldots, \mathcal{N}$.

Some assumptions are presented as follows:

Assumption 6.2. (E, A, α) is regular and impulse-free.

Assumption 6.3. (E, B, α) is $\mathbb{R}$-controllable.

Assumption 6.4. There exists positive constant δ_1 such that $\| f(x) - f(y)\| \le \delta_1 \|x - y\|$ holds.

Assumption 6.5. For the given leader system (6.1), the pseudo-state trajectory $x_0(t)$ is norm-bounded.

Remark 6.1. Assumption 6.4 means that the trajectory of the leader system (6.1) is a convergent or periodic response curve. Under this hypothesis, the study of an event-triggered SMC problem for a class of FOD-NMASs is meaningful.

6.3 Main results

Design the following event-triggered condition (ETC) for agent i:

$$t_{k+1}^i = \inf\{t > t_k^i | \mathcal{F}_i(t) > 0\}, \tag{6.3}$$

where $\mathcal{F}_i(t) = \|\hat{e}_i(t)\|^2 + q_i \|\hat{e}_{0i}(t)\|^2 - \phi_i(t)$, and function $\phi_i(t)$ satisfies $0 \le \phi_i(t) \le \xi_i \|e_i(t)\|^2$ with $\xi_i > 0$. Here $\hat{e}_i(t) = x_i(t_k^i) - x_i(t)$ denotes the sampled error for each agent i, while $\hat{e}_{0i}(t) = x_0(t_k^i) - x_0(t)$ denotes the sampled error for the leader system.

Remark 6.2. In ETC (6.3), we introduce a nonnegative threshold function $\phi_i(t)$, which is beneficial to exclude the possibility of Zeno behavior by incorporating the specific information of MASs in the following part. Generally speaking, there exist two sampling methods in some existing literature on the problem of MAS control:

(1) Similar to the sampling method used in this chapter, the current sampled signal $x_i(t_k^i)$ is obtained by utilizing a zero-order holder for agent i. This approach has been widely used in the literature [35–37]. However, such an event-triggered mechanism requires continuous measurements of the signal from the neighbors of each individual agent, which is more beneficial to explore the Zeno behavior;

(2) Based on the network communication information, by constructing a combined measurement at first, and using a zero-order holder for the combined measurement, the current sampled signal can be also derived; see the literature [31,38,39]. This method is more suitable to discuss the Zeno behavior for linear systems.

6.3.1 Event-triggered SMC strategy

In this section, we construct the following sliding surface $s_i(t)$:

$$\begin{cases} s_i(t) = GE(x_i(t) - x_0(t)) - {}_{t_0}D_t^{-\alpha}z_i(t), \\ z_i(t) = GA(x_i(t) - x_0(t)) + GB\hat{u}_{il}(t), \end{cases} \tag{6.4}$$

where $G \in \mathbb{R}^{m \times n}$ is such that GB is nonsingular, and

$$\hat{u}_{il}(t) = K\left\{ \sum_{j=1}^{\mathcal{N}} p_{ij}(x_i(t_k^i) - x_j(t_{k'}^j)) + q_i(x_i(t_k^i) - x_0(t_k^i)) \right\}$$

denotes the control law with $K \in \mathbb{R}^{m \times n}$.

Let $e_i(t) = x_i(t) - x_0(t)$ denote the tracking error for agent i. If the trajectory of the FOD-NMASs arrives to the designed sliding manifold, we have

$$_{t_0}^{C}D_t^{\alpha}s_i(t) = GE\,{}_{t_0}^{C}D_t^{\alpha}e_i(t) - GAe_i(t) - GB\hat{u}_{il}(t) = 0, \tag{6.5}$$

and $s_i(t) = 0$ holds, which means that the nonlinear function $f(x_i(t))$ can be compensated by the control law $\hat{u}_{il}(t)$. Therefore, when the state trajectories of FOD-NMASs (6.1) and (6.2) fully reach the fractional-order integral sliding manifold (6.4), by choosing ${}_{t_0}^{C}D_t^{\alpha}s_i(t) = s_i(t) = 0$, we get the following sliding mode dynamics:

$$E\,{}_{t_0}^{C}D_t^{\alpha}e_i(t) = Ae_i(t) + B\hat{u}_{il}(t). \tag{6.6}$$

Therefore, the asymptotic stability problem of sliding mode dynamics (6.6) is equivalent to the consensus problem of FOD-NMASs (6.1) and (6.2).

Theorem 6.1. *Assume that Assumptions 6.1–6.3 hold, and the ETC shown in (6.3) is considered, then the leader-following exponential consensus of FOD-NMASs can be achieved by the proposed control strategy if there exists a matrix H satisfying*

$$\mathcal{A} - \left(\mu_i - \frac{\omega_{1i}\mu_i}{2} - \frac{\omega_{2i}q_i}{2}\right)\mathcal{H} + \varrho I_n < 0, \tag{6.7}$$

and $H^T E = E^T H \geq 0$ holds for each agent i, where $\mathcal{A} \triangleq \frac{1}{2}(H^T A + A^T H)$, $\mathcal{H} \triangleq H^T BB^T H$, μ_i are the eigenvalues of $\mathcal{J} \triangleq \mathcal{Y}^T \mathcal{L}\mathcal{Y}$, with $\mathcal{Y}$ being an orthogonal matrix, $\mathcal{L} = \hat{P} + \mathcal{Q}$, $\varrho > \frac{\lambda_{\max}\{\mathcal{H}\}}{2\omega}\xi$, $\frac{1}{2\omega} = \max_i\left\{\frac{\mu_i}{2\omega_{1i}}, \frac{1}{2\omega_{2i}}\right\}$, $\xi = \max_i\{\xi_i\}$, and the gain matrix is $K = -B^T H$.

Proof. Considering the sliding mode dynamics (6.6) with controller $\hat{u}_{il}(t)$ for agent i, one has

$$E_{t_0}^{\mathcal{C}} D_t^\alpha e_i(t) = A e_i(t) + BK\left\{ \sum_{j=1}^{\mathcal{N}} p_{ij}(x_i(t_k^i) - x_j(t_{k'}^j)) + q_i(x_i(t_k^i) - x_0(t_k^i)) \right\}. \tag{6.8}$$

Since $x_i(t_k^i) = x_i(t) + \hat{e}_i(t)$, $x_j(t_{k'}^j) = x_j(t) + \hat{e}_j(t)$, and $x_0(t_k^i) = x_0(t) + \hat{e}_{0i}(t)$, we obtain

$$E_{t_0}^{\mathcal{C}} D_t^\alpha e_i(t) = A e_i(t) + BK\left\{ \sum_{j=1}^{\mathcal{N}} p_{ij}(\hat{e}_i(t) - \hat{e}_j(t) + e_i(t) - e_j(t)) \right. \tag{6.9}$$
$$\left. + q_i(\hat{e}_i(t) - \hat{e}_{0i}(t) + e_i(t)) \right\}.$$

According to the properties of the Kronecker product, one has

$$(I_{\mathcal{N}} \otimes E)_{t_0}^{\mathcal{C}} D_t^\alpha e(t) = (I_{\mathcal{N}} \otimes A)e(t) + (\mathcal{L} \otimes BK)(e(t) \tag{6.10}$$
$$+ \hat{e}_0(t)) - (\mathcal{Q} \otimes BK)\hat{e}_0(t),$$

where $e(t) = \mathrm{col}\{e_i(t)\}$, $\hat{e}(t) = \mathrm{col}\{\hat{e}_i(t)\}$, and $\hat{e}_0(t) = \mathrm{col}\{e_{0i}(t)\}$.

Consider the following Lyapunov function:

$$V(t) = \frac{1}{2} e^{\mathrm{T}}(t)(I_{\mathcal{N}} \otimes H^{\mathrm{T}} E)e(t), \tag{6.11}$$

then

$$_{t_0}^{\mathcal{C}} D_t^\alpha V(t) \le e^{\mathrm{T}}(t)(I_{\mathcal{N}} \otimes H^{\mathrm{T}} E)_{t_0}^{\mathcal{C}} D_t^\alpha e(t)$$
$$= e^{\mathrm{T}}(t)(I_{\mathcal{N}} \otimes H^{\mathrm{T}})$$
$$\times \left[(I_{\mathcal{N}} \otimes A)e(t) + (\mathcal{L} \otimes BK)(e(t) + \hat{e}(t)) - (\mathcal{Q} \otimes BK)\hat{e}_0(t) \right]$$
$$= e^{\mathrm{T}}(t)(I_{\mathcal{N}} \otimes \mathcal{A} - \mathcal{L} \otimes \mathcal{H})e(t)$$
$$- e^{\mathrm{T}}(t)(\mathcal{L} \otimes \mathcal{H})\hat{e}(t) + e^{\mathrm{T}}(t)(\mathcal{Q} \otimes \mathcal{H})\hat{e}_0(t). \tag{6.12}$$

In this chapter, since the matrix $\mathcal{L} = \hat{\mathcal{P}} + \mathcal{Q}$ is symmetric and positive definite, using the singular value decomposition method, we can find an orthogonal matrix $\mathcal{Y}$ satisfying $\mathcal{L} = \mathcal{Y}\mathcal{J}\mathcal{Y}^{\mathrm{T}}$, where $\mathcal{J}$ is a diagonal matrix. Let $\theta(t) = (\mathcal{Y} \otimes I_{\mathcal{N}})e(t)$, $\psi(t) = (\mathcal{Y} \otimes I_{\mathcal{N}})\hat{e}(t)$, and $\varphi(t) = (\mathcal{Y} \otimes I_{\mathcal{N}})\hat{e}_0(t)$, then (6.12) is equivalent to

$$
\begin{aligned}
{}^{C}_{t_0}D^{\alpha}_t V(t) &\leq \boldsymbol{\theta}^{\mathrm{T}}(t)(\boldsymbol{I}_{\mathcal{N}} \otimes \mathcal{A} - \mathcal{J} \otimes \mathcal{H})\boldsymbol{\theta}(t) \\
&\quad - \boldsymbol{\theta}^{\mathrm{T}}(t)(\mathcal{J} \otimes \mathcal{H})\boldsymbol{\psi}(t) + \boldsymbol{\theta}^{\mathrm{T}}(t)(\mathcal{Q} \otimes \mathcal{H})\boldsymbol{\varphi}(t) \\
&= \sum_{i=1}^{\mathcal{N}} \left\{ \boldsymbol{\theta}^{\mathrm{T}}_i(t)(\mathcal{A} - \mu_i\mathcal{H})\boldsymbol{\theta}_i(t) - \mu_i\boldsymbol{\theta}^{\mathrm{T}}_i(t)\mathcal{H}\boldsymbol{\psi}(t) + q_i\boldsymbol{\theta}^{\mathrm{T}}_i(t)\mathcal{H}\boldsymbol{\varphi}(t) \right\}.
\end{aligned}
$$

$$(6.13)$$

Since $2\boldsymbol{a}^{\mathrm{T}}\boldsymbol{b} \leq \omega\boldsymbol{a}^{\mathrm{T}}\boldsymbol{a} + \frac{1}{\omega}\boldsymbol{b}^{\mathrm{T}}\boldsymbol{b}$ with $\omega > 0$, there exist some positive constants ω_{1i} and ω_{2i} satisfying

$$
\begin{cases}
-\boldsymbol{\theta}^{\mathrm{T}}_i(t)\mathcal{H}\boldsymbol{\psi}(t) \leq \frac{\omega_{1i}}{2}\boldsymbol{\theta}^{\mathrm{T}}_i(t)\mathcal{H}\boldsymbol{\theta}(t) + \frac{1}{2\omega_{1i}}\boldsymbol{\psi}^{\mathrm{T}}_i(t)\mathcal{H}\boldsymbol{\psi}(t), \\
\boldsymbol{\theta}^{\mathrm{T}}_i(t)\mathcal{H}\boldsymbol{\varphi}(t) \leq \frac{\omega_{2i}}{2}\boldsymbol{\theta}^{\mathrm{T}}_i(t)\mathcal{H}\boldsymbol{\theta}(t) + \frac{1}{2\omega_{2i}}\boldsymbol{\varphi}^{\mathrm{T}}_i(t)\mathcal{H}\boldsymbol{\varphi}(t).
\end{cases}
$$

$$(6.14)$$

Combining (6.13) and (6.14), we have

$$
\begin{aligned}
{}^{C}_{t_0}D^{\alpha}_t V(t) &\leq \sum_{i=1}^{\mathcal{N}} \boldsymbol{\theta}^{\mathrm{T}}_i(t)\left(\mathcal{A} - \left(\mu_i - \frac{\omega_{1i}\mu_i}{2} - \frac{\omega_{1i}q_i}{2}\right)\mathcal{H}\right)\boldsymbol{\theta}_i(t) \\
&\quad + \sum_{i=1}^{\mathcal{N}} \left(\frac{\mu_i}{2\omega_{1i}}\boldsymbol{\psi}^{\mathrm{T}}_i(t)\mathcal{H}\boldsymbol{\psi}_i(t) + \frac{q_i}{2\omega_{2i}}\boldsymbol{\varphi}^{\mathrm{T}}_i(t)\mathcal{H}\boldsymbol{\varphi}_i(t)\right).
\end{aligned}
$$

$$(6.15)$$

According to (6.7), we have that $\mathcal{A} - \left(\mu_i - \frac{\omega_{1i}\mu_i}{2} - \frac{\omega_{2i}q_i}{2}\right)\mathcal{H} < -\varrho\boldsymbol{I}_n$, and then

$$
{}^{C}_{t_0}D^{\alpha}_t V(t) \leq \sum_{i=1}^{\mathcal{N}} -\varrho\|\boldsymbol{\theta}_i(t)\|^2 + \sum_{i=1}^{\mathcal{N}} \left(\frac{\mu_i\lambda_{\max}\{\mathcal{H}\}}{2\omega_{1i}}\|\boldsymbol{\psi}(t)\|^2 + \frac{q_i\lambda_{\max}\{\mathcal{H}\}}{2\omega_{2i}}\|\boldsymbol{\varphi}(t)\|^2\right).
$$

$$(6.16)$$

Since $\sum_{i=1}^{\mathcal{N}} \|\boldsymbol{\theta}_i(t)\|^2 = \sum_{i=1}^{\mathcal{N}} \|\boldsymbol{e}_i(t)\|^2$, $\sum_{i=1}^{\mathcal{N}} \|\boldsymbol{\psi}_i(t)\|^2 = \sum_{i=1}^{\mathcal{N}} \|\hat{\boldsymbol{e}}_i(t)\|^2$, $\sum_{i=1}^{\mathcal{N}} \|\boldsymbol{\varphi}_i(t)\|^2 = \sum_{i=1}^{\mathcal{N}} \|\hat{\boldsymbol{e}}_{0i}(t)\|^2$, and $\frac{1}{2\omega} = \max_i\left\{\frac{\mu_i}{2\omega_{1i}}, \frac{1}{2\omega_{2i}}\right\}$, we obtain

$$
{}^{C}_{t_0}D^{\alpha}_t V(t) \leq \sum_{i=1}^{\mathcal{N}} -\varrho\|\boldsymbol{e}_i(t)\|^2 + \frac{1}{2\omega}\sum_{i=1}^{\mathcal{N}} \left(\|\hat{\boldsymbol{e}}_i(t)\|^2 + q_i\|\hat{\boldsymbol{e}}_{0i}(t)\|^2\right). \tag{6.17}
$$

According to the definition of the trigger function $\mathcal{F}_i(t)$, then

$$
\|\hat{\boldsymbol{e}}_i(t)\|^2 + q_i\|\hat{\boldsymbol{e}}_{0i}(t)\|^2 \leq \phi_i(t) \leq \xi_i\|\boldsymbol{e}_i(t)\|^2. \tag{6.18}
$$

Substituting (6.18) into (6.17), then

$$\begin{aligned}
{}^{C}_{t_0}D^{\alpha}_t V(t) &\le \sum_{i=1}^{\mathcal{N}} -\varrho \|e_i(t)\|^2 + \frac{\lambda_{\max}\{\mathcal{H}\}}{2\omega} \sum_{i=1}^{\mathcal{N}} \left(\xi_i \|e_i(t)\|^2 \right) \\
&\le -\varrho \|e(t)\|^2 + \frac{\lambda_{\max}\{\mathcal{H}\}}{2\omega} \xi \|e(t)\|^2 \\
&\le \left\{ -\varrho + \frac{\lambda_{\max}\{\mathcal{H}\}}{2\omega} \xi \right\} \|e(t)\|^2 .
\end{aligned} \tag{6.19}$$

Based on [40, Theorem 5.1], the latter implies that the dynamics (6.6) is Mittag-Leffler stable, i.e., it is also asymptotically stable. By Lemma 2.8, then

$$V(t) \le V(t_0) E_\alpha \left[-\eta (t - t_0)^\alpha \right], \tag{6.20}$$

where $\eta \triangleq \varrho - \frac{\lambda_{\max}\{\mathcal{H}\}}{2\omega}\xi$. Furthermore, according to [41], we know that

$$\|e(t)\|^2 \le \frac{2}{\lambda_{\mathrm{m}}\{H^T E\}} V(t) \tag{6.21}$$

is feasible. Combining (6.20) with (6.21) leads to

$$\|e(t)\| \le \sqrt{\frac{2}{\lambda_{\mathrm{m}}\{H^T E\}} V(t_0)} \left(E_\alpha \left[-\eta (t - t_0)^\alpha \right] \right)^{\frac{1}{2}}. \tag{6.22}$$

Based on Definition 6.2, the exponential consensus of FOD-NMASs (6.1) and (6.2) can be achieved under the presented ETC. $\qquad\square$

Remark 6.3. In Theorem 6.1, we provide the following method to obtain the feasible solution H:

Step 1 By selecting some appropriate parameters ϱ, ω_{1i}, and ω_{2i}, obtain matrix H by solving the Riccati inequality (6.7) via Matlab$^{\circledR}$ toolbox;

Step 2 using matrix H in Step 1, verify whether the condition $H^T E = E^T H \ge 0$ holds or not. If this condition holds, then H is a feasible solution; if not, go to the Step 1.

Remark 6.4. Noting the trigger function in ETC (6.3), there is a specific form of the trigger condition which relies on the network information, namely

$$t^i_{k+1} = \inf\{t > t^i_k | \mathcal{F}_i(t) > 0\},$$

where $\mathcal{F}_i(t) = \|\hat{e}_i(t)\|^2 + q_i \|\hat{e}_{0i}(t)\|^2 - \phi_i(t)$, with

$$\phi_i(t) = \sigma_i \left\| \sum_{j=1}^{\mathcal{N}} p_{ij}(x_i(t) - x_j(t)) \right\|^2 + \sigma_i q_i \|x_i(t) - x_0(t)\|^2 \ge 0, \quad \sigma_i > 0.$$

Obviously, the parameter ξ can be selected as $\xi = \sigma \lambda_{\max}\{\mathcal{L}^{\mathrm{T}}\mathcal{L} + \mathcal{Q}\}$, with $\sigma = \max_{i}\{\sigma_i\}$. Therefore, similar to the above analysis in Theorem 6.1, the exponential consensus of FOD-NMASs can also be achieved by the proposed ETC.

6.3.2 Reachability analysis

Design the following event-triggered SMC law:

$$\begin{cases} \boldsymbol{u}_i(t) = \hat{\boldsymbol{u}}_{il}(t) + \hat{\boldsymbol{u}}_{iw}(t), \\[2mm] \hat{\boldsymbol{u}}_{il}(t) = \boldsymbol{K}\left\{ \sum_{j=1}^{\mathcal{N}} p_{ij}(\boldsymbol{x}_i(t_k^i) - \boldsymbol{x}_j(t_{k'}^j)) + q_i(\boldsymbol{x}_i(t_k^i) - \boldsymbol{x}_0(t_k^i)) \right\}, \\[2mm] \hat{\boldsymbol{u}}_{iw}(t) = -(\boldsymbol{GB})^{-1}\{\delta_1\|\boldsymbol{G}\|\|\boldsymbol{x}_i(t_k^i)\| + \delta_2 + \delta_3\}\mathrm{sgn}(s_i(t)), \end{cases} \tag{6.23}$$

where $\delta_2 \geq \sup\{\delta_1\|\boldsymbol{G}\|(\|\hat{\boldsymbol{e}}_i(t)\| + \|\boldsymbol{x}_0(t)\|)\} \geq 0$ and $\delta_3 \geq 0$.

Theorem 6.2. *Assume that Assumptions 6.1–6.4 hold. Then the trajectory of FOD-NMASs can be forced to the given integral sliding surface (6.4) in finite time under the presented event-triggered controller (6.23).*

Proof. Construct the following Lyapunov function:

$$V_i(t) = \frac{1}{2}s_i^{\mathrm{T}}(t)s_i(t). \tag{6.24}$$

Then

$$\begin{aligned} {}_{t_0}^{\mathcal{C}}D_t^{\alpha} V_i(t) &\leq s_i^{\mathrm{T}}(t) \cdot {}_{t_0}^{\mathcal{C}}D_t^{\alpha} s_i(t) \\ &= s_i^{\mathrm{T}}(t)\{\boldsymbol{GE}\,{}_{t_0}^{\mathcal{C}}D_t^{\alpha} e_i(t) - \boldsymbol{GA}e_i(t) - \boldsymbol{GB}\hat{\boldsymbol{u}}_{il}(t)\} \\ &= s_i^{\mathrm{T}}(t)\{\boldsymbol{G}(\boldsymbol{f}(\boldsymbol{x}_i(t)) - \boldsymbol{f}(\boldsymbol{x}_0(t))) + \boldsymbol{GB}\hat{\boldsymbol{u}}_{iw}(t)\} \\ &\leq \|s_i^{\mathrm{T}}(t)\|\|\boldsymbol{G}\|\|\boldsymbol{f}(\boldsymbol{x}_i(t)) - \boldsymbol{f}(\boldsymbol{x}_0(t))\| + s_i^{\mathrm{T}}(t)\boldsymbol{GB}\hat{\boldsymbol{u}}_{iw}(t). \end{aligned} \tag{6.25}$$

According to Assumption 6.4, one has

$${}_{t_0}^{\mathcal{C}}D_t^{\alpha} V_i(t) \leq \|s_i^{\mathrm{T}}(t)\|\|\boldsymbol{G}\|\delta_1\|\boldsymbol{x}_i(t) - \boldsymbol{x}_0(t)\| - \{\delta_1\|\boldsymbol{G}\|\|\boldsymbol{x}_i(t_k^i)\| + \delta_2 + \delta_3\}\|s_i(t)\|. \tag{6.26}$$

Since $\|\boldsymbol{x}_i(t) - \boldsymbol{x}_0(t)\| \leq \|\boldsymbol{x}_i(t_k^i)\| + \|\hat{\boldsymbol{e}}_i(t)\| + \|\boldsymbol{x}_0(t)\|$, then

$${}_{t_0}^{\mathcal{C}}D_t^{\alpha} V_i(t) \leq -\delta_3\|s_i(t)\|, \tag{6.27}$$

i.e.,

$${}_{t_0}^{\mathcal{C}}D_t^{\alpha} V_i(t) \leq -\delta_3\sqrt{2V_i(t)}. \tag{6.28}$$

Then, according to Lemma 2.6, we can easily obtain $\lim\limits_{t \to T_s^i} s_i(t) = 0$ when $t \geq T_s$, where the settling time is

$$T_s^i = \left(\frac{\Gamma(\frac{5}{4} - \alpha)\Gamma(1 + \alpha)}{4\delta_3 \Gamma(\frac{5}{4} - \alpha)} \|s_i(0)\| \right)^{\frac{1}{\alpha}}. \tag{6.29}$$

$\square$

Remark 6.5. Noting the controller in (6.23) and replacing $\mathrm{sgn}(s_i(t))$ with $\mathrm{sgn}(s_i(t_k^i))$, we can obtain another approach in designing the controller as follows:

$$\begin{cases} \boldsymbol{u}_i(t) = \hat{\boldsymbol{u}}_{il}(t) + \hat{\boldsymbol{u}}_{iw}(t), \\ \hat{\boldsymbol{u}}_{iw}(t) = -(\boldsymbol{GB})^{-1}\{\delta_1 \|\boldsymbol{G}\| \|\boldsymbol{x}_i(t_k^i)\| + \delta_2 + \delta_3\}\mathrm{sgn}(s_i(t_k^i)). \end{cases} \tag{6.30}$$

Under this control strategy, it can also be proved that the trajectory remains on the practical sliding mode surface.

Remark 6.6. As we know, sliding mode control (SMC), as an effective nonlinear control method, has been widely used in various engineering fields [42,43]. Generally speaking, the design of SMC can be divided into two steps:

Step 1   Design a suitable sliding surface $s_i(t)$ to make the sliding mode dynamic have good dynamic characteristics. As shown in Theorem 6.1, we have presented a novel leader-following exponential consensus control strategy for FOD-NMASs under Assumptions 6.1–6.3;

Step 2   Design a suitable sliding mode controller $\boldsymbol{u}_i(t)$ to guarantee the finite-time reachability of a sliding surface. As shown in Theorem 6.2, due to Assumptions 6.1–6.4 and based on the gain matrix $\boldsymbol{K}$ which was obtained in Theorem 6.1, using the presented event-triggered controller (6.23), the trajectory of FOD-NMASs can be forced to the given integral sliding surface (6.4) in finite time.

6.3.3 Discussion of Zeno behavior

As we know, the fractional-order calculus operator reflects not a local feature or property of the function, but a distributed influence that takes into account a combination of historical information, which is also known as the long memory property. In addition, according to Definition 6.2, we know that the system (6.1) should be a continuous time system rather than a discrete time system, therefore it is appropriate to consider Zeno behavior when we use the event-triggered mechanism to explore the present problem. Seen from the event-triggered mechanism, how to avoid the Zeno behavior is a crucial research. It requires the time interval between two adjacent trigger moments be a positive constant. Based on this consideration, we propose the following theorem.

Theorem 6.3. *Under the assumptions and the conditions in Theorems 6.1 and 6.2, if $\mathcal{F}_i(t) > 0$ for all $t \geq t_0$, then the Zeno behavior can be avoided for each agent i.*

Proof. According to Theorems 6.1 and 6.2, the error trajectories $e_i(t)$ of the tracking error system (6.6) are norm-bounded. And considering Assumption 6.4, it means that the pseudo-state trajectories $x_i(t)$ of system (6.2) are norm-bounded. Furthermore, combining with Assumption 6.5, it can be obtained that ${}^{C}_{t_0}D_t^\alpha x_i(t)$ is also norm-bounded for each agent i. Therefore, from $\hat{e}_i(t) = x_i(t_k^i) - x_i(t)$, one has that

$$
\begin{aligned}
\|\hat{e}_i(t)\| &= \|x_i(t_k^i) - x_i(t)\| \\
&= \left\| {}_{t_0}D_{t_k^i}^{-\alpha}\left({}^{C}_{t_0}D_{t_k^i}^{\alpha}x_i(t_k^i)\right) - {}_{t_0}D_t^{-\alpha}\left({}^{C}_{t_0}D_t^{\alpha}x_i(t)\right)\right\| \\
&= \left\| \frac{1}{\Gamma(\alpha)}\int_{t_0}^{t_k^i}\frac{{}^{C}_{t_0}D_\tau^{\alpha}x_i(\tau)}{(t_k^i-\tau)^{1-\alpha}}\,\mathrm{d}\tau - \frac{1}{\Gamma(\alpha)}\int_{t_0}^{t}\frac{{}^{C}_{t_0}D_\tau^{\alpha}x_i(\tau)}{(t-\tau)^{1-\alpha}}\,\mathrm{d}\tau\right\| \\
&= \left\| \frac{1}{\Gamma(\alpha)}\int_{t_0}^{t_k^i}\left[\frac{{}^{C}_{t_0}D_\tau^{\alpha}x_i(\tau)}{(t_k^i-\tau)^{1-\alpha}} - \frac{{}^{C}_{t_0}D_\tau^{\alpha}x_i(\tau)}{(t-\tau)^{1-\alpha}}\right]\mathrm{d}\tau \right. \\
&\quad \left. - \frac{1}{\Gamma(\alpha)}\int_{t_k^i}^{t}\frac{{}^{C}_{t_0}D_\tau^{\alpha}x_i(\tau)}{(t-\tau)^{1-\alpha}}\,\mathrm{d}\tau\right\|.
\end{aligned}
\tag{6.31}
$$

Since ${}^{C}_{t_0}D_t^{\alpha}x_i(t)$ is norm-bounded, there exists $M_i \geq 0$ such that $\|{}^{C}_{t_0}D_t^{\alpha}x_i(t)\| \leq M_i$ holds, i.e., there exists a positive constant $\varepsilon \to 0$ such that $\|{}^{C}_{t_0}D_t^{\alpha}x_i(t)\| < M_i + \varepsilon$ holds, and then one has

$$
\begin{aligned}
\|\hat{e}_i(t)\| &< \frac{M_i+\varepsilon}{\Gamma(\alpha)}\left|\int_{t_0}^{t_k^i}[(t_k^i-\tau)^{\alpha-1}-(t-\tau)^{\alpha-1}]\mathrm{d}\tau\right| \\
&\quad + \frac{M_i+\varepsilon}{\Gamma(\alpha)}\left|\int_{t_k^i}^{t}(t-\tau)^{\alpha-1}\mathrm{d}\tau\right| \\
&= \frac{M_i+\varepsilon}{\Gamma(\alpha+1)}\left|(t_k^i-t_0)^\alpha + (t-t_k^i)^\alpha - (t-t_0)^\alpha\right| \\
&\quad + \frac{M_i+\varepsilon}{\Gamma(\alpha+1)}(t-t_k^i)^\alpha \\
&\leq \frac{\gamma_i(M_i+\varepsilon)}{\Gamma(\alpha+1)}(t-t_k^i)^\alpha,
\end{aligned}
\tag{6.32}
$$

where $\gamma_i \geq 2$. According to (6.32), we have $\|\hat{e}_i(t)\| < \frac{\gamma_i(M_i+\varepsilon)}{\Gamma(\alpha+1)}(t-t_k^i)^\alpha$. Similarly, it can also be obtained that $\|\hat{e}_{0i}(t)\| < \frac{\gamma_0(M_0+\varepsilon)}{\Gamma(\alpha+1)}(t-t_k^i)^\alpha$ with $\gamma_0 \geq 2$ holds for the given leader system.

On the other hand, when $t = t^i_{k+1}$, we have $\mathcal{F}_i(t) = 0$, that is to say, $\|\hat{e}_i(t^i_{k+1})\|^2 + q_i\|\hat{e}_{0i}(t^i_{k+1})\|^2 = \phi_i(t^i_{k+1})$ with $\phi_i(t) = \sigma_i \left\| \sum\limits_{j=1}^{N} p_{ij}(x_i(t) - x_j(t)) \right\|^2 + \sigma_i q_i \|x_i(t) - x_0(t)\|^2 \geq 0$, and then

$$\phi_i(t^i_{k+1}) < \left[\frac{\gamma_i(M_i + \varepsilon)}{\Gamma(\alpha + 1)}(T^i_k)^\alpha\right]^2 + q_i\left[\frac{\gamma_0(M_0 + \varepsilon)}{\Gamma(\alpha + 1)}(T^i_k)^\alpha\right]^2, \tag{6.33}$$

where $T^i_k = t^i_{k+1} - t^i_k$.

Case 1 If $\inf\limits_{k \geq 1}\{T^i_k\} > 0$, it is seen that Zeno behavior will not happen;

Case 2 If $\inf\limits_{k \geq 1}\{T^i_k\} = 0$, it can be obtained that $\phi_i(t^i_{k+1}) < 0$, however, it contradicts with $\phi_i(t^i_{k+1}) \geq 0$ for any $t \geq t_0$.

Therefore, Zeno behavior does not occur. $\qquad\square$

Remark 6.7. Noticing (6.32) and (6.33) and by introducing a sufficiently small positive constant ε to transform them into strict inequalities to verify whether (Case 2) holds or not, we can easily obtain a criterion so that Zeno behavior will not happen.

Remark 6.8. To our knowledge, there exists another method to avoid Zeno behavior in the existing literature, namely by giving a specific mathematical form of inter-execution intervals T_k. However, as mentioned in Remark 2.1, some results obtained from Lemma 2.4 are incorrect and cannot guarantee that $T_k > 0$ holds. For example, by utilizing the Gronwall–Bellman inequality, Feng et al. gave the inter-execution intervals as follows [44]:

$$T_k = t_{k+1} - t_k > \left(\frac{\Gamma(1 + \alpha)\ln(\sqrt{\rho}\|\hat{x}_{t_k}\|)}{\|A\| + \beta(\|\hat{x}_{t_k}\|)}\right)^{\frac{1}{\alpha}}, \tag{6.34}$$

where $\beta(\|\hat{x}_{t_k}\|) \geq 0$, $\rho > 0$, $k = 0, 1, 2, \ldots$ It is worth pointing out that $T_k > 0$ will not hold when $\|\hat{x}_{t_k}\| \to 0$. To overcome this problem, Cheng et al. [45] utilized Laplace transform to prove that the proposed event-triggered control method is feasible and Zeno behavior can also be avoided.

6.4 Numerical examples

Example 6.1. Consider the following FOD-NMASs with 7 followers:

$$\begin{cases} E^{\mathcal{C}}_{t_0}D^\alpha_t x_0(t) = Ax_0(t) + f(x_0(t)), \\ E^{\mathcal{C}}_{t_0}D^\alpha_t x_i(t) = Ax_i(t) + f(x_i(t)) + Bu_i(t), \end{cases} \tag{6.35}$$

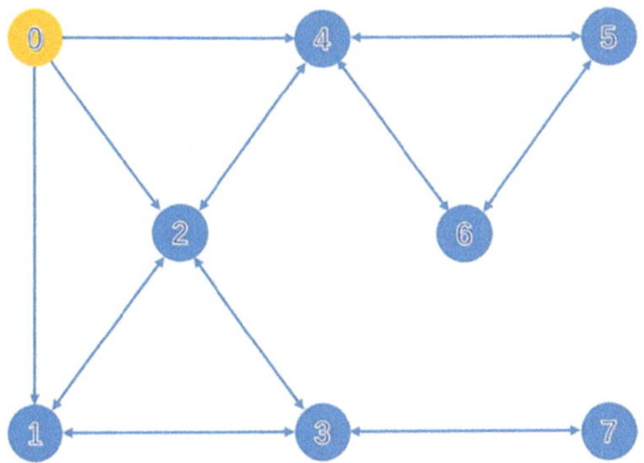

FIGURE 6.1   Interaction topology graph with seven followers.

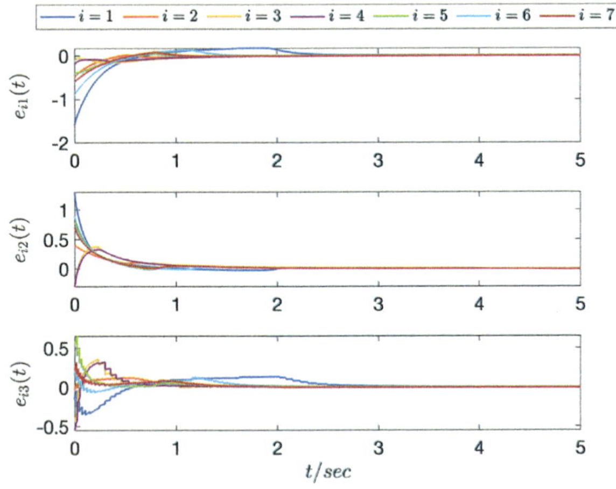

FIGURE 6.2   Tracking errors of agents when $\delta_1 = 0.2$, $\delta_2 = 0.25$.

where $\alpha = 0.9$, $\boldsymbol{E} = \begin{bmatrix} 1 & 0 & 0 \\ 0 & 1 & 0 \\ 0 & 0 & 0 \end{bmatrix}$, $\boldsymbol{A} = \begin{bmatrix} -0.400 & 1.566 & -3.564 \\ 2.132 & -3.950 & -0.968 \\ 1 & 1 & -1 \end{bmatrix}$, $\boldsymbol{B} = \begin{bmatrix} 0.25 & 0 \\ 0 & 0.25 \\ 0 & 0 \end{bmatrix}$, and $\boldsymbol{f}(\boldsymbol{x}_i(t)) = \begin{bmatrix} -0.1\sin(\boldsymbol{x}_{i1}(t)) \\ 0.1\cos(\boldsymbol{x}_{i2}(t)) \\ 0.2\cos(\boldsymbol{x}_{i1}(t))\sin(\boldsymbol{x}_{i2}(t)) \end{bmatrix}$.

From Fig. 6.1, the adjacency and weight matrices are

$$\mathcal{P} = \begin{bmatrix} 0 & 1 & 1 & 0 & 0 & 0 & 0 \\ 1 & 0 & 1 & 1 & 0 & 0 & 0 \\ 1 & 1 & 0 & 0 & 0 & 0 & 1 \\ 0 & 1 & 0 & 0 & 1 & 1 & 0 \\ 0 & 0 & 0 & 1 & 0 & 1 & 0 \\ 0 & 0 & 0 & 1 & 1 & 0 & 0 \\ 0 & 0 & 1 & 0 & 0 & 0 & 0 \end{bmatrix}, \mathcal{Q} = \begin{bmatrix} 1 & 0 & 0 & 0 & 0 & 0 & 0 \\ 0 & 1 & 0 & 0 & 0 & 0 & 0 \\ 0 & 0 & 0 & 0 & 0 & 0 & 0 \\ 0 & 0 & 0 & 1 & 0 & 0 & 0 \\ 0 & 0 & 0 & 0 & 0 & 0 & 0 \\ 0 & 0 & 0 & 0 & 0 & 0 & 0 \\ 0 & 0 & 0 & 0 & 0 & 0 & 0 \end{bmatrix}.$$

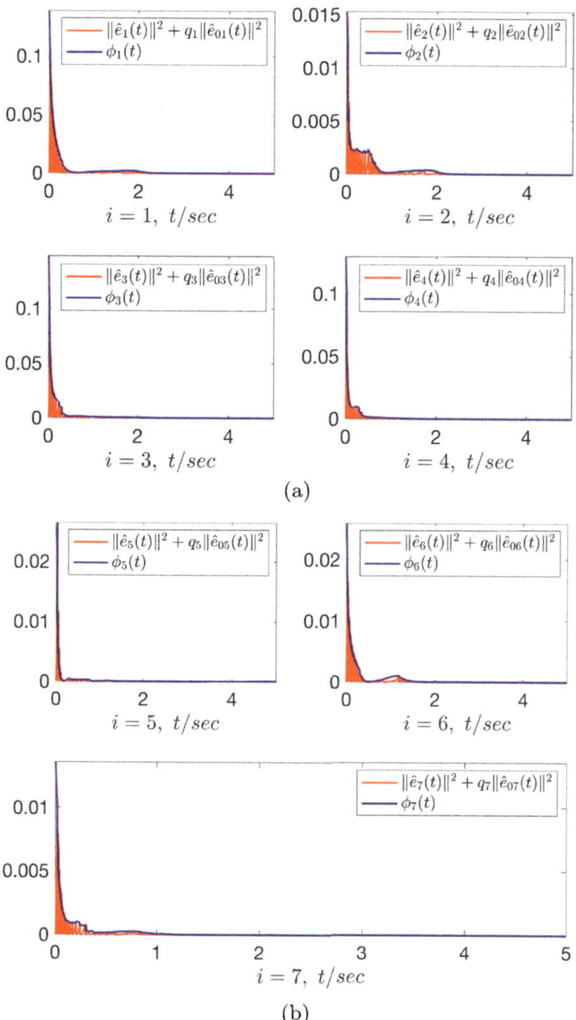

FIGURE 6.3   The evolution of $\|\hat{e}_i(t)\|^2 + q_i\|\hat{e}_{0i}(t)\|^2$.

Then matrix $\mathcal{L}$ is

$$\mathcal{L} = \begin{bmatrix} 3 & -1 & -1 & 0 & 0 & 0 & 0 \\ -1 & 4 & -1 & -1 & 0 & 0 & 0 \\ -1 & -1 & 3 & 0 & 0 & 0 & -1 \\ 0 & -1 & 0 & 4 & -1 & -1 & 0 \\ 0 & 0 & 0 & -1 & 2 & -1 & 0 \\ 0 & 0 & 0 & -1 & -1 & 2 & 0 \\ 0 & 0 & -1 & 0 & 0 & 0 & 1 \end{bmatrix}.$$

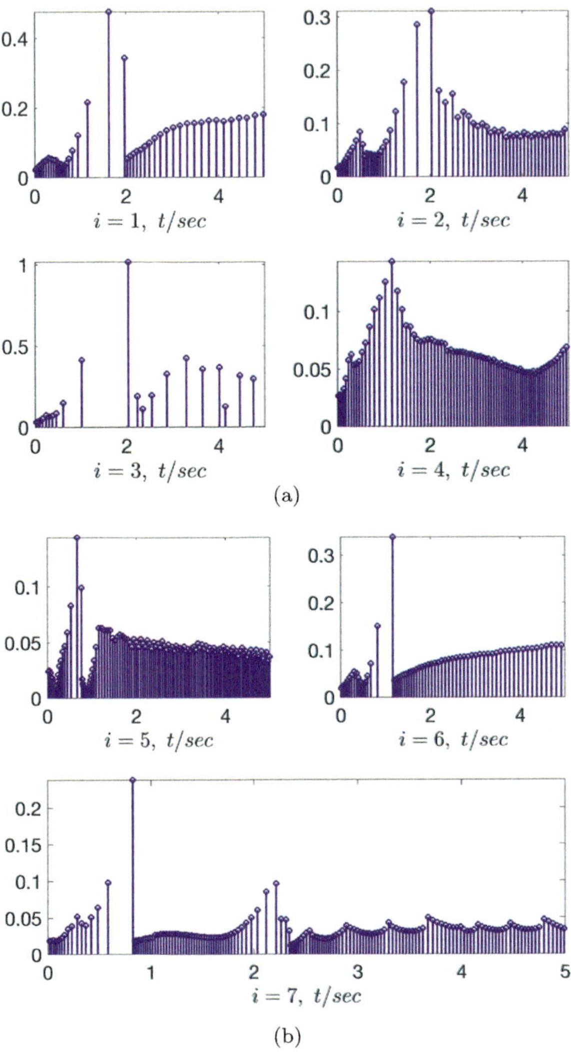

FIGURE 6.4 The event intervals of leader–follower systems.

The initial states of the leader and follower systems are chosen as $x_1(0) = [-0.8, 1.1, 0.42]^{\mathrm{T}}$, $x_2(0) = [0.3, 0.2, 0.54]^{\mathrm{T}}$, $x_3(0) = [0.7, -0.5, 0.13]^{\mathrm{T}}$, $x_4(0) = [0.6, -0.5, 0.02]^{\mathrm{T}}$, $x_5(0) = [0.4, -0.6, 1.10]^{\mathrm{T}}$, $x_6(0) = [-0.1, 0.7, 0.73]^{\mathrm{T}}$, $x_7(0) = [0.2, 0.5, 0.79]^{\mathrm{T}}$, and $x_0(0) = [0.8, -0.2, 0.57]^{\mathrm{T}}$. In ETC (6.3), if we choose $\phi_i(t) = \sigma_i \left\| \sum_{j=1}^{\mathcal{N}} p_{ij}(x_i(t) - x_j(t)) \right\|^2 + \sigma_i q_i \|x_i(t) - x_0(t)\|^2$, then, according to Theorems 6.1 and 6.2, some relevant parameters are chosen as

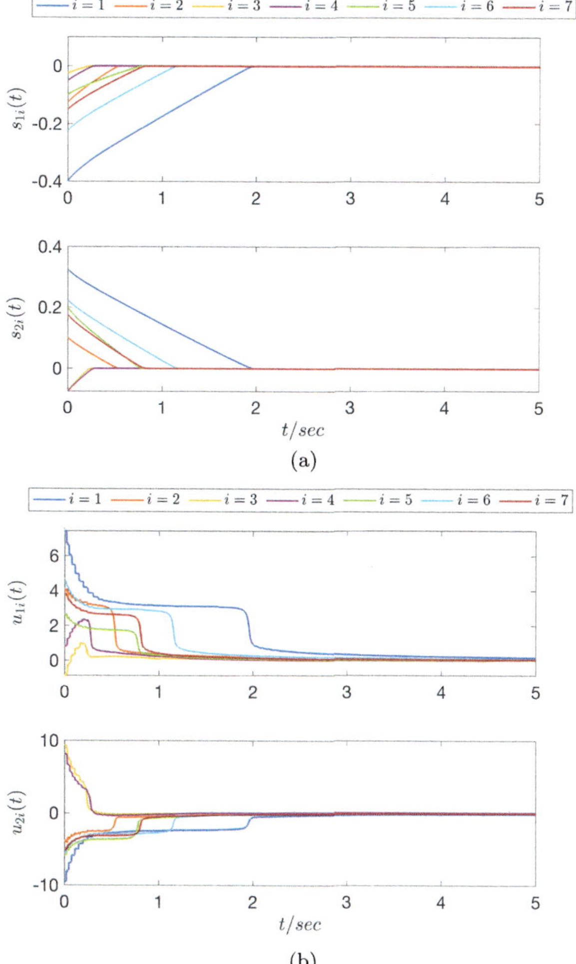

FIGURE 6.5 Response curves of sliding mode surface and event-triggered SMC law.

$\sigma = 0.008$, $\varrho = 0.9$, $\min_{i}\{\mu_i - \frac{\omega_{1i}\mu_i}{2} - \frac{\omega_{2i}q_i}{2}\} = 0.0036$, $\omega = 0.3$, by solving the inequality (6.7) in Theorem 6.1, the corresponding matrices $\boldsymbol{H}$ and $\boldsymbol{K}$ are obtained as follows:

$$\boldsymbol{H} = \begin{bmatrix} 1.659 & -1.676 & 0 \\ -1.676 & 4.631 & 0 \\ -2.146 & 0.471 & 3.534 \end{bmatrix}, \quad \boldsymbol{K} = \begin{bmatrix} -0.415 & 0.419 & 0 \\ 0.419 & -1.158 & 0 \end{bmatrix}.$$

Letting $\delta_1 = 0.2$, $\delta_2 = 0.25$, and $\delta_3 = 0.01$ in controller (6.23), several simulation results are obtained as presented in Figs. 6.2 to 6.6. Fig. 6.2 plots

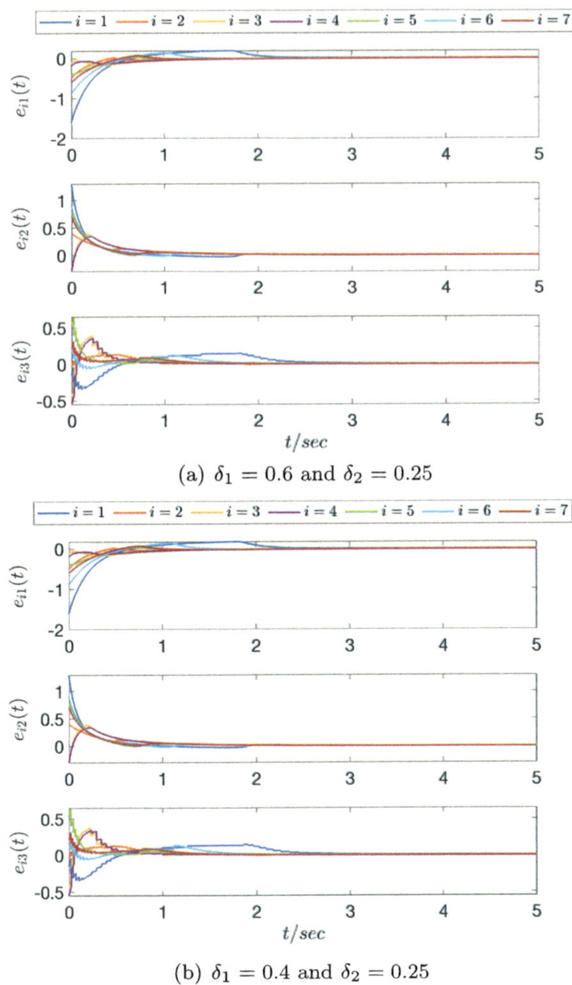

(a) $\delta_1 = 0.6$ and $\delta_2 = 0.25$

(b) $\delta_1 = 0.4$ and $\delta_2 = 0.25$

FIGURE 6.6 Tracking error response curves under different parameters.

the tracking errors of each agent, and shows that the followers can well track the trajectory of the leader. Fig. 6.3 depicts the evolution of ETC (6.3). Meanwhile, the triggering intervals of seven agents are shown in Fig. 6.4. Furthermore, from Fig. 6.4, we know that Zeno behavior will not happen. In Fig. 6.5, the curves of sliding mode surface and controller are presented, respectively. By changing the parameters δ_1 and δ_2 in controller (6.23), some simulation results are listed in Table 6.1 and Fig. 6.6. Combining with Fig. 6.2, it is obvious that the growth rate of sampling numbers for δ_1 is less than that for δ_2.

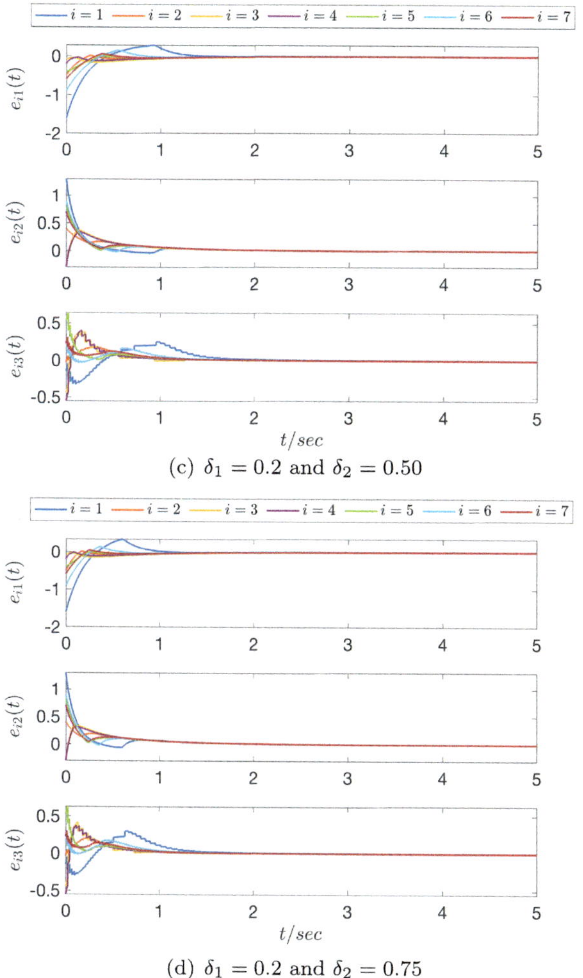

(c) $\delta_1 = 0.2$ and $\delta_2 = 0.50$

(d) $\delta_1 = 0.2$ and $\delta_2 = 0.75$

FIGURE 6.6 *(continued)*

TABLE 6.1 Event-triggering numbers with $\alpha = 0.9$ and for various δ_1, δ_2.

δ_1	δ_2	1	2	3	4	5	6	7	Figure
0.6	0.25	54	64	34	96	96	74	257	6.6(a)
0.4	0.25	53	62	33	90	93	71	246	6.6(b)
0.2	**0.25**	**52**	**59**	**32**	**84**	**90**	**69**	**216**	**6.2**
0.2	0.50	72	104	47	173	177	119	351	6.6(c)
0.2	0.75	129	155	65	263	273	170	446	6.6(d)

Example 6.2. According to the Kirchhoff's laws [46], Zhan and Ma [47] presented the electrical circuit shown in Fig. 6.7, which can be described by the fractional-order descriptor nonlinear system as follows:

$$\begin{cases} e_1 = u_{R_1} + L_1 \frac{d^\alpha i_1}{dt^\alpha} + R_3 i_3 + L_3 \frac{d^\alpha i_3}{dt^\alpha}, \\ e_2 = R_2 i_2 + L_2 \frac{d^\alpha i_2}{dt^\alpha} + R_3 i_3 + L_3 \frac{d^\alpha i_3}{dt^\alpha} + u_w, \\ i_1 + i_2 - i_3 = 0, \end{cases} \tag{6.36}$$

where L_1, L_2, L_3 denote fractional-order inductances, R_1 denotes a fractional-order nonlinear resistance with the voltage $u_{R_1} = R_1 i_1 - 0.3 \sin(i_1)$, R_2 and R_3 denote linear resistances, e_1 and e_2 denote source voltages, while W denotes the unknown disturbance signal.

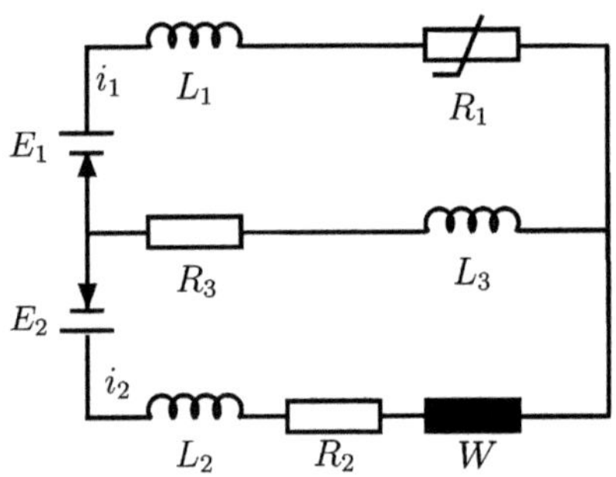

FIGURE 6.7   Fractional-order electrical circuit adopted in [47].

In a parallel circuit, to avoid large deviations between the currents of each circuit, it is necessary to consider the consensus problem of the current in the corresponding circuit network. Consider the following FOD-NMASs described by

$$\begin{cases} E_{t_0}^C D_t^\alpha x_0(t) = A x_0(t) + f(x_0(t)), \\ E_{t_0}^C D_t^\alpha x_i(t) = A x_i(t) + f(x_i(t)) + B u_i(t), \end{cases} \tag{6.37}$$

where $x(t) = [i_1, i_2, i_3]^\mathrm{T}$, $u(t) = [e_1, e_2]^\mathrm{T}$, and the parameters of system (6.37) are

$$E = \begin{bmatrix} L_1 & 0 & L_3 \\ 0 & L_2 & L_3 \\ 0 & 0 & 0 \end{bmatrix}, \quad A = \begin{bmatrix} -R_1 & 0 & -R_3 \\ 0 & -R_2 & -R_3 \\ 1 & 1 & -1 \end{bmatrix},$$

$$B = \begin{bmatrix} 1 & 0 \\ 0 & 1 \\ 0 & 0 \end{bmatrix}, \quad f(x_i(t)) = \begin{bmatrix} -0.3 \sin(x_{i1}(t)) \\ -0.25 \sin(t) \\ 0 \end{bmatrix},$$

with $L_1 = L_2 = L_3 = 1$, $R_2 = 1$, $R_3 = 2$, and $\alpha = 0.8$.

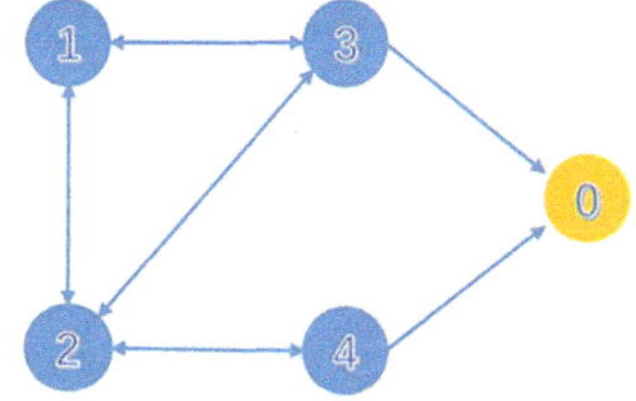

FIGURE 6.8   Interaction topology graph with four followers.

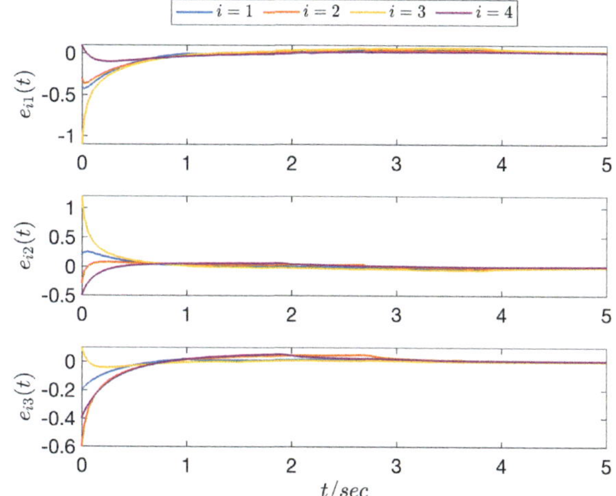

FIGURE 6.9   Tracking errors of agents.

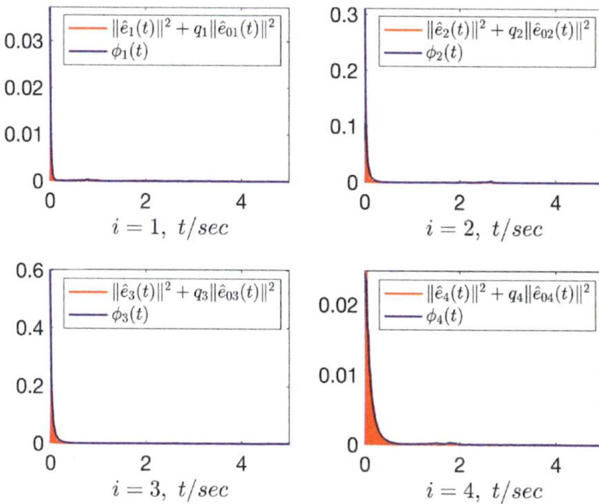

FIGURE 6.10   The evolution of event-triggered function.

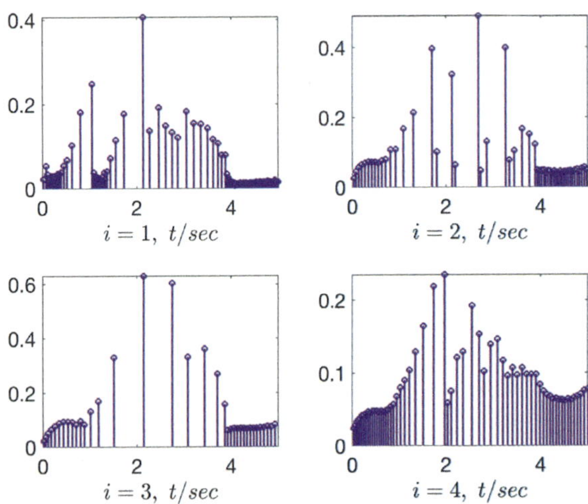

FIGURE 6.11 The event intervals of leader–follower systems.

According to the interaction topology graph shown in Fig. 6.8, choose the following parameters:

$$\mathcal{P} = \begin{bmatrix} 0 & 1 & 1 & 0 \\ 1 & 0 & 1 & 1 \\ 1 & 1 & 0 & 0 \\ 0 & 1 & 0 & 0 \end{bmatrix}, \quad \mathcal{Q} = \begin{bmatrix} 0 & 0 & 0 & 0 \\ 0 & 0 & 0 & 0 \\ 0 & 0 & 0.2 & 0 \\ 0 & 0 & 0 & 0.4 \end{bmatrix}.$$

Then matrix $\mathcal{L}$ is

$$\mathcal{L} = \begin{bmatrix} 2 & -1 & -1 & 0 \\ -1 & 3 & -1 & -1 \\ -1 & -1 & 2.2 & 0 \\ 0 & -1 & 0 & 1.4 \end{bmatrix}.$$

It means that Assumptions 6.1–6.3 are all satisfied. Also the initial states are chosen as $x_1(0) = [0.1, -0.2, 0.1]^{\mathrm{T}}$, $x_2(0) = [0.2, -0.7, -0.5]^{\mathrm{T}}$, $x_3(0) = [-0.6, 0.8, 0.2]^{\mathrm{T}}$, $x_4(0) = [0.6, -0.9, -0.3]^{\mathrm{T}}$, and $x_0(0) = [0.5, -0.4, 0.1]^{\mathrm{T}}$. In controller (6.23), we choose $\sigma = 0.06$, $\delta_1 = 0.3$, $\delta_2 = 0.5$, and $\delta_3 = 0.01$. In Theorems 6.1 and 6.2, choose $\varrho = 3.15$, $\min_i\{\mu_i - \frac{\omega_{1i}\mu_i}{2} - \frac{\omega_{2i}q_i}{2}\} = 0.13$. Moreover, by solving the inequality (6.7) in Theorem 6.1, the corresponding matrices H and K are obtained as follows:

$$H = \begin{bmatrix} 1.9369 & -0.7000 & 1.2369 \\ -0.7000 & 1.7800 & 1.0800 \\ -2.1550 & -2.2161 & 2.8530 \end{bmatrix},$$

$$K = \begin{bmatrix} -1.9369 & 0.7000 & -1.2390 \\ 0.7000 & -1.7800 & -1.0800 \end{bmatrix}.$$

Several simulation results are demonstrated in Figs. 6.9 to 6.11. From Fig. 6.9, we know that the followers can well track the trajectory of the leader and, according to Fig. 6.11, Zeno behavior will not happen under the presented control strategy.

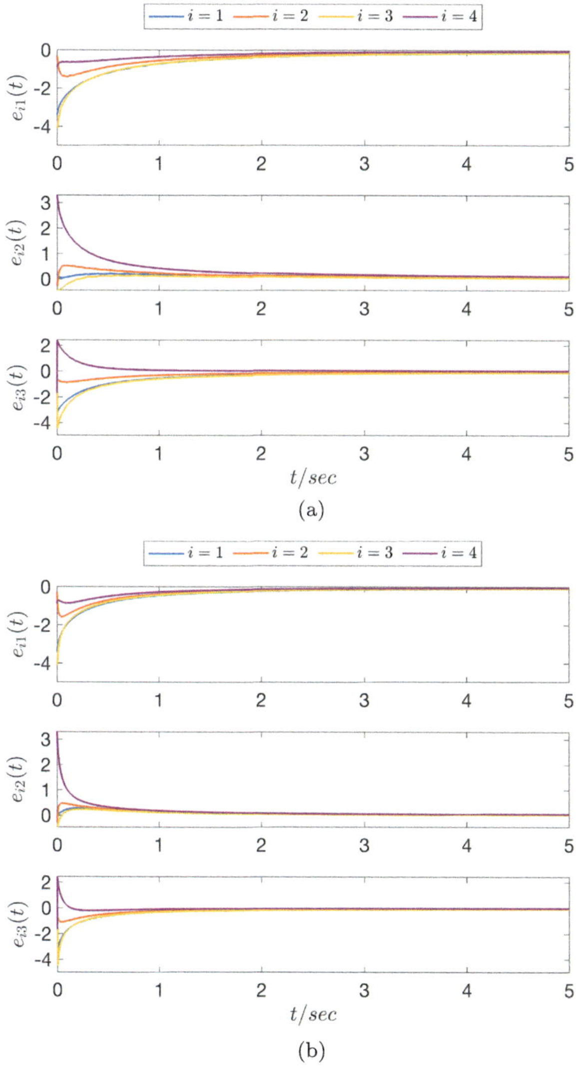

FIGURE 6.12   Tracking errors of agents: (a) Chapter 6, (b) those in [30].

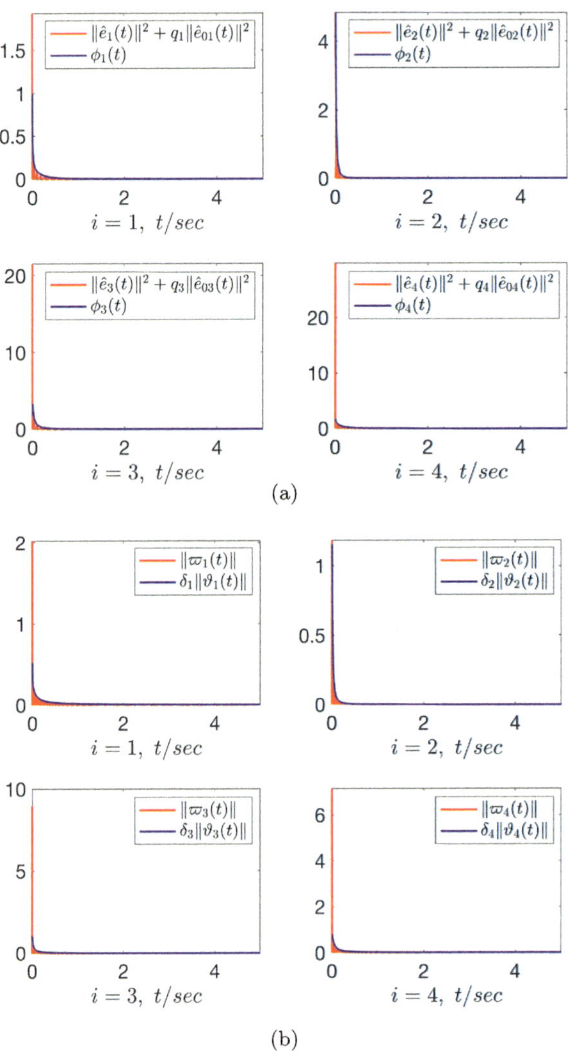

FIGURE 6.13   The evolution of event-triggered function: (a) Chapter 6, (b) that in [30].

Example 6.3. In [30], the authors considered the exponential consensus problem of FOD-LMASs, therefore, based on Example 6.2, we present a simulation result to compare with the control strategy in [30]. The initial states of the leader and follower systems are chosen as $x_1(0) = [0.1, -1.2, 0.3]^T$, $x_2(0) = [3.2, -1.7, 1.5]^T$, $x_3(0) = [-0.6, -1.8, 2.2]^T$, $x_4(0) = [2.6, 1.9, -0.9]^T$, and $x_0(0) = [3.5, -1.4, 0.8]^T$. In controller (6.23), choose $\sigma = 0.06$, $\delta_1 = \delta_2 = 0$, and $\delta_3 = 0.01$. Fig. 6.12 shows the tracking errors of each agent, both in our work and in [30]. Fig. 6.13 depicts the evolu-

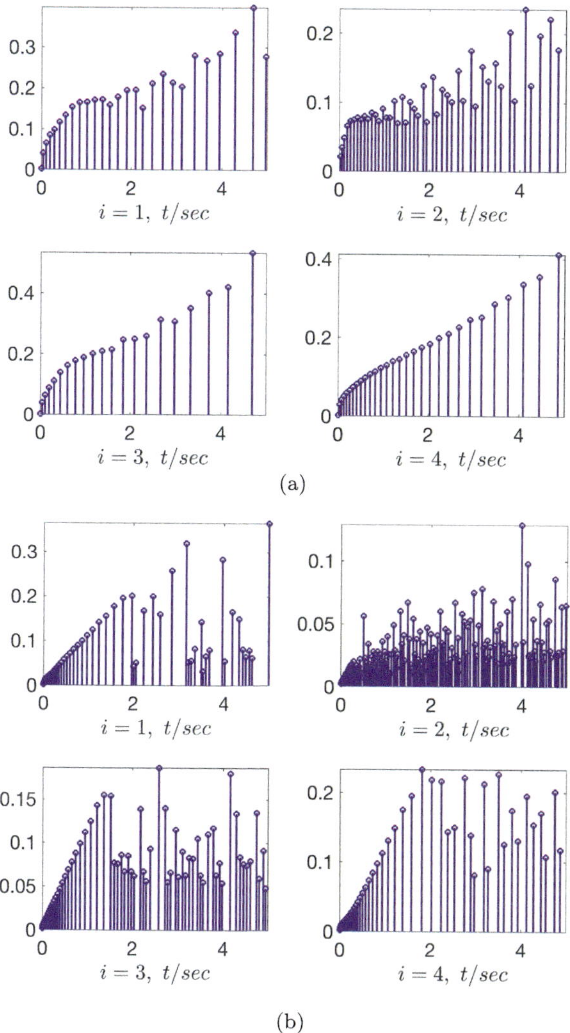

(a)

(b)

FIGURE 6.14 The event intervals of leader–follower systems: (a) Chapter 6, (b) those in [30].

tion of different ETCs. Meanwhile, the triggering intervals of four agents are shown in Fig. 6.14, and Zeno behavior will not happen for these two control methods. Furthermore, from Fig. 6.15, compared with the control strategy proposed in the literature [30], the controller designed in this chapter can save more resources. Lastly, it is obviously seen from Table 6.2 that the sampling numbers under the presented event-triggered mechanism are smaller than those in [30].

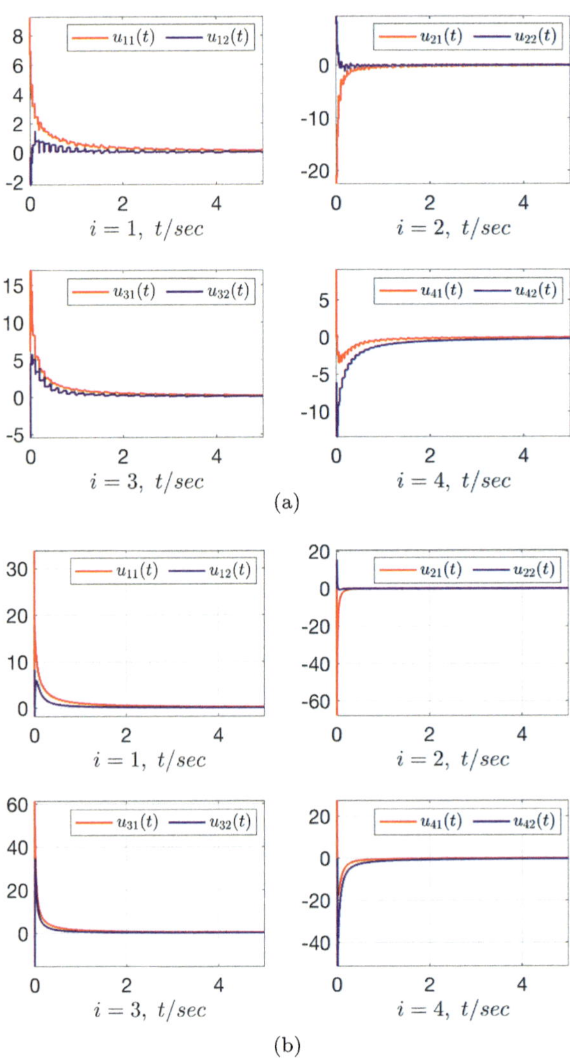

FIGURE 6.15   Event-triggered controller signals: (a) Chapter 6, (b) those in [30].

TABLE 6.2   Event-triggering numbers compared to methods in [30].

Agents		1	2	3	4
$\alpha = 0.90$	Our strategy	48	91	34	44
	[30]	55	182	65	54
$\alpha = 0.85$	Our strategy	34	60	26	35
	[30]	47	249	69	52
$\alpha = 0.80$	**Our strategy**	27	45	21	30
	[30]	55	184	72	53

6.5 Conclusion

The exponential consensus problem of FOD-NMASs has been explored by presenting an event-triggered SMC strategy. First, by designing a fractional-order sliding surface and a controller, the exponential consensus for each agent has been achieved under the Mittag-Leffler stability. Second, the finite-time reachability of state trajectories to the predefined sliding surface has been guaranteed by the proposed event-triggered SMC strategy for each agent. Meanwhile, a condition has also been obtained to avoid the Zeno behavior, which verified the validity of the presented event-triggered mechanism. Finally, three simulation examples have been given to illustrate the proposed method. Besides, the consensus control problem of heterogeneous/homogeneous FO-MASs will be included in further research.

References

[1] A. Dorri, S.S. Kanhere, R. Jurdak, Multi-agent systems: a survey, IEEE Access 6 (2018) 28573–28593.

[2] D. Ye, M. Zhang, A.V. Vasilakos, A survey of self-organization mechanisms in multi-agent systems, IEEE Transactions on Systems, Man, and Cybernetics: Systems 47 (3) (2016) 441–461.

[3] P.G. Balaji, D. Srinivasan, An introduction to multi-agent systems, in: Innovations in Multi-Agent Systems and Applications – 1, 2010, pp. 1–27.

[4] A. Nedic, A. Ozdaglar, P.A. Parrilo, Constrained consensus and optimization in multi-agent networks, IEEE Transactions on Automatic Control 55 (4) (2010) 922–938.

[5] G. Dudek, M.R.M. Jenkin, E. Milios, D. Wilkes, A taxonomy for multi-agent robotics, Autonomous Robots 3 (1996) 375–397.

[6] J. Qin, Q. Ma, Y. Shi, L. Wang, Recent advances in consensus of multi-agent systems: a brief survey, IEEE Transactions on Industrial Electronics 64 (6) (2016) 4972–4983.

[7] Y. Dong, J. Chen, J. Cao, Fixed-time consensus of nonlinear multi-agent systems with stochastically switching topologies, International Journal of Control 95 (10) (2022) 2828–2839.

[8] W. He, G. Chen, Q.-L. Han, F. Qian, Network-based leader-following consensus of nonlinear multi-agent systems via distributed impulsive control, Information Sciences 380 (2017) 145–158.

[9] Z. Zhang, S. Chen, Y. Zheng, Fully distributed scaled consensus tracking of high-order multiagent systems with time delays and disturbances, IEEE Transactions on Industrial Informatics 18 (1) (2021) 305–314.

[10] M. Rehan, A. Jameel, C.K. Ahn, Distributed consensus control of one-sided Lipschitz nonlinear multiagent systems, IEEE Transactions on Systems, Man, and Cybernetics: Systems 48 (8) (2017) 1297–1308.

[11] H. Zhou, S. Sui, S. Tong, Fuzzy adaptive finite-time consensus control for high-order nonlinear multiagent systems based on event-triggered, IEEE Transactions on Fuzzy Systems 30 (11) (2022) 4891–4904.

[12] X.-R. Yang, G.-P. Liu, Necessary and sufficient consensus conditions of descriptor multiagent systems, IEEE Transactions on Circuits and Systems. I, Regular Papers 59 (11) (2012) 2669–2677.

[13] J. Xi, M. He, H. Liu, J. Zheng, Admissible output consensualization control for singular multi-agent systems with time delays, Journal of the Franklin Institute 353 (16) (2016) 4074–4090.

[14] X. Liu, Y. Xie, F. Li, W. Gui, Admissible consensus for homogeneous descriptor multiagent systems, IEEE Transactions on Systems, Man, and Cybernetics: Systems 51 (2) (2019) 965–974.

[15] I. Podlubny, An introduction to fractional derivatives, fractional differential equations, to methods of their solution and some of their applications, Mathematics in Science and Engineering 198 (340) (1999), 0924–34008.

[16] X. Meng, Z. Wu, C. Gao, B. Jiang, H.R. Karimi, Finite-time projective synchronization control of variable-order fractional chaotic systems via sliding mode approach, IEEE Transactions on Circuits and Systems. II, Express Briefs 68 (7) (2021) 2503–2507.

[17] G. Ren, Y. Yu, Robust consensus of fractional multi-agent systems with external disturbances, Neurocomputing 218 (2016) 339–345.

[18] W. Yu, Y. Li, G. Wen, Xin Yu, J. Cao, Observer design for tracking consensus in second-order multi-agent systems: fractional order less than two, IEEE Transactions on Automatic Control 62 (2) (2016) 894–900.

[19] Z. Yu, H. Jiang, C. Hu, J. Yu, Leader-following consensus of fractional-order multi-agent systems via adaptive pinning control, International Journal of Control 88 (9) (2015) 1746–1756.

[20] J. Shen, J. Cao, Necessary and sufficient conditions for consensus of delayed fractional-order systems, Asian Journal of Control 14 (6) (2012) 1690–1697.

[21] G. Wen, Y. Zhang, Z. Peng, Y. Yu, A. Rahmani, Observer-based output consensus of leader-following fractional-order heterogeneous nonlinear multi-agent systems, International Journal of Control 93 (10) (2020) 2516–2524.

[22] Z. Gao, H. Zhang, Y. Wang, Y. Mu, Time-varying output formation-containment control for homogeneous/heterogeneous descriptor fractional-order multi-agent systems, Information Sciences 567 (2021) 146–166.

[23] H. Pan, X. Yu, L. Guo, Admissible leader-following consensus of fractional-order singular multiagent system via observer-based protocol, IEEE Transactions on Circuits and Systems. II, Express Briefs 66 (8) (2018) 1406–1410.

[24] X. Zhang, S. Chen, J.-X. Zhang, Adaptive sliding mode consensus control based on neural network for singular fractional order multi-agent systems, Applied Mathematics and Computation 434 (2022) 127442.

[25] C. Peng, F. Li, A survey on recent advances in event-triggered communication and control, Information Sciences 457 (2018) 113–125.

[26] X. Yin, D. Yue, S. Hu, Adaptive periodic event-triggered consensus for multi-agent systems subject to input saturation, International Journal of Control 89 (4) (2016) 653–667.

[27] X. Meng, B. Jiang, H.R. Karimi, C. Gao, An event-triggered mechanism to observer-based sliding mode control of fractional-order uncertain switched systems, ISA Transactions 135 (2023) 115–129.

[28] Q. Zhang, J. Zhang, H. Wang, Data-driven minimum entropy control for stochastic nonlinear systems using the cumulant-generating function, IEEE Transactions on Automatic Control (2022).

[29] Q. Zhang, H. Wang, A novel data-based stochastic distribution control for non-Gaussian stochastic systems, IEEE Transactions on Automatic Control 67 (3) (2021) 1506–1513.

[30] H. Zhang, Z. Gao, Y. Wang, Y. Cai, Leader-following exponential consensus of fractional-order descriptor multiagent systems with distributed event-triggered strategy, IEEE Transactions on Systems, Man, and Cybernetics: Systems 52 (6) (2021) 3967–3979.

[31] P. Xiao, Z. Gu, Adaptive event-triggered consensus of fractional-order nonlinear multiagent systems, IEEE Access 10 (2021) 213–220.

[32] L. Wang, J. Dong, Event-based distributed adaptive fuzzy consensus for nonlinear fractional-order multiagent systems, IEEE Transactions on Systems, Man, and Cybernetics: Systems 52 (9) (2021) 5901–5912.

[33] Z. Gao, H. Zhang, Y. Cai, Y. Mu, Finite-time event-triggered output consensus of heterogeneous fractional-order multiagent systems with intermittent communication, IEEE Transactions on Cybernetics 53 (4) (2021) 2164–2176.

[34] D. Yao, H. Li, R. Lu, Y. Shi, Distributed sliding-mode tracking control of second-order nonlinear multiagent systems: an event-triggered approach, IEEE Transactions on Cybernetics 50 (9) (2020) 3892–3902.

[35] T. Hu, Z. He, X. Zhang, S. Zhong, Leader-following consensus of fractional-order multi-agent systems based on event-triggered control, Nonlinear Dynamics 99 (2020) 2219–2232.

[36] F. Wang, Y. Yang, Leader-following consensus of nonlinear fractional-order multi-agent systems via event-triggered control, International Journal of Systems Science 48 (3) (2017) 571–577.

[37] L. Luo, W. Mi, S. Zhong, Adaptive consensus control of fractional multi-agent systems by distributed event-triggered strategy, Nonlinear Dynamics 100 (2) (2020) 1327–1341.

[38] G. Ren, Y. Yu, C. Xu, X. Hai, Consensus of fractional multi-agent systems by distributed event-triggered strategy, Nonlinear Dynamics 95 (2019) 541–555.

[39] X. Yuan, L. Mo, Y. Yu, Observer-based quasi-containment of fractional-order multi-agent systems via event-triggered strategy, International Journal of Systems Science 50 (3) (2019) 517–533.

[40] Y. Li, Y.Q. Chen, I. Podlubny, Stability of fractional-order nonlinear dynamic systems: Lyapunov direct method and generalized Mittag-Leffler stability, Computers & Mathematics with Applications 59 (5) (2010) 1810–1821.

[41] R. Olfati-Saber, R.M. Murray, Consensus problems in networks of agents with switching topology and time-delays, IEEE Transactions on Automatic Control 49 (9) (2004) 1520–1533.

[42] B. Jiang, Z. Wu, Z. Liu, B. Li, Adaptive sliding mode security control of wheeled mobile manipulators with Markov switching joints against adversarial attacks, Control Engineering Practice 137 (2023) 105558.

[43] B. Jiang, C. Gao, Decentralized adaptive sliding mode control of large-scale semi-Markovian jump interconnected systems with dead-zone input, IEEE Transactions on Automatic Control 67 (3) (2021) 1521–1528.

[44] T. Feng, Y.-E. Wang, L. Liu, B. Wu, Observer-based event-triggered control for uncertain fractional-order systems, Journal of the Franklin Institute 357 (14) (2020) 9423–9441.

[45] Y. Cheng, T. Hu, Y. Li, S. Zhong, Consensus of fractional-order multi-agent systems with uncertain topological structure: a Takagi–Sugeno fuzzy event-triggered control strategy, Fuzzy Sets and Systems 416 (2021) 64–85.

[46] T. Kaczorek, Positivity and reachability of fractional electrical circuits, Acta Mechanica et Automatica 5 (2) (2011) 42–51.

[47] T. Zhan, S. Ma, Reduced-order observer design with unknown input for fractional order descriptor nonlinear systems, Transactions of the Institute of Measurement and Control 41 (13) (2019) 3705–3713.

Leader–follower sliding mode formation control of fractional-order multi-agent systems: a dynamic event-triggered mechanism

7.1 Introduction

In recent years, the formation control problem [1–3] has become one of the most popular and important topics in the field of multi-agent system research, and has been widely used in practical applications, such as flight formation control, satellite formation control, multi-robot collaborative formation control, etc. The core purpose of formation control is to design a controller such that each agent can be driven to its desired state, such as position, displacement, velocity, etc., thereby the target formation can be achieved. In addition, some performance indexes are often used for evaluating the formation control case, such that position deviation, velocity deviation, drift, energy consumption, and efficiency, etc. All of the above indexes have a wide range of applications in formation control, with different indexes being applied to different environments and tasks. Compared to other existing methods, these indexes allow a quantitative assessment of the performance of formation control algorithms and systems, thus providing a comprehensive assessment of the strengths and weaknesses of formation control performance. At the same time, these indexes can be compared with other relevant control methods to guide the selection of control strategies and algorithms and to obtain better control results. In some of the existing literature [4–9], formation control methods can be broadly classified into three categories based on leader information, i.e.,

Sliding Mode Control of Fractional-order Systems
https://doi.org/10.1016/B978-0-44-331696-8.00011-2

leaderless mode, virtual leader mode, and leader–follower mode. In [10], the author used the state feedback control method to study the robust leaderless time-varying formation control problem of the tracking-oriented of the UAV swarm system under the directional topology. By using a special matrix decomposition method, the leaderless time-varying formation control problem could be transformed into the asymptotic stability problem of a low-dimensional closed-loop control system, thereby the corresponding sufficiency criterion was obtained. For example, in [8], the authors studied the leader–follower formation control problem of surface vehicles with prescribed performance and collision avoidance. To guarantee that the closed-loop system was always bounded and avoid collisions between each follower and leader, a distributed adaptive formation controller was designed. More works about formation control can be found in survey articles [11,12].

Recently, fractional calculus theory has been widely used in the fields of physics, automatic control, and biochemistry because of its memory and genetic characteristics. As we know, fractional-order control systems are more suitable for describing complex dynamical behavior than integer-order systems [13–15], such as the unmanned ground vehicle driving on the muddy road and the unmanned aerial vehicle flying in extreme environment. The reason why fractional-order systems work better than integer-order systems in multi-intelligent systems is that they can describe the complex relationships between multiple intelligences more flexibly and can better solve nonlinear, nonstationary and nonuniform problems, thus improving the stability and performance of the system. For example, in an MAS formation control system, multiple MASs fly in a certain formation pattern and need to work together to accomplish some tasks, such as aerial photography, rescue, and detection. For this problem, traditional integer-order control methods often encounter the following problems:

(1) State variables are not of integer order. Traditional integer-order control methods can only handle common integer order systems such as first, second, and third order, but in practical applications, MAS are often high order, nonlinear and nonsmooth, and cannot simply be modeled and controlled by integer-order systems.

(2) Time lag between formation members. Due to factors such as communication and transmission delays between MASs, there is a certain amount of time lag between formation members, which can lead to instability or performance degradation in traditional integer-order control methods.

(3) Lack of flexibility. Traditional integer-order control methods cannot be flexibly adjusted and optimized for different mission requirements and environments, making it difficult to implement advanced control strategies such as adaptive and nonlinear control.

In contrast, fractional-order control methods can better cope with the above problems, with the following advantages:

(1) Fractional calculus better describes the complex relationships between formation members and provides a better mathematical way to capture noninteger order equations of state.
(2) Fractional order control methods include weights for historical information, allowing better handling of time lags and nonsmoothness.
(3) Fractional order parameters can be adjusted in real time to achieve better control performance and adapt control to different mission requirements and environments.

Therefore, focusing on the fractional-order multi-agent system, some results on the formation control problem have been reported in [16–18]. In [16], the authors considered the time-varying formation control problem for a class of FO-MASs by designing an observer-based distributed control strategy. In addition, the authors realized the objective of formation control by transforming the time-varying formation control problem into the asymptotic stability problem of the new system, some sufficient conditions were obtained by adopting the stability theory of fractional-order systems with other methods. In [18], the authors studied the formation control problem for a class of FO-MASs in the case of relative damping and nonuniform time delays, where nonuniform time delays include symmetric and asymmetric time delays. Meanwhile, two formation control algorithms of fractional-order multi-agents under undirected and directed network topologies were proposed, respectively.

In the last few years, the event-triggered control has become a research hotspot due to its advantages such as smaller execution times, less energy consumption, and the ability to ensure the required stability of the sampled-data system, which has attracted more and more scholars' attention, and some effective event-triggered control strategies have been put forward; see [19–23] and the references therein. For example, for a class of discrete-time networked control systems under stochastic cyber-attacks, the authors in [22] considered the security tracking controller design problem by utilizing dynamic event-triggered communication strategy. The simulation results are remarkable. More recently, several formation control strategies, including time-varying formation control, robust formation control, circle formation control, etc., have been developed for MASs based on the event-triggered mechanism [24–28]. For example, focusing on integer-order control systems, the authors studied the formation tracking control problem of MASs with communication constraints in [28]. The authors designed a new role "coordinator" in the original leader–follower coordination mechanism to ensure the dissemination of environmental information in the entire communication network, thus forming a negative feedback connection from the follower to the leader. In addition, considering the situation that the multi-agent cannot pass through the original stratum normally, a scale factor is designed to enlarge or reduce the formation target according to the demand. In [27], the authors explored the state formation control problem of MASs with a general linear network topology. A novel fully distributed asynchronous hybrid self-triggered and event-triggered control strategy was adopted to make sure that the multi-agent system could achieve the given state target. Compared with previous research results, this control strategy presented in [27] can save more communication resources within the same inter-event time interval. However, it should be pointed out that the considered MASs in the aforementioned results are not considered as FO-MASs. To the best of the authors' knowledge, by far, there are no available results on the leader–follower formation control for FO-MASs by using the event-triggered sliding mode protocol, which motivates us to consider this formation control problem.

On the other hand, with the improvement of the event-triggered control method, a problem arises naturally, i.e., how to get a better balance between further reducing the event triggers and ensuring the system performance? One of the solutions to this problem is to add a nonnegative threshold parameter that can be dynamically adjusted according to dynamic rules, which stimulates the explore of dynamic event-triggered mechanism; see, for example, [29–33] and the references therein. In [33], by constructing a performance function, the authors get a trade-off between reducing the event trigger number and ensuring the system performance. Recently, we have witnessed a few works on the DETM for the formation control problem of MASs. In [34], the authors investigated the formation control problem of networked MASs constrained by communication resources, where a dynamic parameter that depended on the communication of its neighbors for each agent system was introduced in the trigger condition. In [35], the time-varying output formation control problem for a class of heterogeneous MASs was studied under DETM. Meanwhile, in [36], the authors investigated the bipartite time-varying output formation tracking problems for a class of homogeneous/heterogeneous MASs under switching communication networks by presenting an observer-based distributed dynamic event-triggered control strategy. Generally speaking, the research on formation control problems for FO-MASs has the following main difficulties:

(1) Improving the performance of formation control. Fractional-order sliding mode control is an excellent control method with good adaptive capability, robustness, and antidisturbance ability. The fractional-order sliding mode control method based on dynamic event-triggered mechanism can better control the position and speed of agents in the formation and improve the performance of the formation control system.

(2) Reducing the communication resources. Traditional formation control methods usually require all intelligences to communicate with each other at each time step, which causes significant communication overhead and time delay. The dynamic event-triggered mechanism allows only some of the intelligences in the formation to communicate, reducing communication overhead and improving real-time performance.

However, there are no available results for the dynamic event-triggered sliding mode formation control problem of FO-MASs, which also motivated us to explore this formation control problem. Motivated by the above analysis, we will explore the leader–follower formation control problem for a class of FO-MASs with general linear system dynamics by adopting the dynamic event-triggered SMC method. The main contributions of this chapter are summarized as follows:

(1) This chapter first explores the leader–follower sliding mode formation control strategy based on a DETM for a class of linear FO-MASs. Under the Mittag-Leffler stability of fractional-order systems, such a control protocol can guarantee the asymptotic stability of the closed-loop system and the finite-time reachability of the sliding-mode surface. In addition, by introducing a dynamic event-triggered mechanism, a better balance can be achieved between reducing communication resources and maintaining better formation performance of FO-MASs;

(2) When implementing the DETM-based controller, an auxiliary variable is designed for each agent to avoid the continuous sampling of other agents states. And then, the updated law of each dynamic parameter only depends on the values of the combined measurement and the sampling error at the trigger time.

(3) Due to the memory nature of the fractional-order calculus operator, we propose a new condition to avoid the occurrence of Zeno behavior, which is different from existing event-triggered schemes.

7.2 Problem statement

7.2.1 Graph theory

In this part, some necessary definitions from graph theory are introduced for the following analysis. Let $\hat{\mathcal{G}} = \{\mathcal{V}_0\} \cup \mathcal{G}$ denote the interaction topology, where $\mathcal{V}_0$ presents the leader and the subgraph $\mathcal{G} = \{\mathcal{V}, \mathcal{E}, \mathcal{P}\}$ is the information exchange among the followers. The vertex set in the subgraph $\mathcal{G}$ is represented by $\mathcal{V} = \{1, 2, \ldots, \mathcal{N}\}$, where the ith agent is denotes by a node $i \in \mathcal{I}$, the set of the edges is represented as $\mathcal{E} \subseteq \mathcal{V} \times \mathcal{V}$ in the subgraph $\mathcal{G}$, and the weighted adjacency matrix of the subgraph $\mathcal{G}$ is denoted by $\mathcal{P} = [p_{ij}] \in \mathbb{R}^{\mathcal{N} \times \mathcal{N}}$ with $p_{ij} \geq 0$. Particularly, if agent i can get information from agent j in the subgraph $\mathcal{G}$, then $p_{ij} > 0$; otherwise $p_{ij} = 0$. The

Laplacian matrix of $\mathcal{G}$ is denoted as $\hat{\mathcal{P}} = [\hat{p}_{ij}] \in \mathbb{R}^{\mathcal{N} \times \mathcal{N}}$ with

$$\hat{p}_{ij} = \begin{cases} \displaystyle\sum_{j=1,\, j \neq i}^{\mathcal{N}} p_{ij}, & i = j, \\ -p_{ij}, & i \neq j. \end{cases}$$

Let $\mathcal{L} = \hat{\mathcal{P}} + \mathcal{Q}$ with $\mathcal{Q} = \text{diag}\{q_1, q_2, \ldots, q_{\mathcal{N}}\}$, where q_i denotes the weight of $(\mathcal{V}_i, \mathcal{V}_0)$.

Assumption 7.1. The topology $\mathcal{G}$ among followers is undirected and connected.

7.2.2 Fractional-order multi-agent dynamics

In this chapter, we consider the fractional-order multi-agent systems (FO-MASs) on the topology $\hat{\mathcal{G}}$, and the dynamics of the leader and followers are given respectively as follows:

$$_{t_0}^{C}D_t^{\alpha} x_0(t) = A x_0(t), \tag{7.1}$$

$$_{t_0}^{C}D_t^{\alpha} x_i(t) = A x_i(t) + B(u_i(t) + w_i(t)), \quad i \in V, \tag{7.2}$$

where $_{t_0}^{C}D_t^{\alpha}$ denotes the Caputo fractional derivative operator with order $\alpha \in (0, 1)$, $x_0(t) \in \mathbb{R}^n$ is the state of the leader system while $x_i(t) \in \mathbb{R}^n$, $u_i(t) \in \mathbb{R}^m$, and $w_i(t) \in \mathbb{R}^m$ are state, control input, and disturbance of the ith agent, respectively. And the disturbance $w_i(t)$ is such that $\|w_i(t)\| \leq \delta_i$ with $\delta_i > 0$. Moreover, A, B are known constant matrices.

Assumption 7.2. For the given leader system (7.1), the state trajectory $x_0(t)$ is norm-bounded.

Assumption 7.3. The matrix BB^{T} is nonsingular, i.e., $\lambda_{\min}\{BB^{\mathrm{T}}\} > 0$.

Definition 7.1 ([34]). The FO-MASs (7.1) and (7.2) are said to achieve the desired formation if their solution satisfies

$$\lim_{t \to +\infty} \|x_i(t) - x_0(t) - f_i\| = 0, \tag{7.3}$$

where $f_i \in \mathbb{R}^n$ denotes the formation variable of agent i.

For given the formation variable f_i shown in Definition 7.1, this chapter aims to design a sliding mode formation control protocol for FO-MASs (7.1) and (7.2) such that all followers can reach the desired position with tiny steady-state errors under the given leader information. At the same time, the above agents can be guaranteed to avoid the collision.

7.2.3 Dynamic event-triggered mechanism

Let the combined measurement and the measurement error be given by $v_i(t) = \sum_{j=1}^{\mathcal{N}} p_{ij}(z_i(t) - z_j(t)) + q_i z_i(t)$ and $\hat{e}_i = v_i(t_k^i) - v_i(t)$, respectively. In this chapter, the DETM is presented as follows:

$$\text{DETM}: \quad t_{k+1}^i = \inf\{t > t_k^i \mid \varphi_i(t) > 0\}, \tag{7.4}$$

where $\varphi_i(t) = \theta_i(\hat{e}_i^T(t)\Gamma\hat{e}_i(t) - \beta_i v_i^T(t_k^i)\Gamma v_i(t_k^i)) - \eta_i(t)$, with $\Gamma = GBB^T G$, and the auxiliary variable $\eta_i(t)$ satisfying

$$\,_{t_0}^{C}D_t^{\alpha}\eta_i(t) = -\rho_i\eta_i(t) + \xi_i(\beta_i v_i^T(t_k^i)\Gamma v_i(t_k^i) - \hat{e}_i^T(t)\Gamma\hat{e}_i(t)), \tag{7.5}$$

with $\theta_i, \beta_i, \rho_i, \xi_i \geq 0$, and $\eta_i(t_0) > 0$.

Remark 7.1. Generally speaking, noting the presented DETM (7.4), there exist two other event-triggered mechanisms by choosing some appropriate parameters, i.e., the time-triggered mechanism (TTM) and the static event-triggered mechanism (SETM).

Case (a) If we let $\beta_i = 0$ in (7.4) and $\rho_i = 0$, $\xi_i = 0$ with an initial value $\eta_i(t_0) = 0$ in (7.5), then DETM (7.4) will become a time-dependent communication mechanism;

Case (b) If we let $\rho_i = 0$, $\xi_i = 0$ with an initial value $\eta_i(t_0) > 0$ in (7.5), then the threshold $\eta_i(t)$ DETM (7.4) will become a positive constant, i.e., called the static event-triggered mechanism.

Lemma 7.1. *For the given positive scalars θ_i, β_i, ρ_i, ξ_i, and $\eta_i(t_0)$ in DETC (7.4), the auxiliary variable satisfies $\eta_i(t) > 0$ for each agent $i = 1, 2, \ldots, \mathcal{N}$.*

Proof. According to DETC (7.4), when $t \in [t_k^i, t_{k+1}^i)$, we have

$$\hat{e}_i^T(t)\Gamma\hat{e}_i(t) - \beta_i v_i^T(t_k^i)\Gamma v_i(t_k^i) \leq \frac{1}{\theta_i}\eta_i(t)$$

and, combining with (7.5),

$$\,_{t_0}^{C}D_t^{\alpha}\eta_i(t) \geq -\left(\rho_i + \frac{\xi_i}{\theta_i}\right)\eta_i(t). \tag{7.6}$$

Using the Laplace transform on (7.6), we have that

$$\hat{\eta}_i(s) \geq \frac{s^{\alpha-1}}{s^{\alpha} + (\rho_i + \frac{\xi_i}{\theta_i})}\eta_i(t_0), \tag{7.7}$$

then taking the inverse Laplace transform of (7.7), one can easily obtain

$$\eta_i(t) \geq \eta_i(t_0)E_{\alpha}\left[-\left(\rho_i + \frac{\xi_i}{\theta_i}\right)(t - t_0)^{\alpha}\right]. \tag{7.8}$$

According to [37, Property 1], as $\rho_i + \frac{\xi_i}{\theta_i} > 0$, the Mittag-Leffler function $E_\alpha[-(\rho_i + \frac{\xi_i}{\theta_i})(t - t_0)^\alpha] \in (0, 1)$, therefore, we can easily obtain that $\eta_i(t) > 0$ holds for $t \in [t_0, +\infty)$. The proof is completed. $\qquad\square$

7.3 Main results

7.3.1 Dynamic event-triggered SMC strategy

In this part, construct the following surface function $s_i(t)$:

$$s_i(t) = z_i(t) - {}_{t_0}D_t^{-\alpha}\{A(z_i(t) + f_i) - B\hat{u}_{il}(t)\}, \tag{7.9}$$

where ${}_{t_0}D_t^{-\alpha}$ denotes the fractional-order integral operator, $z_i(t) = x_i(t) - x_0(t) - f_i$ denotes the formation error, and $\hat{u}_{il}(t) = Kv_i(t_k^i) + Hf_i$ denotes the state-feedback control law, in which $K, H \in \mathbb{R}^{m \times n}$ are gain matrices to be designed. From $s_i(t) = 0$ and ${}_{t_0}^{C}D_t^\alpha s_i(t) = 0$, we have

$$_{t_0}^{C}D_t^\alpha z_i(t) = A(z_i(t) + f_i) - B\hat{u}_{il}(t) = Az_i(t) - BKv_i(t_k^i) + (A - BH)f_i. \tag{7.10}$$

Using the Kronecker product, one has

$$_{t_0}^{C}D_t^\alpha z(t) = (I_N \otimes A)z(t) - (I_N \otimes BK)v(t_k) + (I_N \otimes (A - BH))F, \tag{7.11}$$

where $z(t) = [z_1^T(t), z_2^T(t), \ldots, z_N^T(t)]^T$, $v(t_k) = [v_1^T(t_{k_1}^1), v_2^T(t_{k_2}^2), \ldots, v_n^T(t_{k_N}^N)]^T$, and $F = [f_1^T, f_1^T, \ldots, f_N^T]^T$.

Remark 7.2. Noting Definition 7.1, since $z_i(t) = x_i(t) - x_0(t) - f_i$, Eq. (7.3) is equivalent to $\lim\limits_{t \to +\infty} \|z_i(t) - z_j(t)\| = 0$, therefore, the formation control problem of FO-MASs (7.1) and (7.2) can be transformed into the asymptotic stability problem of the closed-loop system (7.11) under the presented DETM (7.4). Particularly, notice from (7.3) that if let $f_i = f_j = 0$ hold for each $i, j \in V$, then we can obtain that $\lim\limits_{t \to +\infty} \|x_i(t) - x_0(t)\| = 0$, which implies that all the followers will achieve a leader-following consensus on states. Some special cases of leader-following consensus frameworks have been discussed; see, for example, [38–41] and the references therein.

Theorem 7.1. *For the FO-MASs (7.1) and (7.2) under the DETM (7.4), if the following conditions hold simultaneously, then it can achieve specified state formation:*

    (i) *For any agent $i \in V$, there exists a gain matrix H of an appropriate dimension satisfying*

$$(A - BH)f_i = 0; \tag{7.12}$$

(ii) *For given scalars $c_0 \in [0, \frac{1}{2})$ and $\rho_0 \in (0, +\infty)$, there exists a matrix $\boldsymbol{G} = \boldsymbol{G}^{\mathrm{T}} > \boldsymbol{0}$ satisfying the following inequality:*

$$\boldsymbol{G}\boldsymbol{A} + \boldsymbol{A}^{\mathrm{T}}\boldsymbol{G} - c_0\lambda_{\min}\{\mathcal{L}\}\boldsymbol{\Gamma} + \rho_0\boldsymbol{I} < \boldsymbol{0}, \tag{7.13}$$

then the gain matrix $\boldsymbol{K} = \boldsymbol{B}^{\mathrm{T}}\boldsymbol{G}$;

(iii) *There exist some positive scalars c_{0i}, θ_i, β_i, ρ_i, and ξ_i satisfying*

$$\beta_i = \frac{1 - 2c_{0i}}{1 + 2c_{0i}}, \tag{7.14}$$

$$\rho_i > \frac{1 + 2c_{0i} - \xi_i}{\theta_i}, \tag{7.15}$$

in which $c_0 = \min\limits_{i}\{c_{0i}\}$, $i \in \mathcal{V}$.

Proof. If condition (i) holds, then we have

$$(\boldsymbol{I}_{\mathcal{N}} \otimes (\boldsymbol{A} - \boldsymbol{B}\boldsymbol{H}))\boldsymbol{F} = \boldsymbol{0}. \tag{7.16}$$

Substituting (7.16) into (7.12), the dynamic systems can be obtained as follows:

$${}^{C}_{t_0}\!D^{\alpha}_t \boldsymbol{z}(t) = (\boldsymbol{I}_{\mathcal{N}} \otimes \boldsymbol{A})\boldsymbol{z}(t) - (\boldsymbol{I}_{\mathcal{N}} \otimes \boldsymbol{B}\boldsymbol{K})\boldsymbol{v}(t_k). \tag{7.17}$$

According to Lemma 7.1, we have that $\eta_i(t) > 0$ holds. Then based on the dynamic event-triggered mechanism, construct the following Lyapunov function:

$$V(t) = V_1(t) + V_2(t), \tag{7.18}$$

where

$$V_1(t) = \boldsymbol{z}^{\mathrm{T}}(t)(\mathcal{L} \otimes \boldsymbol{G})\boldsymbol{z}(t), \quad V_2(t) = \sum_{i=1}^{\mathcal{N}} \eta_i(t). \tag{7.19}$$

Based on Lemma 2.2, taking the Caputo fractional derivative of $V_1(t)$ along the trajectories of (7.17), one has

$$\begin{aligned}
{}^{C}_{t_0}\!D^{\alpha}_t V_1(t) &\leq 2\boldsymbol{z}^{\mathrm{T}}(t)(\mathcal{L} \otimes \boldsymbol{G}){}^{C}_{t_0}\!D^{\alpha}_t \boldsymbol{z}(t) \\
&= 2\boldsymbol{z}^{\mathrm{T}}(t)(\mathcal{L} \otimes \boldsymbol{G})[(\boldsymbol{I}_{\mathcal{N}} \otimes \boldsymbol{A})\boldsymbol{z}(t) - (\boldsymbol{I}_{\mathcal{N}} \otimes \boldsymbol{B}\boldsymbol{K})\boldsymbol{v}(t_k)] \\
&\leq 2\boldsymbol{z}^{\mathrm{T}}(t)[(\mathcal{L} \otimes \boldsymbol{G}\boldsymbol{A})\boldsymbol{z}(t) - (\mathcal{L} \otimes \boldsymbol{\Gamma})\boldsymbol{v}(t_k)] \\
&\quad - c_0\boldsymbol{v}^{\mathrm{T}}(t)(\boldsymbol{I}_{\mathcal{N}} \otimes \boldsymbol{\Gamma})\boldsymbol{v}(t) + \sum_{i=1}^{\mathcal{N}} c_{0i}\boldsymbol{v}_i^{\mathrm{T}}(t)\boldsymbol{\Gamma}\boldsymbol{v}_i(t).
\end{aligned} \tag{7.20}$$

Since $v(t) = (\mathcal{L} \otimes I_n)z(t)$, then

$$\,^{C}_{t_0}D_t^{\alpha} V_1(t) \leq z^{\mathrm{T}}(t)[\mathcal{L} \otimes (GA + A^{\mathrm{T}}G) - c_0\mathcal{L}\mathcal{L} \otimes \Gamma]z(t)$$
$$+ \sum_{i=1}^{\mathcal{N}} c_{0i} v_i^{\mathrm{T}}(t)\Gamma v_i(t) - 2z^{\mathrm{T}}(t)(\mathcal{L} \otimes \Gamma)v(t_k). \tag{7.21}$$

Moreover, recalling that $v_i(t) = v_i(t_k^i) - \hat{e}_i(t)$, one has

$$c_{0i} v_i^{\mathrm{T}}(t)\Gamma v_i(t) = c_{0i}(v_i(t_k^i) - \hat{e}_i(t))^{\mathrm{T}}\Gamma(v_i(t_k^i) - \hat{e}_i(t))$$
$$\leq 2c_{0i} v_i^{\mathrm{T}}(t_k^i)\Gamma v_i(t_k^i) + 2c_{0i}\hat{e}_i^{\mathrm{T}}(t)\Gamma \hat{e}_i(t) \tag{7.22}$$

and

$$-2z^{\mathrm{T}}(t)(\mathcal{L} \otimes \Gamma)v(t_k) = -2z^{\mathrm{T}}(t)(\mathcal{L} \otimes I_n)(I_{\mathcal{N}} \otimes \Gamma)v(t_k)$$
$$= -2v^{\mathrm{T}}(t)(I_{\mathcal{N}} \otimes \Gamma)v(t_k)$$
$$= \sum_{i=1}^{\mathcal{N}} \left\{ -2(\hat{e}_i(t) - v_i(t_k^i))^{\mathrm{T}}\Gamma v(t_k) \right\} \tag{7.23}$$
$$\leq \sum_{i=1}^{\mathcal{N}} \left\{ \hat{e}_i^{\mathrm{T}}(t)\Gamma \hat{e}_i(t) - v_i^{\mathrm{T}}(t_k^i)\Gamma v_i(t_k^i) \right\}.$$

Combining (7.5), (7.22), and (7.23), one gets

$$\,^{C}_{t_0}D_t^{\alpha} V_1(t) \leq z^{\mathrm{T}}(t)[\mathcal{L} \otimes (GA + A^{\mathrm{T}}G) - c_0\mathcal{L}\mathcal{L} \otimes \Gamma]z(t)$$
$$+ \sum_{i=1}^{\mathcal{N}} \left\{ (1 + 2c_{0i})\hat{e}_i^{\mathrm{T}}(t)\Gamma \hat{e}_i(t) - (1 - 2c_{0i})v_i^{\mathrm{T}}(t_k^i)\Gamma v_i(t_k^i) \right\} \tag{7.24}$$

and

$$\,^{C}_{t_0}D_t^{\alpha} V_2(t) = \sum_{i=1}^{\mathcal{N}} \,^{C}_{t_0}D_t^{\alpha} \eta_i(t)$$
$$= \sum_{i=1}^{\mathcal{N}} \left\{ -\rho_i \eta_i(t) + \xi_i(\beta_i v_i^{\mathrm{T}}(t_k^i)\Gamma v_i(t_k^i) - \hat{e}_i^{\mathrm{T}}(t)\Gamma \hat{e}_i(t)) \right\}. \tag{7.25}$$

Therefore,

$$\,^{C}_{t_0}D_t^{\alpha} V(t) = \,^{C}_{t_0}D_t^{\alpha} V_1(t) + \,^{C}_{t_0}D_t^{\alpha} V_2(t)$$
$$\leq z^{\mathrm{T}}(t)[\mathcal{L} \otimes (GA + A^{\mathrm{T}}G) - c_0\mathcal{L}\mathcal{L} \otimes \Gamma]z(t)$$
$$+ \sum_{i=1}^{\mathcal{N}} \left\{ -\rho_i \eta_i(t) + (1 + 2c_{0i} - \xi_i)\hat{e}_i^{\mathrm{T}}(t)\Gamma \hat{e}_i(t) \right.$$

$$- (1 - 2c_{0i} - \xi_i \beta_i) \boldsymbol{v}_i^{\mathrm{T}}(t_k^i) \boldsymbol{\Gamma} \boldsymbol{v}_i(t_k^i) \Big\}. \tag{7.26}$$

From (7.14), it can be obtained that $\frac{1-2c_{0i}-\xi_i\beta_i}{1+2c_{0i}-\xi_i} = \beta_i$, and then, according to DETC, we have

$$_{t_0}^{C}D_t^{\alpha} V(t) \le z^{\mathrm{T}}(t)[\mathcal{L} \otimes (\boldsymbol{G}\boldsymbol{A} + \boldsymbol{A}^{\mathrm{T}}\boldsymbol{G}) - c_0 \mathcal{L}\mathcal{L} \otimes \boldsymbol{\Gamma}]z(t)$$

$$+ \sum_{i=1}^{\mathcal{N}} \left\{ - \rho_i \eta_i(t) + \frac{1 + 2c_{0i} - \xi_i}{\theta_i} \eta_i(t) \right\}. \tag{7.27}$$

Hence, due to conditions (ii) and (iii), the latter implies that $_{t_0}^{C}D_t^{\alpha} V(t) < 0$, which also implies that FO-MASs (7.1) and (7.2) are asymptotically stable, i.e., the FO-MASs (7.1) and (7.2) will achieve leader–follower formation specified by $\boldsymbol{F}$ under the presented DETM (7.4). This completes the proof. $\qquad\square$

Remark 7.3. As mentioned in Remark 7.1, a general framework has been presented in Theorem 7.1 to design an event-triggered formation protocol, which includes SETM and TTM cases. Particularly, letting $f_i = 0$ hold for each agent i, the relevant results of the leader-following consensus control problem of the FO-MASs (7.1) and (7.2) can be derived from Theorem 7.1.

7.3.2 Reachability analysis

Based on the variable structure control theory [42], we design the following event-triggered control law:

$$\begin{cases} \boldsymbol{u}_i(t) = -\hat{\boldsymbol{u}}_{il}(t) + \hat{\boldsymbol{u}}_{iw}(t), \\ \hat{\boldsymbol{u}}_{iw}(t) = -(\gamma + \delta_i)\mathrm{sgn}(\boldsymbol{B}^{\mathrm{T}}\boldsymbol{s}_i(t)), \end{cases} \tag{7.28}$$

where $\gamma > 0$ and $\delta_i > 0$.

Theorem 7.2. *Under the presented dynamic event-triggered controller (7.28) for each agent i, the state trajectory of FO-MASs (7.1) and (7.2) can be forced to the given integral sliding surface (7.9) in finite time, i.e., the settling time is*

$$T_i^s = \left\{ \frac{\Gamma(\frac{5}{4} - \alpha)\Gamma(1 + \alpha)}{2\gamma_0 \Gamma(\frac{5}{4} - \alpha)} \hat{V}_i^{\frac{1}{2}}(t_0) \right\}^{\frac{1}{2}}, \tag{7.29}$$

where $\gamma_0 = \sqrt{2}\gamma \lambda_{\min}^{\frac{1}{2}}\{\boldsymbol{B}\boldsymbol{B}^{\mathrm{T}}\}$.

Proof. Construct the following Lyapunov function:

$$\hat{V}_i(t) = \frac{1}{2}\boldsymbol{s}_i^{\mathrm{T}}(t)\boldsymbol{s}_i(t). \tag{7.30}$$

Then, taking the Caputo fractional derivative of $\hat{V}_i(t)$ and using Assumption 7.3, one has

$$\begin{aligned}
{}^C_{t_0}D^\alpha_t \hat{V}_i(t) &\leq s_i^{\mathrm{T}}(t){}^C_{t_0}D^\alpha_t s_i(t) = s_i^{\mathrm{T}}(t)\Big(B\hat{u}_{iw}(t) + Bw_i(t)\Big) \\
&\leq -(\gamma + \delta_i)\|B^{\mathrm{T}}s_i(t)\| + \delta_i\|B^{\mathrm{T}}s_i(t)\| = -\gamma\|B^{\mathrm{T}}s_i(t)\| \qquad (7.31)\\
&\leq -\gamma_0 \hat{V}_i^{\frac{1}{2}}(t),
\end{aligned}$$

in which $\gamma_0 = \gamma(\lambda_{\min}\{BB^{\mathrm{T}}\})^{\frac{1}{2}} > 0$. According to Lemma 2.6, the setting time can be obtained as

$$T_i^s = \left\{ \frac{\Gamma(\frac{5}{4} - \alpha)\Gamma(1 + \alpha)}{2\gamma_0\Gamma(\frac{5}{4} - \alpha)} \hat{V}_i^{\frac{1}{2}}(t_0) \right\}^{\frac{1}{2}}.$$

This completes the proof. $\qquad\qquad\square$

7.3.3 Discussion of Zeno behavior

Seen from the dynamic event-triggered control strategy, how to avoid the Zeno behavior is still a crucial research. It requires the time interval between two adjacent trigger moments be a positive constant. Based on this consideration, we present the following result to avoid the occurrence of Zeno behavior.

Theorem 7.3. *Under the assumptions and stability conditions in Theorems 7.1 and 7.2, the Zeno behavior can be avoided for each agent i.*

Proof. According to Theorems 7.1 and 7.2, the state error trajectories $z_i(t)$ are norm-bounded. Then Assumption 7.2 implies that the state trajectories $x_i(t)$ are norm-bounded, which also means that the combined measurements $v_i(t)$ are norm-bounded. Therefore, it can be easily obtained that ${}^C_{t_0}D^\alpha_t v_i(t)$ are also norm-bounded, i.e., there exist some positive constants M_i such that $\|{}^C_{t_0}D^\alpha_t v_i(t)\| \leq M_i$ holds.

Besides, since $\hat{e}_i(t) = v_i(t_k^i) - v_i(t)$, then

$$\begin{aligned}
\|\hat{e}_i(t)\| &= \|v_i(t_k^i) - v_i(t_0) - v_i(t) + v_i(t_0)\| \\
&= \left\| \frac{1}{\Gamma(\alpha)} \int_{t_0}^{t_k^i} \frac{{}^C_{t_0}D^\alpha_\tau v_i(\tau)}{(t_k^i - \tau)^{1-\alpha}} d\tau - \frac{1}{\Gamma(\alpha)} \int_{t_0}^{t} \frac{{}^C_{t_0}D^\alpha_\tau v_i(\tau)}{(t - \tau)^{1-\alpha}} d\tau \right\| \\
&= \left\| \frac{1}{\Gamma(\alpha)} \int_{t_0}^{t_k^i} \left[\frac{1}{(t_k^i - \tau)^{1-\alpha}} - \frac{1}{(t - \tau)^{1-\alpha}} \right] {}^C_{t_0}D^\alpha_\tau v_i(\tau) d\tau \right.\\
&\qquad \left. - \frac{1}{\Gamma(\alpha)} \int_{t_k^i}^{t} \frac{{}^C_{t_0}D^\alpha_\tau v_i(\tau)}{(t - \tau)^{1-\alpha}} d\tau \right\|
\end{aligned}$$

$$\leq \frac{M_i}{\Gamma(\alpha)} \left| \int_{t_0}^{t_k^i} \left[\frac{1}{(t_k^i - \tau)^{1-\alpha}} - \frac{1}{(t - \tau)^{1-\alpha}} \right] d\tau \right|$$

$$+ \frac{M_i}{\Gamma(\alpha)} \left| \int_{t_k^i}^{t} \frac{1}{(t - \tau)^{1-\alpha}} d\tau \right|$$

$$= \frac{M_i}{\Gamma(\alpha + 1)} \left| (t_k^i - t_0)^\alpha + (t - t_k^i)^\alpha + (t - t_0)^\alpha \right| + \frac{M_i}{\Gamma(\alpha + 1)} (t - t_k^i)^\alpha$$

$$\leq \frac{\gamma_i M_i}{\Gamma(\alpha + 1)} (t - t_k^i)^\alpha, \tag{7.32}$$

where $\gamma_i \geq 2$.

According to DETC, when $t = t_{k+1}^i$, one has

$$\theta_i (\hat{e}_i^{\mathrm{T}}(t_{k+1}^i) \mathbf{\Gamma} \hat{e}_i(t_{k+1}^i) - \beta_i v_i^{\mathrm{T}}(t_k^i) \mathbf{\Gamma} v_i(t_k^i)) \geq \eta_i(t_{k+1}^i), \tag{7.33}$$

with $\mathbf{\Gamma} = \mathbf{G} \mathbf{B} \mathbf{B}^{\mathrm{T}} \mathbf{G}$, that is to say,

$$\|\mathbf{B}^{\mathrm{T}} \mathbf{G} \hat{e}_i(t_{k+1}^i)\|^2 \geq \frac{1}{\theta_i} \eta_i(t_{k+1}^i) + \beta_i \|\mathbf{B}^{\mathrm{T}} \mathbf{G} v_i(t_k^i)\|^2. \tag{7.34}$$

On the other hand, there exists a small positive scalar ε satisfying

$$\|\mathbf{B}^{\mathrm{T}} \mathbf{G} \hat{e}_i(t_{k+1}^i)\| \leq \|\mathbf{B}^{\mathrm{T}} \mathbf{G}\| \|\hat{e}_i(t_{k+1}^i)\|$$
$$\leq (\|\mathbf{B}^{\mathrm{T}} \mathbf{G}\| + \varepsilon) \|\hat{e}_i(t_{k+1}^i)\|. \tag{7.35}$$

Combining (7.34) and (7.35), one has

$$\|\hat{e}_i(t_{k+1}^i)\|^2 \geq \frac{1}{(\|\mathbf{B}^{\mathrm{T}} \mathbf{G}\| + \varepsilon)^2} \left[\frac{1}{\theta_i} \eta_i(t_{k+1}^i) + \beta_i \|\mathbf{B}^{\mathrm{T}} \mathbf{G} v_i(t_k^i)\|^2 \right]. \tag{7.36}$$

Therefore, from (7.32) and (7.36), it can be obtained that the inter-event time intervals $T_k^i = t_{k+1}^i - t_k^i$ satisfy

$$T_k^i \geq \left\{ \frac{\Gamma(\alpha + 1)}{\gamma_i M_i (\|\mathbf{B}^{\mathrm{T}} \mathbf{G}\| + \varepsilon)} \left[\frac{1}{\theta_i} \eta_i(t_{k+1}^i) + \beta_i \|\mathbf{B}^{\mathrm{T}} \mathbf{G} v_i(t_k^i)\|^2 \right]^{\frac{1}{2}} \right\}^{\frac{1}{\alpha}}. \tag{7.37}$$

Meanwhile, according to Lemma 7.1, the auxiliary variable is such that $\eta_i(t) > 0$ holds, therefore, it can easily be obtained that $T_k^i > 0$ holds for each agent i. Obviously, the DETM (7.4) can completely avoid Zeno behavior. $\qquad \square$

Remark 7.4. In the design of an event-triggered mechanism, how to avoid Zeno behavior is always a problem that is worth exploring. Different from the integer-order control system, due to the memory nature of the fractional-order calculus operator, some results obtained from Lemma 2.4 are incorrect and can not strictly guarantee the inter-execution intervals to satisfy $T_k > 0$. To the best of the authors' knowledge, there exist currently two methods to derive a mathematical expression of a positive lower bound of T_k. The first is to use the Gronwall–Bellman inequality to give the minimum positive lower bound of T_k, which was shown in [38,43]. The second is as described in Theorem 7.3 in this chapter, where exploiting the boundedness of the fractional-order derivative of state signals and the specific triggering mechanism, a positive lower bound of T_k can be also obtained.

Remark 7.5. It can be seen from Theorems 7.1–7.3 that for the considered FO-MASs (7.1) and (7.2), and the sliding surface function (7.9), the state trajectories of agents can converge to a desired motion under the control protocol (7.28) and the settling time is upper-bounded as in (7.29). Fig. 7.1 describes an overview of the presented control strategy.

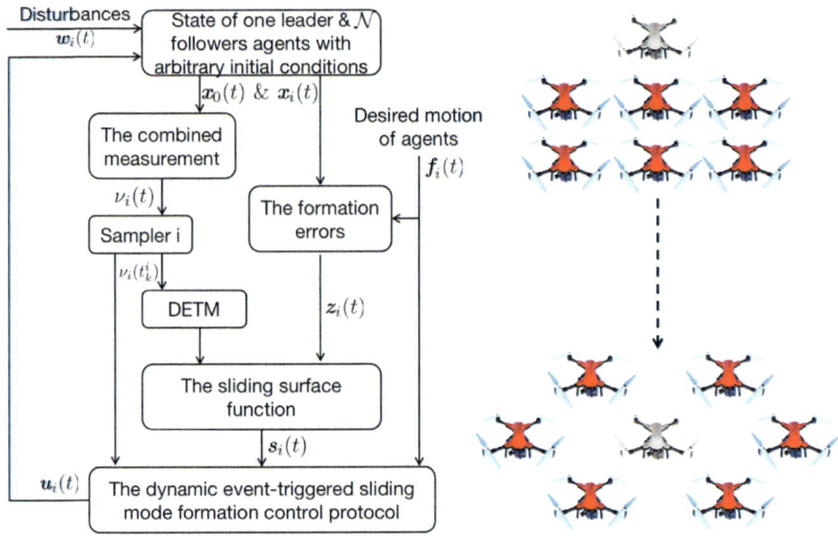

FIGURE 7.1　An overview of the proposed DETM sliding mode control protocol.

7.4 Simulation results

In this section, some simulations are given to verify the presented results.

Example 7.1. In Fig. 7.1, we treat each unmanned aerial vehicle (UAV) as a prime point and consider an FO-MASs consisting of a leader and six followers, and the dynamic equations are as follows:

$$\begin{cases} {}_{t_0}^{C}D_t^{\alpha}x_0(t) = Ax_0(t), \\ {}_{t_0}^{C}D_t^{\alpha}x_i(t) = Ax_i(t) + B(u_i(t) + w_i(t)), \end{cases} \tag{7.38}$$

where $\alpha = 0.9$, $A = \begin{bmatrix} 0.1898 & -0.8483 \\ 0.8752 & -0.1167 \end{bmatrix}$, $B = \begin{bmatrix} 0.25 & 0 \\ 0 & 0.5 \end{bmatrix}$, $w_i(t) = \begin{bmatrix} e^{-t}\sin(it) \\ e^{-t}\cos(it) \end{bmatrix}$ with $i \in \{1, 2, \ldots, 6\}$. The state formation targets of six agents are chosen as $f_1 = [-1/3, \sqrt{3}/3]^{\mathrm{T}}$, $f_2 = [1/3, \sqrt{3}/3]^{\mathrm{T}}$, $f_3 = [2/3, 0]^{\mathrm{T}}$, $f_4 = [1/3, -\sqrt{3}/3]^{\mathrm{T}}$, $f_5 = [-1/3, -\sqrt{3}/3]^{\mathrm{T}}$, and $f_6 = [-2/3, 0]^{\mathrm{T}}$. Choose matrix $H = \begin{bmatrix} 0.7592 & -3.3932 \\ 1.7504 & -0.2334 \end{bmatrix}$. It is clear that $(A + BH)f_i = 0$ holds for any $i = 1, 2, \ldots, 6$. According to the given formation vectors, those agents can be driven to achieve a regular hexagon formation in the 2D plane as time progresses. The interaction topology is described with a directed graph shown in Fig. 7.2, from which the adjacency and weight matrices are:

$$\mathcal{P} = \begin{bmatrix} 0 & 0 & 0 & 0 & 1 & 1 \\ 0 & 0 & 1 & 0 & 0 & 0 \\ 0 & 1 & 0 & 1 & 1 & 0 \\ 0 & 0 & 1 & 0 & 0 & 0 \\ 1 & 0 & 1 & 0 & 0 & 0 \\ 1 & 0 & 0 & 0 & 0 & 0 \end{bmatrix}, \quad \mathcal{Q} = \begin{bmatrix} 1 & 0 & 0 & 0 & 0 & 0 \\ 0 & 0 & 0 & 0 & 0 & 0 \\ 0 & 0 & 1 & 0 & 0 & 0 \\ 0 & 0 & 0 & 0 & 0 & 0 \\ 0 & 0 & 0 & 0 & 0 & 0 \\ 0 & 0 & 0 & 0 & 0 & 0 \end{bmatrix}.$$

Then the matrix of $\mathcal{L}$ can be obtained as follows:

$$\mathcal{L} = \begin{bmatrix} 3 & 0 & 0 & 0 & -1 & -1 \\ 0 & 1 & -1 & 0 & 0 & 0 \\ 0 & -1 & 4 & -1 & -1 & 0 \\ 0 & 0 & -1 & 1 & 0 & 0 \\ -1 & 0 & -1 & 0 & 2 & 0 \\ -1 & 0 & 0 & 0 & 0 & 1 \end{bmatrix}.$$

The initial states of each agent system are chosen as $x_1(0) = [4, 5]^{\mathrm{T}}$, $x_2(0) = [2, -4]^{\mathrm{T}}$, $x_3(0) = [-2, 4]^{\mathrm{T}}$, $x_4(0) = [3, -2]^{\mathrm{T}}$, $x_5(0) = [-3, -3]^{\mathrm{T}}$, $x_6(0) = [-4, 1]^{\mathrm{T}}$, and $x_0(0) = [0, -1]^{\mathrm{T}}$. Then the states of the open-loop system (7.38) are shown in Fig. 7.3.

As mentioned in Remark 7.1, to verify the effectiveness and feasibility of the proposed formation control protocol, the following three different triggering mechanisms are considered in this simulation:

(1) *Time-triggered mechanism* (TTM). In this case, according to Theorem 7.1, the relevant parameters in (7.4) and (7.5) are chosen as $c_0 = 0.5$, $\rho_0 = 0.01$, $\beta_i = \frac{1-2c_0}{1+2c_0}$, $\xi_i = 0$, $\theta_i = 2$, $\rho_i = 0$, and $\eta_i(0) = 0$ for each agent i;

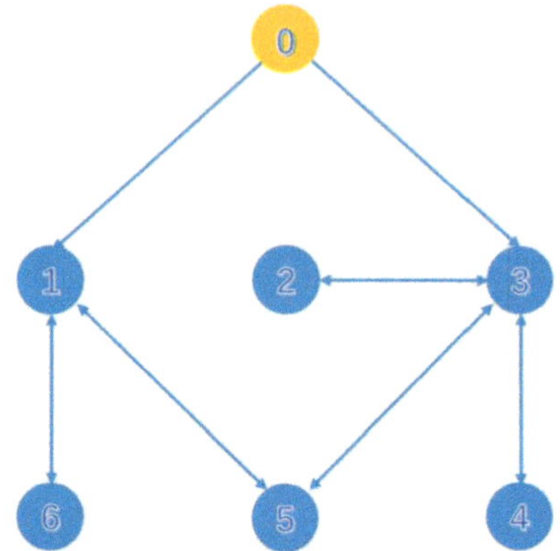

FIGURE 7.2   Interaction topology graph with six followers.

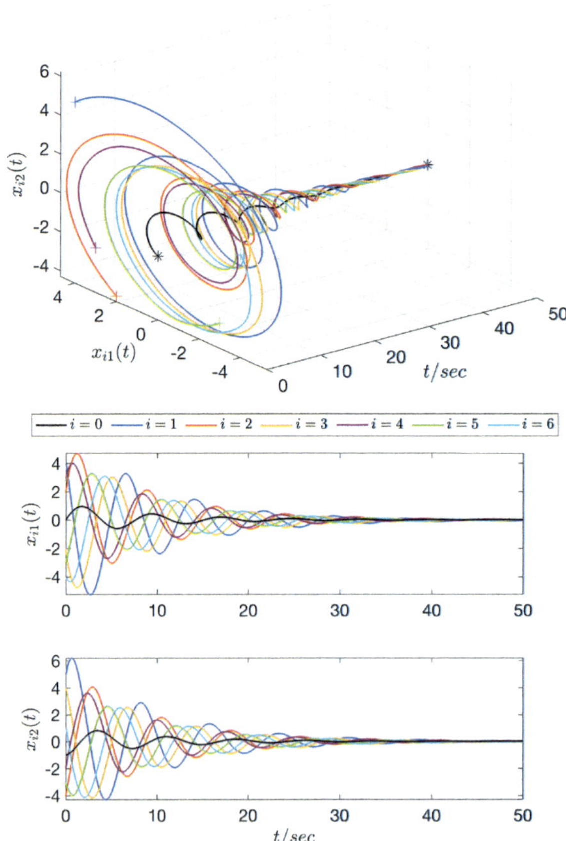

FIGURE 7.3   State trajectories of the open-loop system (7.38).

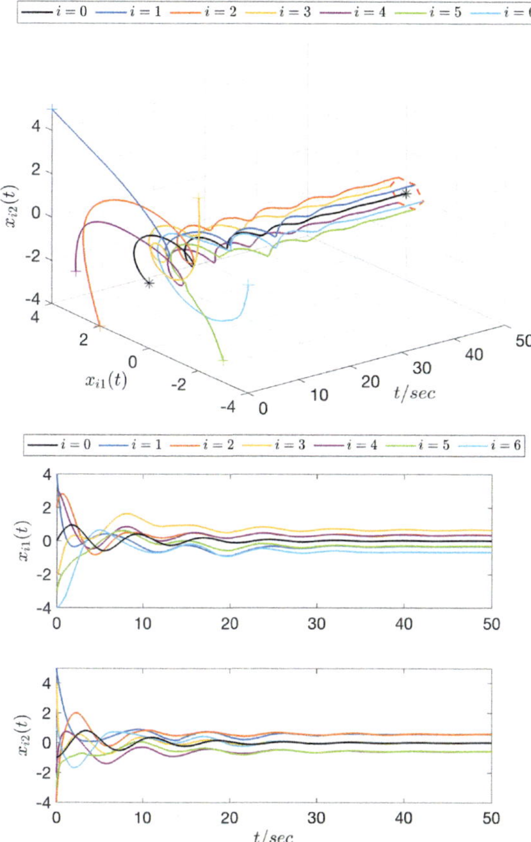

FIGURE 7.4 State trajectories of each agent i with $t \in (0, 50)$ s under DETM (7.4).

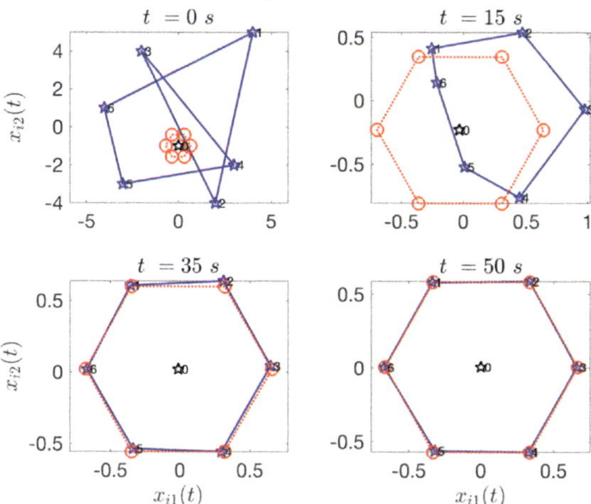

FIGURE 7.5 Position snapshots of each agent i under DETM (7.4): (1) $t = 0$ s; (2) $t = 15$ s; (3) $t = 35$ s; and (4) $t = 50$ s.

(2) *Static event-triggered mechanism* (SETM). In this case, according to Theorem 7.1, the relevant parameters in (7.4) and (7.5) are chosen as $c_0 = 0.495$, $\rho_0 = 0.01$, $\beta_i = \frac{1-2c_0}{1+2c_0}$, $\xi_i = 0$, $\theta_i = 1.8 + 0.2i$, $\rho_i = 0$, and $\eta_i(0) = 0.4$ in Case (2a) while $\eta_i(0) = 0$ in Case (2b), for each agent i;

(3) *Dynamic event-triggered mechanism* (DETM). In this case, according to Theorem 7.1, the relevant parameters in (7.4) and (7.5) are chosen as $c_0 = 0.495$, $\rho_0 = 0.01$, $\beta_i = \frac{1-2c_0}{1+2c_0}$, $\xi_i = 1.3 - 0.1i$, $\theta_i = 1.8 + 0.2i$, $\rho_i = \frac{1+2c_0-\xi_i}{\theta_i} + 0.1$, as well as $\eta_1(0) = 0.4$, $\eta_2(0) = 1.2$, $\eta_3(0) = 2.2$, $\eta_4(0) = 1.1$, $\eta_5(0) = 0.9$, and $\eta_6(0) = 1.8$.

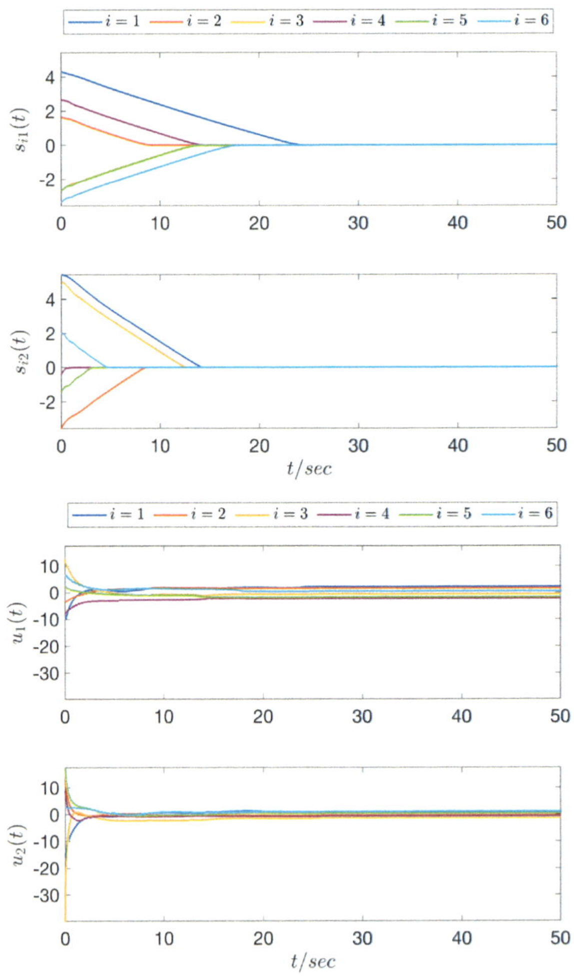

FIGURE 7.6   Response curves of sliding mode surfaces $s_i(t)$ and controllers $u_i(t)$ for each agent i under DETM (7.4).

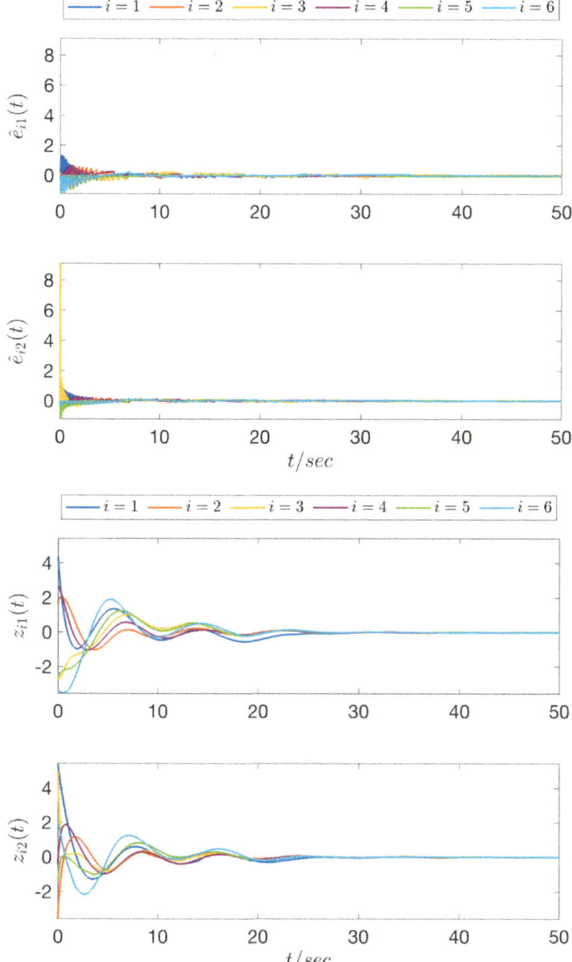

FIGURE 7.7   Time evolutions of measurement error $\hat{e}_i(t)$ and formation error $z_i(t)$ for each agent i under DETM (7.4).

Focusing on the above three different triggering mechanisms, there exists a matrix $G = \begin{bmatrix} 2.8404 & -0.5499 \\ -0.5499 & 2.6936 \end{bmatrix}$ such that the inequality (7.13) in Theorem 7.1 holds, and then the gain matrix $K = \begin{bmatrix} 0.7101 & -0.1375 \\ -0.2750 & 1.3468 \end{bmatrix}$ can be also obtained.

In addition, let $\gamma = 0.005$ and $\delta_i = 1$, then some comparative simulation results are obtained as shown in Figs. 7.4 to 7.14 and Table 7.1, respectively.

First, under the presented DETM (7.4), Fig. 7.4 depicts the state trajectories of each agent i and leader system, and the position snapshots have

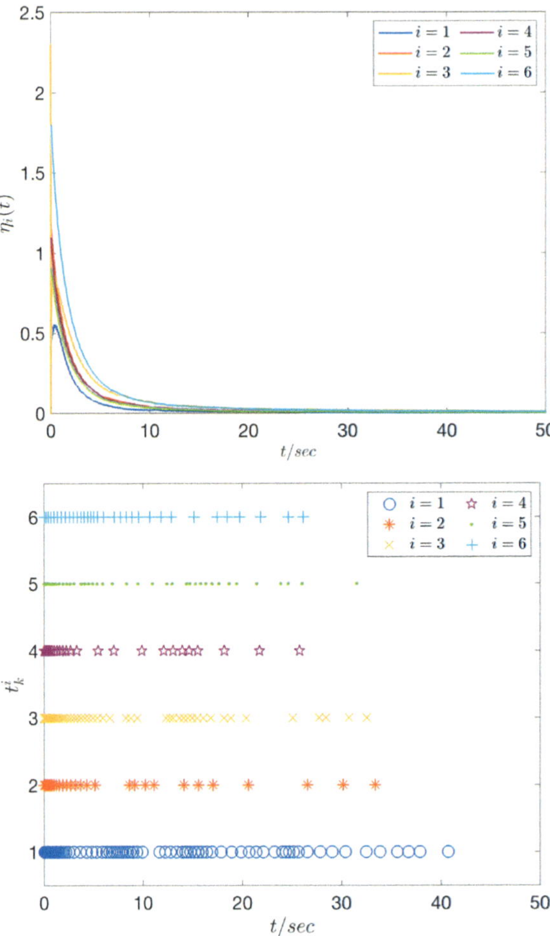

FIGURE 7.8　Time evolutions of dynamic threshold $\eta_i(t)$ and event release instants t_k^i for each agent i under DETM (7.4).

also been shown in Fig. 7.5 for different time instants. Figs. 7.6 to 7.8 illustrate the time evolutions of sliding mode surface, controller, measurement error, formation error, dynamic threshold, and event release instants for each agent i, respectively.

Second, under the presented SETM with $\eta_i = 0.4$, the state trajectories and the position snapshots for each agent i are shown in Figs. 7.9 and 7.10, respectively. Meanwhile, the state trajectories and the position snapshots for each agent i are also depicted in Figs. 7.11 and 7.12 under the presented SETM with $\eta_i = 0$, respectively. Figs. 7.13 and 7.14 illustrate the time evolutions of formation error and event release instants under different SETMs, respectively.

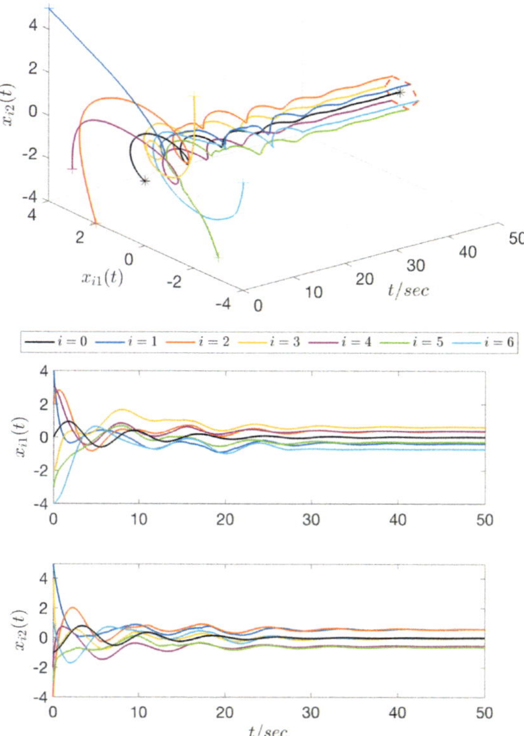

FIGURE 7.9   State trajectories of each agent i with $t \in (0, 50)$ s under SETM with $\eta_i = 0.4$.

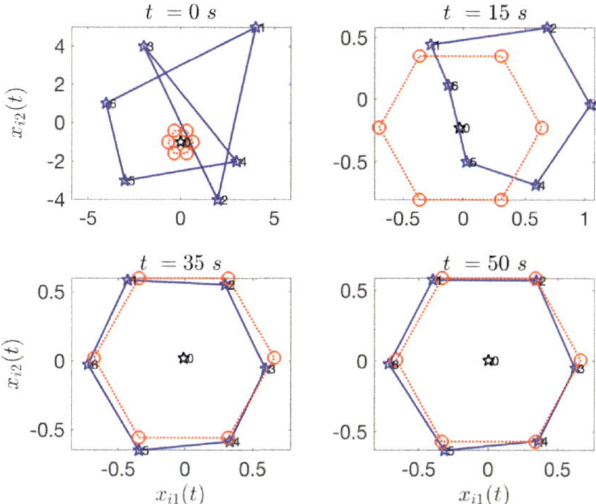

FIGURE 7.10   Position snapshots of each agent i under SETM with $\eta_i = 0.4$: (1) $t = 0$ s; (2) $t = 15$ s; (3) $t = 35$ s; and (4) $t = 50$ s.

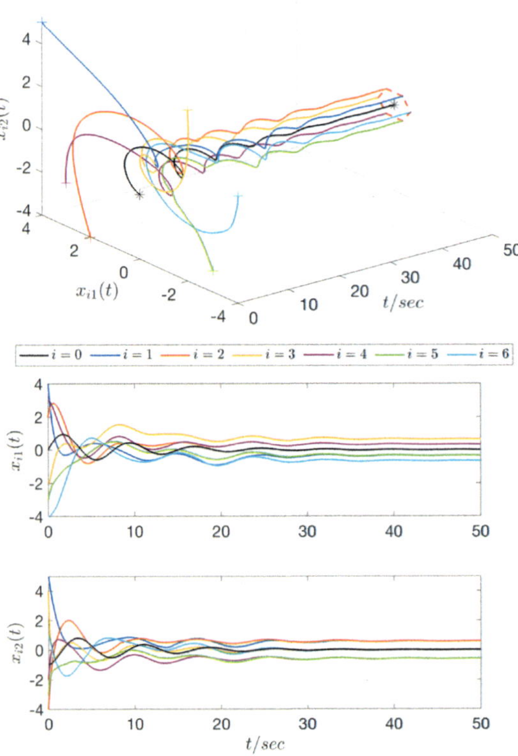

FIGURE 7.11   State trajectories of each agent i with $t \in (0, 50)$ s under SETM with $\eta_i = 0$.

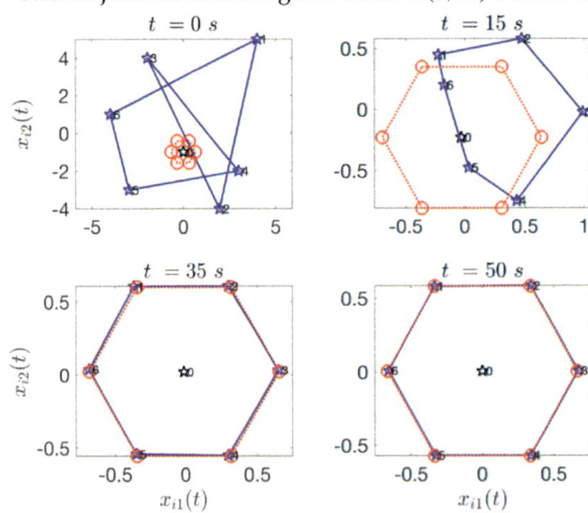

FIGURE 7.12   Position snapshots of each agent i under SETM with $\eta_i = 0$: (1) $t = 0$ s; (2) $t = 15$ s; (3) $t = 35$ s; and (4) $t = 50$ s.

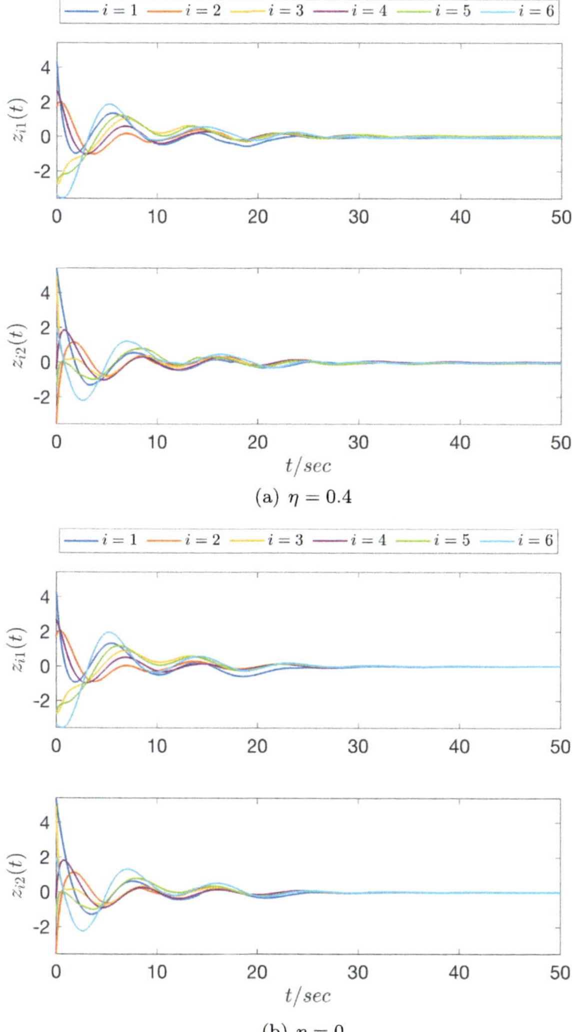

FIGURE 7.13 Time evolutions of formation error $z_i(t)$ for each agent i under different SETMs.

At last, the numbers of trigging times for each agent under TTM, SETM, and DETM are listed in Table 7.1.

Select the position deviation as the performance index. Then, according to the above simulation results, we can get the following comparison result: By choosing an appropriate initial value of the threshold parameter $\eta_i(t)$, compared with SETM, the presented DETM in this chapter can achieve a balance between reducing the number of triggering times and maintaining the formation performance.

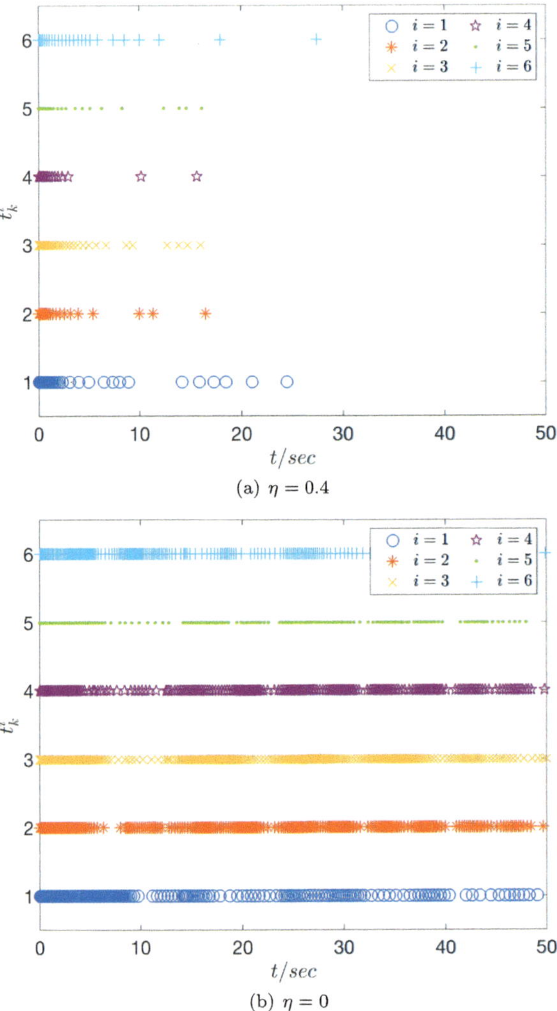

(a) $\eta = 0.4$

(b) $\eta = 0$

FIGURE 7.14 Time evolutions of event release instants t_k^i for each agent i under different SETMs.

TABLE 7.1 Numbers of triggering times for FO-MASs (7.38) under different triggering mechanisms.

Agent i		1	2	3	4	5	6
TTM		5000	5000	5000	5000	5000	5000
SETM	$\eta_i = 0.4$	33	25	40	22	28	24
	$\eta_i = 0$	213	311	376	323	222	199
DETM		76	32	55	29	45	33

7.5 Conclusion

In this chapter, a dynamic event-triggered SMC strategy was presented to study the leader–follower formation control problem for a class of linear FO-MASs. First, a DETM had been designed based on the combined measurement vector, and the threshold parameter could be dynamically adjusted using fractional-order dynamic rules for each agent i. Second, the leader–follower formation control problem has been transformed into the asymptotic stability problem of the new system under the proposed formation protocol and DETM, and the finite-time reachability of state trajectories onto the predefined sliding surface has been guaranteed for each agent i. Furthermore, a condition has also been obtained to exclude the Zeno behavior, which verified the validity of the presented DETM. Finally, a simulation example was given to illustrate the effectiveness and feasibility of the proposed method. In the future, we will continue to explore the results of time-varying formation control of FO-MASs by utilizing the DETM.

References

[1] K.-K. Oh, M.-C. Park, H.-S. Ahn, A survey of multi-agent formation control, Automatica 53 (2015) 424–440.

[2] Y. Xia, X. Na, Z. Sun, J. Chen, Formation control and collision avoidance for multi-agent systems based on position estimation, ISA Transactions 61 (2016) 287–296.

[3] H. Zamani, K. Khandani, V.J. Majd, Fixed-time sliding-mode distributed consensus and formation control of disturbed fractional-order multi-agent systems, ISA Transactions 138 (2023) 37–48.

[4] X. Dong, B. Yu, Z. Shi, Y. Zhong, Time-varying formation control for unmanned aerial vehicles: theories and applications, IEEE Transactions on Control Systems Technology 23 (1) (2014) 340–348.

[5] H. Rezaee, F. Abdollahi, Pursuit formation of double-integrator dynamics using consensus control approach, IEEE Transactions on Industrial Electronics 62 (7) (2014) 4249–4256.

[6] C. Yuan, H. He, C. Wang, Cooperative deterministic learning-based formation control for a group of nonlinear uncertain mechanical systems, IEEE Transactions on Industrial Informatics 15 (1) (2018) 319–333.

[7] Y. Yu, J. Guo, C.K. Ahn, Z. Xiang, Neural adaptive distributed formation control of nonlinear multi-UAVs with unmodeled dynamics, IEEE Transactions on Neural Networks and Learning Systems (2022).

[8] S. He, M. Wang, S.-L. Dai, F. Luo, Leader–follower formation control of USVs with prescribed performance and collision avoidance, IEEE Transactions on Industrial Informatics 15 (1) (2018) 572–581.

[9] H. Zhi, L. Chen, C. Li, Y. Guo, Leader–follower affine formation control of second-order nonlinear uncertain multi-agent systems, IEEE Transactions on Circuits and Systems. II, Express Briefs 68 (12) (2021) 3547–3551.

[10] Y. Kang, D. Luo, B. Xin, J. Cheng, T. Yang, S. Zhou, Robust leaderless time-varying formation control for nonlinear unmanned aerial vehicle swarm system with communication delays, IEEE Transactions on Cybernetics (2022) 1–14.

[11] G.-P. Liu, S. Zhang, A survey on formation control of small satellites, Proceedings of the IEEE 106 (3) (2018) 440–457.

[12] Q. Ouyang, Z. Wu, Y. Cong, Z. Wang, Formation control of unmanned aerial vehicle swarms: a comprehensive review, Asian Journal of Control 25 (1) (2023) 570–593.

[13] I. Podlubny, Fractional-order systems and $PI^\lambda D^\mu$-controllers, IEEE Transactions on Automatic Control 44 (1) (1999) 208–214.

[14] J. Liu, N. Zhou, K. Qin, B. Chen, Y. Wu, K.-S. Choi, Distributed optimization for consensus performance of delayed fractional-order double-integrator multi-agent systems, Neurocomputing 522 (2023) 105–115.

[15] X. Meng, C. Gao, B. Jiang, Z. Wu, Finite-time synchronization of variable-order fractional uncertain coupled systems via adaptive sliding mode control, International Journal of Control, Automation, and Systems 20 (2022) 1535–1543.

[16] Y. Gong, G. Wen, Z. Peng, T. Huang, Y. Chen, Observer-based time-varying formation control of fractional-order multi-agent systems with general linear dynamics, IEEE Transactions on Circuits and Systems. II, Express Briefs 67 (1) (2019) 82–86.

[17] J. Bai, G. Wen, A. Rahmani, Y. Yu, Distributed formation control of fractional-order multi-agent systems with absolute damping and communication delay, International Journal of Systems Science 46 (13) (2015) 2380–2392.

[18] J. Liu, P. Li, L. Qi, W. Chen, K. Qin, Distributed formation control of double-integrator fractional-order multi-agent systems with relative damping and nonuniform time-delays, Journal of the Franklin Institute 356 (10) (2019) 5122–5150.

[19] C. Peng, F. Li, A survey on recent advances in event-triggered communication and control, Information Sciences 457 (2018) 113–125.

[20] J. Bai, H. Wu, J. Cao, Topology identification for fractional complex networks with synchronization in finite time based on adaptive observers and event-triggered control, Neurocomputing 505 (2022) 166–177.

[21] A.K. Behera, B. Bandyopadhyay, M. Cucuzzella, A. Ferrara, X. Yu, A survey on event-triggered sliding mode control, IEEE Journal of Emerging and Selected Topics in Industrial Electronics 2 (3) (2021) 206–217.

[22] J. Liu, Y. Dong, L. Zha, E. Tian, X. Xiangpeng, Event-based security tracking control for networked control systems against stochastic cyber-attacks, Information Sciences 612 (2022) 306–321.

[23] Y. Li, F. Song, J. Liu, X. Xie, E. Tian, Decentralized event-triggered synchronization control for complex networks with nonperiodic DoS attacks, International Journal of Robust and Nonlinear Control 21 (3) (2021) 1633–1653.

[24] X. Li, X. Dong, Q. Li, Z. Ren, Event-triggered time-varying formation control for general linear multi-agent systems, Journal of the Franklin Institute 356 (17) (2019) 10179–10195.

[25] B. Yan, P. Shi, C.-C. Lim, Robust formation control for nonlinear heterogeneous multiagent systems based on adaptive event-triggered strategy, IEEE Transactions on Automation Science and Engineering 19 (4) (2022) 2788–2800.

[26] M. Yu, H. Wang, G. Xie, K. Jin, Event-triggered circle formation control for second-order-agent system, Neurocomputing 275 (2018) 462–469.

[27] W. Li, H. Zhang, Y. Cai, Y. Wang, Fully distributed formation control of general linear multiagent systems using a novel mixed self- and event-triggered strategy, IEEE Transactions on Systems, Man, and Cybernetics: Systems 52 (9) (2022) 5736–5745.

[28] X. Liu, S.S. Ge, Ch.-H. Goh, Y. Li, Event-triggered coordination for formation tracking control in constrained space with limited communication, IEEE Transactions on Cybernetics 49 (3) (2018) 1000–1011.

[29] X. Ge, Q.-L. Han, X.-M. Zhang, D. Ding, Dynamic event-triggered control and estimation: a survey, International Journal of Automation and Computing 18 (6) (2021) 857–886.

[30] X. Ge, Q.-L. Han, L. Ding, Y.-L. Wang, X.-M. Zhang, Dynamic event-triggered distributed coordination control and its applications: a survey of trends and techniques, IEEE Transactions on Systems, Man, and Cybernetics: Systems 50 (9) (2020) 3112–3125.

[31] X. Yi, K. Liu, D.V. Dimarogonas, K.H. Johansson, Dynamic event-triggered and self-triggered control for multi-agent systems, IEEE Transactions on Automatic Control 64 (8) (2018) 3300–3307.

[32] B. Jiang, Z. Wu, H.R. Karimi, A distributed dynamic event-triggered mechanism to HMM-based observer design for H_∞ sliding mode control of Markov jump systems, Automatica 142 (2022) 110357.

[33] Z. Ai, L. Peng, G. Zong, K. Shi, Impulsive control for nonlinear systems under DoS attacks: a dynamic event-triggered method, IEEE Transactions on Circuits and Systems. II, Express Briefs 69 (9) (2022) 3839–3843.

[34] X. Ge, Q.-L. Han, Distributed formation control of networked multi-agent systems using a dynamic event-triggered communication mechanism, IEEE Transactions on Industrial Electronics 64 (10) (2017) 8118–8127.

[35] W. Song, J. Feng, H. Zhang, W. Wang, Dynamic event-triggered formation control for heterogeneous multiagent systems with nonautonomous leader agent, IEEE Transactions on Neural Networks and Learning Systems (2022).

[36] H. Zhang, W. Li, J. Zhang, Y. Wang, J. Sun, Fully distributed dynamic event-triggered bipartite formation tracking for multiagent systems with multiple nonautonomous leaders, IEEE Transactions on Neural Networks and Learning Systems (2022).

[37] M. Shahvali, M.-B. Naghibi-Sistani, J. Askari, Dynamic event-triggered control for a class of nonlinear fractional-order systems, IEEE Transactions on Circuits and Systems. II, Express Briefs 69 (4) (2021) 2131–2135.

[38] H. Zhang, Z. Gao, Y. Wang, Y. Cai, Leader-following exponential consensus of fractional-order descriptor multiagent systems with distributed event-triggered strategy, IEEE Transactions on Systems, Man, and Cybernetics: Systems 52 (6) (2021) 3967–3979.

[39] A. Dorri, S.S. Kanhere, R. Jurdak, Multi-agent systems: a survey, IEEE Access 6 (2018) 28573–28593.

[40] K. Chen, J. Wang, Y. Zhang, Z. Liu, Leader-following consensus for a class of nonlinear strict-feedback multiagent systems with state time-delays, IEEE Transactions on Systems, Man, and Cybernetics: Systems 50 (7) (2018) 2351–2361.

[41] Z. Yang, S. Zheng, F. Liu, Y. Xie, Adaptive output feedback control for fractional-order multi-agent systems, ISA Transactions 96 (2020) 195–209.

[42] B. Jiang, C. Gao, Decentralized adaptive sliding mode control of large-scale semi-Markovian jump interconnected systems with dead-zone input, IEEE Transactions on Automatic Control 67 (3) (2021) 1521–1528.

[43] X. Meng, B. Jiang, H.R. Karimi, C. Gao, An event-triggered mechanism to observer-based sliding mode control of fractional-order uncertain switched systems, ISA Transactions 135 (2023) 115–129.

An event-triggered mechanism to observer-based sliding mode control of fractional-order uncertain switched systems

8.1 Introduction

Switched systems as a special type of hybrid systems, which include a series of subsystems and a switching rule directing the switching signal among those subsystems, have received great attention during the past few decades; see [1,2] and the references therein. Focusing on the continuous switched systems, the authors in [3] addressed the adaptive neural network observer design control problem on the basis of quantized output signals. In [4], by designing a nonfragile observer, the exponential stability of uncertain switched systems via the average dwell time approach was investigated. Meanwhile, with the development of fractional calculus theory [5], it has witnessed many applications of fractional order systems in mathematical modeling of physical systems; see, for instance, [6–9] and the references therein. Paralleled to the above theoretical results, some practical systems described by fractional-order switched systems have also been explored. In [10], the authors explored the exponential stability problem for a class of fractional-order switched systems by utilizing the mode-dependent average impulsive interval method. Meanwhile, the numerical simulation is carried out by taking the switching capacity DC–DC power converter as an example to verify the effectiveness and feasibility of the proposed method. In [11], the authors considered the impulsive stabilization problem for a class of fractional-order uncertain switched systems, and applied the theoretical results to blind source separation. In [12], the authors proposed a modeling and analysis approach for the DC–DC converter model which was described by a fractional-order switched system.

Sliding mode control (SMC) as one of the momentous methods to investigate the stabilization of complex nonlinear systems has been widely applied to engineering fields due to its advantages, such as good and fast transient response, and easy realization [13,14]. Different from existing time-triggered control strategies, the execution of control missions in the event-triggered control strategy depends only on the given triggered condition. Recently, some crucial superiorities of the event-triggered control strategy have been illustrated in [15–17]. In [12], a sufficient condition based on the LMI was also proposed to ensure the asymptotic stability for a class of fractional-order uncertain systems under the event-triggered mechanism. In [18], the guaranteed cost problem of fractional-order switched systems has been considered by utilizing the finite-time event-triggered control strategy. And then, by solving the strict LMI, some sufficient conditions were derived. In [19,20], the authors obtained some sufficient conditions for the robust and asymptotic stability of the augmented system by adopting the LMI method, and applied the theoretical results to the vehicle dynamics simulation experiment.

For integer-order dynamic systems, the event-triggered mechanism has been well studied in the literature. In [21], the authors proposed the event-triggered SMC method for the discrete-time 2D Roesser systems. In [22], the authors considered the problem of stabilization for linear uncertain systems with time delays by presenting the event-triggered SMC strategy. In [23], the authors investigated the stability problem for singular systems with disturbance via the event-triggered SMC method. For uncertain nonlinear control systems, a robust nonsingular fast terminal SMC strategy has been presented based on the event-triggered mechanism in [24]. By presenting the dynamic event-triggered SMC strategy, the authors in [25] considered the stabilization problem for a class of interval type-2 (IT2) fuzzy system subjects to fading channels. It should be pointed out that the problem of fractional-order control systems via the event-triggered SMC approach has not been fully addressed in the existing literature due to the following issues: (a) controller design and (b) the minimal inter-event time. For the issue (a), the problem is similar to that of time-triggered control design. Besides, Zeno behavior refers to infinitely many triggers within a limited time. For the designed event triggering conditions, those conditions are continuously satisfied, and the controller cannot effectively adjust to the triggering. In this way, the convergence of the system may not be guaranteed. Therefore, for the issue (b), in order to avoid Zeno behavior, it must be guaranteed that the minimal inter-event time has a positive lower bound. To our knowledge, utilizing an event-triggered mechanism to study the finite-time stability problem of the fractional-order uncertain switched system is still an open question, which also motivates us to explore this problem. Meanwhile, how to solve the issue (b) for such a system is also a challenge that motivates us to attempt this research.

Based on the above analysis, this chapter will deal with the problem of observer-based event-triggered SMC for a class of fractional-order uncertain switched systems. A set of sliding mode dynamic equations is obtained based on the proposed fractional-order integral sliding surface function. Meanwhile, an event-triggered SMC law is also designed to guarantee the finite-time stability of the sliding mode dynamics and reachability of the sliding surface. The main contributions of this chapter are threefold:

First, by introducing a fractional-order observer, the event-triggered SMC problem of uncertain switched systems is considered for the first time in the presence of arbitrary switching signals. Here, we design a piecewise output signal depending on the last sampled observer state during the triggered interval.

Second, some sufficient conditions in terms of strict LMIs are derived to ensure the finite-time stability of the obtained sliding mode dynamics via an average dwell-time approach. In addition, the advantages of the presented results are compared with the existing literature; see, for instance, [7,4].

Third, compared with existing works, a corrected result is presented to avoid the Zeno behavior for the fractional-order switched systems under the presented event-triggered condition (ETC). At the same time, to avoid the occurrence of infinitely many triggers within a finite time, a positive lower bound on the inter-execution intervals is also given by utilizing the Gronwall–Bellman inequality, which is one of the main significant theoretic contributions of this chapter.

8.2 Problem statement

Definition 8.1 ([4]). For given switching signal $\sigma(t)$ and parameters $T_2 > T_1 > 0$, let N_σ denote the number of switchings of $\sigma(t)$ over the interval (T_1, T_2). Then T_σ is said to be the average dwell time if

$$N_\sigma \leq N_0 + \frac{T_2 - T_1}{T_\sigma}$$

holds for any $T_\sigma > 0$ and $N_0 \geq 0$.

Definition 8.2 ([26]). A closed-loop system with variable $x(t)$ is said to be finite-time stable with respect to $(c_1, c_2, T_f, \sigma(t))$, with $c_2 > c_1 > 0$ and $T_f = T_2 - T_1 > 0$, if

$$\|x(T_1)\|^2 \leq c_1 \quad \Longrightarrow \quad \|x(t)\|^2 < c_2, \ \forall t \in [T_1, T_2].$$

Remark 8.1. In [27], Polyakov and Fridman have systematically given some stability notions for time-invariant systems. According to the standard definition of system stability, it can be seen that Lyapunov, asymptotic, and finite time stabilities are the three basic notions that the former concept is based on.

Consider the following fractional-order uncertain switched systems:

$$\begin{cases} {}^{C}_{t_0}D_t^{\alpha}x(t) = (A_{\sigma(t)} + \Delta A_{\sigma(t)})x(t) + (B_{\sigma(t)} + \Delta B_{\sigma(t)})u(t), \\ y(t) = C_{\sigma(t)}x(t), \end{cases} \tag{8.1}$$

where order $\alpha \in (0,1)$, $x(t) \in \mathbb{R}^n$ denotes the pseudo-state vector, while $u(t) \in \mathbb{R}^m$ and $y(t) \in \mathbb{R}^l$ are control input and output signals, respectively. Also $\sigma(t) \in \{1, 2, \ldots, s\}$ is a piecewise constant function of continuous time t, called the switching signal. Furthermore, $A_{\sigma(t)}$, $B_{\sigma(t)}$, $C_{\sigma(t)}$ are known constant matrices for the switching signal $\sigma(t)$, and $\Delta A_{\sigma(t)}$, $\Delta B_{\sigma(t)}$ are the corresponding admissible parameter uncertainties.

In addition, for $\sigma(t) = i$, $i \in \{1, 2, \ldots, s\}$, the switched system (8.1) is rewritten as

$$\begin{cases} {}^{C}_{t_0}D_t^{\alpha}x(t) = (A_i + \Delta A_i)x(t) + (B_i + \Delta B_i)u(t), \\ y(t) = C_i x(t). \end{cases} \tag{8.2}$$

Besides, it is assumed that $\Delta A_i = E_{1i} F_{1i}(t) H_{1i}$, $\Delta B_i = E_{2i} F_{2i}(t) H_{2i}$, where E_{1i}, H_{1i}, E_{2i} and H_{2i} are known real matrices, $F_{1i}(t)$ and $F_{2i}(t)$ are unknown functions satisfying $F_{1i}^{\mathrm{T}}(t) F_{1i}(t) \leq I$ and $F_{2i}^{\mathrm{T}}(t) F_{2i}(t) \leq I$, respectively.

Remark 8.2. Fractional-order switching systems have been used in some practical applications, especially in the field of electrical systems. For example, in [12,6], the authors presented a DC–DC converter model which could be described by fractional-order switched systems, where the expression of the mathematical model was given by

$$\text{State 1}: \begin{cases} {}^{C}_{t_0}D_t^{\alpha}i_L(t) = \frac{1}{L}U_{in}, \\ {}^{C}_{t_0}D_t^{\alpha}v_C(t) = -\frac{1}{RC}v_C(t); \end{cases}$$

$$\text{State 2}: \begin{cases} {}^{C}_{t_0}D_t^{\alpha}i_L(t) = \frac{1}{L}U_{in} - \frac{1}{L}v_C(t), \\ {}^{C}_{t_0}D_t^{\alpha}v_C(t) = \frac{1}{C}i_L(t) - \frac{1}{RC}v_C(t). \end{cases}$$

Meanwhile, another fractional-order electrical circuit was also presented in [8], where the mathematical model was described as

$$\text{State 1}: \begin{cases} {}^{C}_{t_0}D^{\alpha}_t v_C(t) = -\frac{1}{RC} v_C(t) + \frac{1}{C} i_L(t), \\ {}^{C}_{t_0}D^{\alpha}_t i_L(t) = -\frac{1}{L} v_C(t); \end{cases}$$

$$\text{State 2}: \begin{cases} {}^{C}_{t_0}D^{\alpha}_t v_C(t) = -\frac{1}{RC} v_C(t), \\ {}^{C}_{t_0}D^{\alpha}_t i_L(t) = -\frac{1}{L} E. \end{cases}$$

More details about the modeling and application of fractional-order switching systems can be found in [12,6,8,11,10].

8.3 Main results

8.3.1 Fractional-order state observer design

Considering system (8.2), the fractional-order state observer is presented as follows:

$$\begin{cases} {}^{C}_{t_0}D^{\alpha}_t \hat{x}(t) = A_i \hat{x}(t) + B_i u(t) + L_i(y(t) - \hat{y}(t)), \\ \hat{y}(t) = C_i \hat{x}(t_k), \end{cases} \tag{8.3}$$

where $t \in [t_k, t_{k+1})$ is illustrated in (8.6), $\hat{x}(t) \in \mathbb{R}^n$ and $\hat{y}(t) \in \mathbb{R}^l$ denote the observer state and output, respectively; $\hat{x}(t_k)$ denotes the last sampled observer state during the triggered interval; $L_i \in \mathbb{R}^{n \times l}$ is the observer gain matrix.

Then the error system can be obtained as follows:

$${}^{C}_{t_0}D^{\alpha}_t e(t) = (A_i - L_i C_i + \Delta A_i)e(t) + \Delta A_i \hat{x}(t) + \Delta B_i u(t) - L_i C_i \hat{e}(t), \tag{8.4}$$

where $e(t) = x(t) - \hat{x}(t)$ and $\hat{e}(t) = \hat{x}(t) - \hat{x}(t_k)$ are the state estimation and sampled errors, respectively.

Based on the SMC theory, the fractional-order integral sliding surface function can be designed as follows:

$$s(t) = G_i \hat{x}(t) - G_i(A_i + B_i K_i){}_{t_0}D^{-\alpha}_t \hat{x}(t), \tag{8.5}$$

where $G_i \in \mathbb{R}^{m \times n}$ is such that $G_i B_i$ is nonsingular, and $K_i \in \mathbb{R}^{m \times n}$ is selected to ensure that $A_i + B_i K_i$ is Hurwitz.

8.3.2 Event-triggered SMC strategy

Next, we aim to design an event-triggered detection mechanism-based SMC law to ensure the stability of the error system (8.4). Therefore, we present the following ETC:

$$\begin{cases} t_{k+1} = \min\{t_k + \Delta, \mathfrak{t}_{k+1}\}, \\ \mathfrak{t}_{k+1} \triangleq \inf\{t > t_k \mid \|\hat{e}(t)\|^2 \geq \rho\|\hat{x}(t)\|^2 \vee |(s_e)_j(t)| \geq \frac{\sqrt{\rho}}{\sqrt{m}}\|\hat{x}(t)\|\}, \end{cases} \tag{8.6}$$

where $s_e(t) = s(t) - s(t_k)$, $\rho > 0$ is a designed threshold constant, and $\Delta > 0$ is a given constant.

Remark 8.3. With the rapid development of computer communication, network and information processing technology, network control systems have made significant progress. In the traditional time-triggered control strategy, such as the SMC approach, when a system reaches its steady state and no disturbance is imposed on the system, executing control tasks periodically will waste communication resources. Thus, the event-triggered strategy became the focus of attention of many scholars since it can significantly reduce the transmission of data in the network, save network resources, and reduce energy consumption; see [13,16,23] and the references therein. Notice that in ETC (8.6), based on the event-triggering mechanism of [23], for given fractional-order systems, a threshold is introduced to adjust the triggering time interval to ensure that Zeno behavior can be excluded.

Besides, for periodically sampled data, the event-triggered equivalent controller is

$$\hat{u}_{i\text{eq}}(t) = K_i\hat{x}(t_k) - (G_iB_i)^{-1}G_iL_iC_i(e(t) + \hat{e}(t)), \quad t \in [t_k, t_{k+1}). \tag{8.7}$$

Then, substituting (8.7) into (8.3) and (8.4), the sliding mode dynamic can be obtained as follows:

$$\begin{cases} {}^C_{t_0}D^\alpha_t\hat{x}(t) = (A_i + B_iK_i)\hat{x}(t) + \tilde{B}_1L_iC_ie(t) + (\tilde{B}_1L_iC_i - B_iK_i)\hat{e}(t), \\ {}^C_{t_0}D^\alpha_t e(t) = (\Delta A_i + \Delta B_iK_i)\hat{x}(t) + (A_i + \Delta A_i - \Delta B_i\tilde{B}_2L_iC_i - L_iC_i)e(t) \\ \qquad\qquad - (\Delta B_iK_i + \Delta B_i\tilde{B}_2L_iC_i + L_iC_i)\hat{e}(t), \end{cases} \tag{8.8}$$

where $\tilde{B}_1 \triangleq I - B_i(G_iB_i)^{-1}G_i$, $\tilde{B}_2 \triangleq (G_iB_i)^{-1}G_i$.

Remark 8.4. In [7], the finite-time control problem for fractional-order positive impulsive switched systems via an average dwell time method was investigated. However, as mentioned in Remark 2.1, for the given fractional-order switched systems, the authors have not consider whether Lemma 2.4 applies or not. Therefore, to handle this technical concern, a better method is presented for fractional-order switched systems in the following.

Theorem 8.1. *For given positive scalars ρ, T_f, ε, $c_2 > c_1$, and $\beta_2 > \beta_1$, the augmented closed-loop system (8.8) is finite-time stable under ETC (8.6) with* *respect to (c_1, c_2, T_f, σ) if there exist symmetric positive definite matrices $\boldsymbol{P}_i \in \mathbb{R}^{n \times n}$, some positive scalars ε_{ki} $(k = 1, 2, \ldots, 6)$, $\gamma_1 \geq 1 \geq \gamma_2$ and $\hat{\mu}$ such that the following inequalities hold:*

$$
\begin{bmatrix}
\boldsymbol{\Phi}_{11} & \boldsymbol{P}_i \tilde{\boldsymbol{B}}_1 \boldsymbol{L}_i \boldsymbol{C}_i & \boldsymbol{\Phi}_{13} & \boldsymbol{H}_{1i}^{\mathrm{T}} & \boldsymbol{K}_i^{\mathrm{T}} \boldsymbol{H}_{2i}^{\mathrm{T}} & 0 & 0 & 0 & 0 \\
* & \boldsymbol{\Phi}_{22} & -\boldsymbol{P}_i \boldsymbol{L}_i \boldsymbol{C}_i & 0 & 0 & \boldsymbol{H}_{1i}^{\mathrm{T}} & \boldsymbol{C}_i^{\mathrm{T}} \boldsymbol{L}_i^{\mathrm{T}} \tilde{\boldsymbol{B}}_2^{\mathrm{T}} \boldsymbol{H}_{2i}^{\mathrm{T}} & 0 & 0 \\
* & * & -\boldsymbol{I} & 0 & 0 & 0 & 0 & \boldsymbol{K}_i^{\mathrm{T}} \boldsymbol{H}_{2i}^{\mathrm{T}} & \boldsymbol{C}_i^{\mathrm{T}} \boldsymbol{L}_i^{\mathrm{T}} \tilde{\boldsymbol{B}}_2^{\mathrm{T}} \boldsymbol{H}_{2i}^{\mathrm{T}} \\
* & * & * & -\varepsilon_{1i} \boldsymbol{I} & 0 & 0 & 0 & 0 & 0 \\
* & * & * & * & -\varepsilon_{2i} \boldsymbol{I} & 0 & 0 & 0 & 0 \\
* & * & * & * & * & -\varepsilon_{3i} \boldsymbol{I} & 0 & 0 & 0 \\
* & * & * & * & * & * & -\varepsilon_{4i} \boldsymbol{I} & 0 & 0 \\
* & * & * & * & * & * & * & -\varepsilon_{5i} \boldsymbol{I} & 0 \\
* & * & * & * & * & * & * & * & -\varepsilon_{6i} \boldsymbol{I}
\end{bmatrix} < 0,
\tag{8.9}
$$

$$
\beta_1 \gamma_1 \leq \varepsilon \lambda_{\min}(\boldsymbol{P}_i) \leq \varepsilon \lambda_{\max}(\boldsymbol{P}_i) \leq \beta_2 \gamma_2,
\tag{8.10}
$$

$$
\hat{\mu} \geq \mu \triangleq \max_{i \neq j} \left\{ \frac{\lambda_{\max}(\boldsymbol{P}_i)}{\lambda_{\min}(\boldsymbol{P}_j)} \right\} \geq 1,
\tag{8.11}
$$

$$
\ln \frac{\gamma_1}{\gamma_2} > \frac{\varepsilon(\alpha T_f - \alpha + 1)}{\Gamma(1 + \alpha)},
\tag{8.12}
$$

where

$$
\boldsymbol{\Phi}_{11} = \mathrm{sym}\{\boldsymbol{P}_i(\boldsymbol{A}_i + \boldsymbol{B}_i \boldsymbol{K}_i)\} + \rho \boldsymbol{I}, \quad \boldsymbol{\Phi}_{13} = \boldsymbol{P}_i(\tilde{\boldsymbol{B}}_1 \boldsymbol{L}_i \boldsymbol{C}_i - \boldsymbol{B}_i \boldsymbol{K}_i),
$$

$$
\boldsymbol{\Phi}_{22} = \mathrm{sym}\{\boldsymbol{P}_i(\boldsymbol{A}_i - \boldsymbol{L}_i \boldsymbol{C}_i)\} + (\varepsilon_{1i} + \varepsilon_{3i})\boldsymbol{P}_i \boldsymbol{E}_{1i} \boldsymbol{E}_{1i}^{\mathrm{T}} \boldsymbol{P}_i
$$

$$
+ (\varepsilon_{2i} + \varepsilon_{4i} + \varepsilon_{5i} + \varepsilon_{6i})\boldsymbol{P}_i \boldsymbol{E}_{2i} \boldsymbol{E}_{2i}^{\mathrm{T}} \boldsymbol{P}_i,
$$

and the average dwell time is

$$
T_\sigma > T_\sigma^*,
\tag{8.13}
$$

where $T_\sigma^* \triangleq T_f \left[\ln \hat{\mu} + \frac{\varepsilon(1-\alpha)}{\Gamma(1+\alpha)} \right] \left[\ln \frac{\gamma_1}{\gamma_2} - \frac{\varepsilon(\alpha T_f - \alpha + 1)}{\Gamma(1+\alpha)} \right]^{-1}.$

Proof. Construct the following Lyapunov function:

$$
V_i(t) = \hat{\boldsymbol{x}}^{\mathrm{T}}(t) \boldsymbol{P}_i \hat{\boldsymbol{x}}(t) + \boldsymbol{e}^{\mathrm{T}}(t) \boldsymbol{P}_i \boldsymbol{e}(t).
\tag{8.14}
$$

Apparently there exist positive scalars β_1 and β_2 $(\beta_2 > \beta_1)$ satisfying

$$
\beta_1 \|\boldsymbol{\zeta}(t)\|^2 \leq V_i(t) \leq \beta_2 \|\boldsymbol{\zeta}(t)\|^2,
$$

where $\boldsymbol{\zeta}^{\mathrm{T}}(t) = [\hat{\boldsymbol{x}}^{\mathrm{T}}(t), \boldsymbol{e}^{\mathrm{T}}(t)]$.

Then, combining with Lemma 2.2, one has

$$
\begin{aligned}
{}_{t_0}^{\mathcal{C}} D_t^\alpha V_i(t) \leq\; & 2\hat{x}^{\mathrm{T}}(t) P_i\, {}_{t_0}^{\mathcal{C}} D_t^\alpha \hat{x}(t) + 2 e^{\mathrm{T}}(t) P_i\, {}_{t_0}^{\mathcal{C}} D_t^\alpha e(t) \\
=\; & 2\hat{x}^{\mathrm{T}}(t) P_i\{(A_i + B_i K_i)\hat{x}(t) + \tilde{B}_1 L_i C_i e(t) \\
& + (\tilde{B}_1 L_i C_i - B_i K_i)\hat{e}(t)\} + 2 e^{\mathrm{T}}(t) P_i\{(\Delta A_i + \Delta B_i K_i)\hat{x}(t) \\
& + (A_i + \Delta A_i - \Delta B_i \tilde{B}_2 L_i C_i - L_i C_i) e(t) \\
& - (\Delta B_i K_i + \Delta B_i \tilde{B}_2 L_i C_i + L_i C_i)\hat{e}(t)\}.
\end{aligned}
$$

Meanwhile, according to Lemma 1 in [28], there exist positive constants $\varepsilon_{ji}\ (j = 1, 2, \ldots, 6)$ such that

$$
\begin{aligned}
2 e^{\mathrm{T}}(t) P_i \Delta A_i \hat{x}(t) =\; & 2 e^{\mathrm{T}}(t) P_i E_{1i} F_{1i}(t) H_{1i} \hat{x}(t) \\
\leq\; & \varepsilon_{1i} e^{\mathrm{T}}(t) P_i E_{1i} E_{1i}^{\mathrm{T}} P_i e(t) \\
& + \frac{1}{\varepsilon_{1i}} \hat{x}^{\mathrm{T}}(t) H_{1i}^{\mathrm{T}} H_{1i} \hat{x}(t), \\
2 e^{\mathrm{T}}(t) P_i \Delta B_i K_i \hat{x}(t) =\; & 2 e^{\mathrm{T}}(t) P_i E_{2i} F_{2i}(t) H_{2i} K_i \hat{x}(t) \\
\leq\; & \varepsilon_{2i} e^{\mathrm{T}}(t) P_i E_{2i} E_{2i}^{\mathrm{T}} P_i e(t) \\
& + \frac{1}{\varepsilon_{2i}} \hat{x}^{\mathrm{T}}(t) K_i^{\mathrm{T}} H_{2i}^{\mathrm{T}} H_{2i} K_i \hat{x}(t), \\
2 e^{\mathrm{T}}(t) P_i \Delta A_i e(t) =\; & 2 e^{\mathrm{T}}(t) P_i E_{1i} F_{1i}(t) H_{1i} e(t) \\
\leq\; & \varepsilon_{3i} e^{\mathrm{T}}(t) P_i E_{1i} E_{1i}^{\mathrm{T}} P_i e(t) \\
& + \frac{1}{\varepsilon_{3i}} e^{\mathrm{T}}(t) H_{1i}^{\mathrm{T}} H_{1i} e(t), \\
-2 e^{\mathrm{T}}(t) P_i \Delta B_i \tilde{B}_2 L_i C_i e(t) =\; & -2 e^{\mathrm{T}}(t) P_i E_{2i} F_{2i}(t) H_{2i} \tilde{B}_2 L_i C_i e(t) \\
\leq\; & \varepsilon_{4i} e^{\mathrm{T}}(t) P_i E_{2i} E_{2i}^{\mathrm{T}} P_i e(t) \\
& + \frac{1}{\varepsilon_{4i}} e^{\mathrm{T}}(t) C_i^{\mathrm{T}} L_i^{\mathrm{T}} \tilde{B}_2^{\mathrm{T}} H_{2i}^{\mathrm{T}} H_{2i} \tilde{B}_2 L_i C_i e(t), \\
-2 e^{\mathrm{T}}(t) P_i \Delta B_i K_i \hat{e}(t) =\; & -2 e^{\mathrm{T}}(t) P_i E_{2i} F_{2i}(t) H_{2i} K_i \hat{e}(t) \\
\leq\; & \varepsilon_{5i} e^{\mathrm{T}}(t) P_i E_{2i} E_{2i}^{\mathrm{T}} P_i^{\mathrm{T}} e(t) \\
& + \frac{1}{\varepsilon_{5i}} \hat{e}^{\mathrm{T}}(t) K_i^{\mathrm{T}} H_{2i}^{\mathrm{T}} H_{2i} K_i \hat{e}(t), \\
-2 e^{\mathrm{T}}(t) P_i \Delta B_i \tilde{B}_2 L_i C_i \hat{e}(t) =\; & -2 e^{\mathrm{T}}(t) P_i E_{2i} F_{2i}(t) H_{2i} \tilde{B}_2 L_i C_i \hat{e}(t) \\
\leq\; & \varepsilon_{6i} e^{\mathrm{T}}(t) P_i E_{2i} E_{2i}^{\mathrm{T}} P_i^{\mathrm{T}} e(t) \\
& + \frac{1}{\varepsilon_{6i}} \hat{e}^{\mathrm{T}}(t) C_i^{\mathrm{T}} L_i^{\mathrm{T}} \tilde{B}_2^{\mathrm{T}} H_{2i}^{\mathrm{T}} H_{2i} \tilde{B}_2 L_i C_i \hat{e}(t).
\end{aligned}
$$

Therefore, considering ETC (8.6), one can obtain

$$\substack{C \\ t_0} D_t^\alpha V_i(t) \le \varpi^{\mathrm{T}}(t)\boldsymbol{\Psi}\varpi(t),$$

where $\varpi^{\mathrm{T}}(t) = [\hat{x}^{\mathrm{T}}(t), e^{\mathrm{T}}(t), \hat{e}^{\mathrm{T}}(t)]$, and

$$\boldsymbol{\Psi} = \begin{bmatrix} \boldsymbol{\Psi}_{11} & P_i \tilde{B}_1 L_i C_i & \boldsymbol{\Psi}_{13} \\ * & \boldsymbol{\Psi}_{22} & P_i L_i C_i \\ * & * & \boldsymbol{\Psi}_{33} \end{bmatrix},$$

with

$$\boldsymbol{\Psi}_{11} = \mathrm{sym}\{P_i(A_i + B_i K_i)\} + \frac{1}{\varepsilon_{1i}} H_{1i}^{\mathrm{T}} H_{1i} + \frac{1}{\varepsilon_{2i}} K_i^{\mathrm{T}} H_{2i}^{\mathrm{T}} H_{2i} K_i + \rho I,$$

$$\boldsymbol{\Psi}_{13} = P_i \tilde{B}_1 L_i C_i - P_i B_i K_i,$$

$$\boldsymbol{\Psi}_{22} = \mathrm{sym}\{P_i(A_i - L_i C_i)\} + (\varepsilon_{1i} + \varepsilon_{3i}) P_i E_{1i} E_{1i}^{\mathrm{T}} P_i$$

$$+ (\varepsilon_{2i} + \varepsilon_{4i} + \varepsilon_{5i} + \varepsilon_{6i}) P_i E_{2i} E_{2i}^{\mathrm{T}} P_i + \frac{1}{\varepsilon_{3i}} H_{1i}^{\mathrm{T}} H_{1i}$$

$$+ \frac{1}{\varepsilon_{4i}} C_i^{\mathrm{T}} L_i^{\mathrm{T}} \tilde{B}_2^{\mathrm{T}} H_{2i}^{\mathrm{T}} H_{2i} \tilde{B}_2 L_i C_i, \quad \text{and}$$

$$\boldsymbol{\Psi}_{33} = \frac{1}{\varepsilon_{5i}} K_i^{\mathrm{T}} H_{2i}^{\mathrm{T}} H_{2i} K_i + \frac{1}{\varepsilon_{6i}} C_i^{\mathrm{T}} L_i^{\mathrm{T}} \tilde{B}_2^{\mathrm{T}} H_{2i}^{\mathrm{T}} H_{2i} \tilde{B}_2 L_i C_i - I.$$

Then, by using the Schur complement lemma, we know that (8.9) is equivalent to $\boldsymbol{\Psi} < 0$. Hence, there exists a positive constant ε such that

$$\substack{C \\ t_0} D_t^\alpha V_i(t) < -\varepsilon V_i(t).$$

And according to Lemma 2.5 and Lemma 2 in [29], one has that

$$V_i(t) < V_i(t_0) E_{\alpha,1}[-\varepsilon(t - t_0)^\alpha] \le V_i(t_0) \frac{\lambda}{1 + \varepsilon(t - t_0)^\alpha}.$$

Therefore, based on the above analysis, it is clear that $V_i(t) \to 0$ holds when $t \to +\infty$, that is to say, there exist positive scalars δ_1, δ_2 such that $|(\hat{x})_j(t)| \le \delta_1$ and $|(e)_j(t)| \le \delta_2$ hold. Then, from (8.8), one has that

$$\left| \substack{C \\ t_0} D_t^\alpha (\hat{x})_j(t) \right| = |(A_i + B_i K_i)_j \hat{x}(t) + (\tilde{B}_1 L_i C_i)_j e(t)$$

$$+ (\tilde{B}_1 L_i C_i - B_i K_i)_j (\hat{x}(t) - \hat{x}(t_k))|$$

$$\le \|(A_i + \tilde{B}_1 L_i C_i)_j\| \|\hat{x}(t)\| + \|(\tilde{B}_1 L_i C_i)_j\| \|e(t)\|$$

$$+ \|(\tilde{B}_1 L_i C_i - B_i K_i)_j\| \|\hat{x}(t_k)\|$$

$$\le \|(A_i + \tilde{B}_1 L_i C_i)_j\| \sqrt{n}\delta_1 + \|(\tilde{B}_1 L_i C_i)_j\| \sqrt{n}\delta_2$$

$$+ \|(\tilde{B}_1 L_i C_i - B_i K_i)_j\| \|\hat{x}(t_k)\|$$

$$\triangleq (\hat{\eta})_j(t_k).$$

Denote $q(t) = \hat{x}'(t)$, then there exists a nonnegative continuous function $p(t)$ satisfying

$$_{t_0}D_t^{\alpha-1}(q)_j(t) = \frac{\sqrt{(\hat{\eta})_j(t_k)}}{\sqrt{n}} - (p)_j(t). \tag{8.15}$$

Then, taking the Laplace transform on both sides of (8.15), one has

$$s(Q)_j(s) = \frac{\sqrt{(\hat{\eta})_j(t_k)}}{\sqrt{n}}s^{1-\alpha} - s^{2-\alpha}(P)_j(s), \tag{8.16}$$

where $Q(s) = \mathcal{L}[q(t)]$ and $P(s) = \mathcal{L}[p(t)]$.

From (8.16), it is easily obtained that $\lim\limits_{s \to 0} s(Q)_j(s) = 0$, which is equivalent to $\lim\limits_{t \to +\infty}(q)_j(t) = 0$ via the final value theorem. Namely, we can also obtain that $\lim\limits_{t \to +\infty} q(t) = 0$ holds. Therefore, there exists a positive scalar $\hat{\delta}_1$ such that $\|\hat{x}'(t)\| \le \hat{\delta}_1$ holds for any $t \in [t_0, +\infty)$. Similarly, it can also easily be deduced that $\|e'(t)\| \le \hat{\delta}_2$ holds for any $t \in [t_0, +\infty)$. Therefore, for $\forall t \in [t_0, t_0 + T_f]$, one can get from (8.14) that

$$\|V_i'(t)\| \le 2\sqrt{n}(\delta_1\hat{\delta}_1 + \delta_2\hat{\delta}_2)\lambda_{\max}(P_i) \triangleq \tilde{\delta}. \tag{8.17}$$

Thereupon, for $\forall t \in [\hat{t}_k, \hat{t}_{k+1})$, from (8.17) and Definition 2.6 we get

$$\begin{aligned}
{}_{\hat{t}_k}^{C}D_t^{\alpha}V_i(t) &< -\varepsilon V_i(t) - \frac{1}{\Gamma(1-\alpha)}\int_{\hat{t}_0}^{\hat{t}_k}\frac{V_i'(\tau)}{(t-\tau)^{\alpha}}d\tau \\
&\le \varepsilon V_i(t) + \frac{1}{\Gamma(1-\alpha)}\int_{\hat{t}_0}^{\hat{t}_k}\frac{\|V_i'(\tau)\|}{(t-\tau)^{\alpha}}d\tau \\
&\le \varepsilon V_i(t) + \frac{\tilde{\delta}}{\Gamma(1-\alpha)}\int_{\hat{t}_0}^{\hat{t}_k}\frac{1}{(t-\tau)^{\alpha}}d\tau \\
&\le \varepsilon V_i(t) + \Upsilon_1(T_f),
\end{aligned} \tag{8.18}$$

where $\Upsilon_1(T_f) \triangleq \frac{\tilde{\delta}}{\Gamma(2-\alpha)}T_f^{1-\alpha} > 0$.

Taking the fractional-order integral of (8.18) from $\hat{t}_k$ to t, we have

$$V_i(t) - V_i(\hat{t}_k) \le {}_{\hat{t}_k}D_t^{-\alpha}(\varepsilon V_i(t) + \Upsilon_1(T_f)) = \frac{1}{\Gamma(\alpha)}\int_{\hat{t}_k}^{t}\frac{\varepsilon V_i(\tau) + \Upsilon_1(T_f)}{(t-\tau)^{1-\alpha}}d\tau.$$

The latter can be also rewritten as

$$\varepsilon V_i(t) + \Upsilon_1(T_f) \le \varepsilon V_i(\hat{t}_k) + \Upsilon_1(T_f) + \frac{\varepsilon}{\Gamma(\alpha)}\int_{\hat{t}_k}^{t}\frac{\varepsilon V_i(\tau) + \Upsilon_1(T_f))}{(t-\tau)^{1-\alpha}}d\tau.$$

Meanwhile, from Lemma 2.9, one has

$$\varepsilon V_i(t) + \Upsilon_1(T_f) \le (\varepsilon V_i(\hat{t}_k) + \Upsilon_1(T_f)) \exp\left(\frac{\varepsilon}{\Gamma(\alpha)} \int_{\hat{t}_k}^t \frac{1}{(t-\tau)^{1-\alpha}} d\tau\right)$$
$$= (\varepsilon V_i(\hat{t}_k) + \Upsilon_1(T_f)) \exp\left(\frac{\varepsilon}{\Gamma(1+\alpha)}(t-\hat{t}_k)^\alpha\right). \tag{8.19}$$

In addition, from (8.11), we have

$$V_i(\hat{t}_k) \le \mu V_j(\hat{t}_k^-).$$

Hence, there exists a scalar $\hat{\mu} \ge \mu$ satisfying

$$\varepsilon V_i(\hat{t}_k) + \Upsilon_1(T_f) \le \hat{\mu}(\varepsilon V_j(\hat{t}_k^-) + \Upsilon_1(T_f)). \tag{8.20}$$

From (8.19), (8.20) and considering the switching instant, $\sigma(\hat{t}_k^-) = j$ and $\sigma(\hat{t}_k^+) = i$, we have

$$\varepsilon V_{\sigma(t)}(t) + \Upsilon_1(T_f) \le (\varepsilon V_{\sigma(\hat{t}_k)}(\hat{t}_k) + \Upsilon_1(T_f)) \exp\left(\frac{\varepsilon}{\Gamma(1+\alpha)}(t-\hat{t}_k)^\alpha\right) \tag{8.21}$$

and

$$\varepsilon V_{\sigma(\hat{t}_k)}(\hat{t}_k) + \Upsilon_1(T_f) \le \hat{\mu}(\varepsilon V_{\sigma(\hat{t}_k^-)}(\hat{t}_k^-) + \Upsilon_1(T_f)). \tag{8.22}$$

According to Definition 8.1, letting $N_\sigma \le N_0 + \frac{t-\hat{t}_0}{T_\sigma}$ and then combining (8.21) and (8.22), we have

$$\varepsilon V_{\sigma(t)}(t) + \Upsilon_1(T_f) \le (\varepsilon V_{\sigma(\hat{t}_k)}(\hat{t}_k) + \Upsilon_1(T_f)) \exp\left(\frac{\varepsilon}{\Gamma(1+\alpha)}(t-\hat{t}_k)^\alpha\right)$$
$$\le \hat{\mu}(\varepsilon V_{\sigma(\hat{t}_k^-)}(\hat{t}_k^-) + \Upsilon_1(T_f)) \exp\left(\frac{\varepsilon}{\Gamma(1+\alpha)}(t-\hat{t}_k)^\alpha\right)$$
$$\vdots$$
$$\le \hat{\mu}^{N_\sigma}(\varepsilon V_{\sigma(\hat{t}_0)}(\hat{t}_0) + \Upsilon_1(T_f))$$
$$\times \exp\left(\frac{\varepsilon}{\Gamma(1+\alpha)}[(t-\hat{t}_k)^\alpha + \cdots + (\hat{t}_1 - \hat{t}_0)^\alpha]\right).$$

According to Lemma 5 in [7], for $\forall t \in [t_0, t_0 + T_f]$, we have

$$\varepsilon V_{\sigma(t)}(t) + \Upsilon_1(T_f) \le (\varepsilon V_{\sigma(\hat{t}_0)}(\hat{t}_0) + \Upsilon_1(T_f))$$
$$\times \exp\left(\frac{T_f}{T_\sigma} \ln \hat{\mu} + \frac{\varepsilon}{\Gamma(1+\alpha)}\left[\left(1 + \frac{T_f}{T_\sigma}\right)^{1-\alpha} + T_f^\alpha\right]\right)$$

$$\leq (\varepsilon V_{\sigma(\hat{t}_0)}(\hat{t}_0) + \Upsilon_1(T_f))$$
$$\times \exp\left(\frac{T_f}{T_\sigma}\ln\hat{\mu} + \frac{\varepsilon}{\Gamma(1+\alpha)}\left[(1-\alpha)\left(1+\frac{T_f}{T_\sigma}\right) + \alpha T_f\right]\right).$$

$$(8.23)$$

In addition, for $\forall t \in [t_0, t_0 + T_f]$, we have

$$\begin{cases} \varepsilon V_{\sigma(t)}(t) + \Upsilon_1(T_f) \geq \beta_1 \gamma_1 \|\zeta(t)\|^2 + \Upsilon_1(T_f), \\ \varepsilon V_{\sigma(\hat{t}_0)}(\hat{t}_0) + \Upsilon_1(T_f) \leq \beta_2 \gamma_2 \|\zeta(\hat{t}_0)\|^2 + \Upsilon_1(T_f). \end{cases} \qquad (8.24)$$

In view of (8.23) and (8.24),

$$\|\zeta(t)\|^2 \leq \left(\frac{\beta_2}{\beta_1}\|\zeta(\hat{t}_0)\|^2 + \frac{\Upsilon_1(T_f)}{\varepsilon\beta_1\gamma_2}\right)\exp\left(-\ln\frac{\gamma_1}{\gamma_2} + \frac{T_f}{T_\sigma}\ln\hat{\mu}\right.$$
$$\left. + \frac{\varepsilon}{\Gamma(1+\alpha)}\left[(1-\alpha)\left(1+\frac{T_f}{T_\sigma}\right) + \alpha T_f\right]\right) - \frac{\Upsilon_1(T_f)}{\varepsilon\beta_1\gamma_1}.$$

$$(8.25)$$

Combining (8.24) with (8.25), we obtain that

$$\|\zeta(t)\|^2 \leq \frac{\beta_2}{\beta_1}\|\zeta(\hat{t}_0)\|^2 + \frac{\Upsilon_1(T_f)}{\varepsilon\beta_1\gamma_2} - \frac{\Upsilon_1(T_f)}{\varepsilon\beta_1\gamma_1}.$$

Therefore, from Definition 8.2, we conclude that the closed-loop system (8.8) is finite-time stable with respect to (c_1, c_2, T_f, σ), with $c_2 = \frac{\beta_2}{\beta_1}c_1 + \frac{\Upsilon_1(T_f)}{\varepsilon\beta_1\gamma_2} - \frac{\Upsilon_1(T_f)}{\varepsilon\beta_1\gamma_1} > c_1 > 0$. $\qquad\square$

Remark 8.5. Notice that the matrices P_i, K_i and L_i in inequality (8.9) are unknown, as are scalars ε_{ki} ($k = 1, 2, \ldots, 6$), which implies that the inequality in (8.9) is nonlinear and cannot be solved directly by using the Matlab$^\circledR$ LMI-Toolbox. Therefore, we need to transform the inequality (8.9) into a strict feasible LMI. To this end, the following Theorem 8.2 is proposed to solve for the feedback matrix K_i and observer gain matrix L_i.

Theorem 8.2. *For a given scalar $\rho > 0$, the sliding mode dynamics (8.8) is finite-time stable under the proposed ETC (8.6) if there exist some matrices X_i, Y_i, Z_i with appropriate dimensions and positive scalars ε_{ki} ($k = 1, 2, \ldots, 6$) such that conditions (8.10)–(8.12) and the following inequality hold:*

$$\begin{bmatrix} \hat{\Phi}_{11} & \tilde{B}_1 Z_i & \hat{\Phi}_{13} & X_i H_{1i}^{\mathrm{T}} & Y_i^{\mathrm{T}} H_{2i}^{\mathrm{T}} & 0 & 0 & 0 & 0 & \rho X_i \\ * & \hat{\Phi}_{22} & -Z_i & 0 & 0 & X H_{1i}^{\mathrm{T}} & Z_i^{\mathrm{T}} \tilde{B}_2^{\mathrm{T}} H_{2i}^{\mathrm{T}} & 0 & 0 & 0 \\ * & * & \lambda^2 I - 2\lambda X_i & 0 & 0 & 0 & 0 & Y_i^{\mathrm{T}} H_{2i}^{\mathrm{T}} & Z_i^{\mathrm{T}} \tilde{B}_2^{\mathrm{T}} H_{2i}^{\mathrm{T}} & 0 \\ * & * & * & -\varepsilon_{1i} I & 0 & 0 & 0 & 0 & 0 & 0 \\ * & * & * & * & -\varepsilon_{2i} I & 0 & 0 & 0 & 0 & 0 \\ * & * & * & * & * & -\varepsilon_{3i} I & 0 & 0 & 0 & 0 \\ * & * & * & * & * & * & -\varepsilon_{4i} I & 0 & 0 & 0 \\ * & * & * & * & * & * & * & -\varepsilon_{5i} I & 0 & 0 \\ * & * & * & * & * & * & * & * & -\varepsilon_{6i} I & 0 \\ * & * & * & * & * & * & * & * & * & -\rho I \end{bmatrix} < 0,$$

$$(8.26)$$

where $\hat{\boldsymbol{\Phi}}_{11} = \text{sym}\{(A_i X_i + B_i Y_i)\}$, $\hat{\boldsymbol{\Phi}}_{13} = \tilde{B}_1 Z_i - B_i Y_i$, $\hat{\boldsymbol{\Phi}}_{22} = \text{sym}\{A_i X_i - Z_i\} + (\varepsilon_{1i} + \varepsilon_{3i}) E_{1i} E_{1i}^{\mathrm{T}} + (\varepsilon_{2i} + \varepsilon_{4i} + \varepsilon_{5i} + \varepsilon_{6i}) E_{2i} E_{2i}^{\mathrm{T}}$. *In addition, the feedback matrix* $K_i = Y_i X_i^{-1}$ *and observer gain matrix* $L_i = Z_i (C_i X_i)^{-1}$ *can be obtained, respectively.*

Proof. Let $X_i = P_i^{-1}$, $Y_i = K_i X_i$, and $Z_i = L_i C_i X_i$. Then, (8.9) is equivalent to

$$\begin{bmatrix} \tilde{\boldsymbol{\Psi}}_{11} & \tilde{B}_1 Z_i & \tilde{\boldsymbol{\Psi}}_{13} & X_i H_{1i}^{\mathrm{T}} & Y_i^{\mathrm{T}} H_{2i}^{\mathrm{T}} & 0 & 0 & 0 & 0 \\ * & \tilde{\boldsymbol{\Psi}}_{22} & -Z_i & 0 & 0 & X_i H_{1i}^{\mathrm{T}} & Z_i^{\mathrm{T}} \tilde{B}_2^{\mathrm{T}} H_{2i}^{\mathrm{T}} & 0 & 0 \\ * & * & -X_i X_i & 0 & 0 & 0 & 0 & Y_i^{\mathrm{T}} H_{2i}^{\mathrm{T}} & Z_i^{\mathrm{T}} \tilde{B}_2^{\mathrm{T}} H_{2i}^{\mathrm{T}} \\ * & * & * & -\varepsilon_{1i} I & 0 & 0 & 0 & 0 & 0 \\ * & * & * & * & -\varepsilon_{2i} I & 0 & 0 & 0 & 0 \\ * & * & * & * & * & -\varepsilon_{3i} I & 0 & 0 & 0 \\ * & * & * & * & * & * & -\varepsilon_{4i} I & 0 & 0 \\ * & * & * & * & * & * & * & -\varepsilon_{5i} I & 0 \\ * & * & * & * & * & * & * & * & -\varepsilon_{6i} I \end{bmatrix} < 0, \tag{8.27}$$

where $\tilde{\boldsymbol{\Psi}}_{11} = \text{sym}\{(A_i + B_i K_i) X_i\} + \rho X_i X_i$, $\tilde{\boldsymbol{\Psi}}_{11} = \tilde{B}_1 Z_i - B_i Y_i$, and $\tilde{\boldsymbol{\Psi}}_{22} = \text{sym}\{(A_i - L_i C_i) X_i\} + (\varepsilon_{1i} + \varepsilon_{3i}) E_{1i} E_{1i}^{\mathrm{T}} + (\varepsilon_{2i} + \varepsilon_{4i} + \varepsilon_{5i} + \varepsilon_{6i}) E_{2i} E_{2i}^{\mathrm{T}}$. Meanwhile, note that there exists a positive scalar λ such that $-X_i X_i \leq \lambda^2 I - 2\lambda X_i$ holds. Hence, by using the Schur complement lemma, one obtains that (8.27) is equivalent to (8.26), i.e., the sliding mode dynamics (8.8) is finite-time stable. $\qquad\square$

8.3.3 Reachability analysis

In this section, the event-triggered controller is presented as follows:

$$\hat{u}(t) = K_i \hat{x}(t_k) - (G_i B_i)^{-1} G_i L_i (y(t) - \hat{y}(t)) - \eta (G_i B_i)^{-1} \text{sgn}(s(t_k)), \tag{8.28}$$

where $\eta \triangleq \|G_i B_i K_i\| \|\hat{e}(t)\| + \iota$ with $\iota > 0$, $t \in [t_k, t_{k+1})$, $k = 0, 1, 2, \ldots$

Different from the traditional SMC strategy in [30,31], in order to deal with $s(t)\text{sgn}(s(t_k))$, the following lemma is proposed.

Lemma 8.1. *If* $\text{sgn}((s)_j(t)) = \text{sgn}((s)_j(t_k))$ *holds for any* $t \in [t_k, t_{k+1})$, *then the observer system trajectory* $\hat{x}(t)$ *will stay out of the region* $\mathcal{B} = \{\hat{x}(t) \mid \|s(t)\| < \sqrt{\rho}\|\hat{x}(t)\|\}$.

Proof. Considering ETC (8.6), when the presented event is triggered, the sliding mode signal will be updated at the triggering instant t_k, then for any $t \in [t_k, t_{k+1})$, one has

$$-\frac{\sqrt{\rho}}{\sqrt{m}}\|\hat{x}(t)\| < |(s)_j(t) - (s)_j(t_k)| < \frac{\sqrt{\rho}}{\sqrt{m}}\|\hat{x}(t)\|.$$

Meanwhile, based on the geometric relations, we know that $\mathrm{sgn}((s)_j(t)) \neq \mathrm{sgn}((s)_j(t_k))$ only when the observer system trajectory $\hat{x}(t)$ falls into the following two bounded regions:

(1) $\{\hat{x}(t)|(s)_j(t) > 0 > (s)_j(t_k) \wedge (s)_j(t) < (s)_j(t_k) + \frac{\sqrt{\rho}}{\sqrt{m}}\|\hat{x}(t)\|\}$;

(2) $\{\hat{x}(t)|(s)_j(t) < 0 < (s)_j(t_k) \wedge (s)_j(t) > (s)_j(t_k) - \frac{\sqrt{\rho}}{\sqrt{m}}\|\hat{x}(t)\|\}$.

Hence, based on the above analysis, when the observer system trajectory $\hat{x}(t)$ stays outside the region $\mathfrak{B} = \{\hat{x}(t)|\|s(t)\| < \sqrt{\rho}\|\hat{x}(t)\|\}$, one has that $\mathrm{sgn}((s)_j(t)) = \mathrm{sgn}((s)_j(t_k))$ holds for $\forall t \in [t_k, t_{k+1})$. $\quad\square$

Theorem 8.3. *The practical sliding mode occurs within the region*

$$\mathfrak{B} = \{\hat{x}(t)|\|s(t)\| < \sqrt{\rho}\|\hat{x}(t)\|\}$$

by the presented sliding mode controller (8.28) and remains there thereafter.

Proof. Construct the Lyapunov function as follows:

$$\hat{V}(t) = \frac{1}{2}s^{\mathrm{T}}(t)s(t).$$

Taking the fractional-order derivative of $\hat{V}(t)$, and considering the event-triggered SMC law (8.28), one has

$$
\begin{aligned}
{}^{C}_{t_0}D^{\alpha}_{t}\hat{V}(t) &= s^{\mathrm{T}}(t){}^{C}_{t_0}D^{\alpha}_{t}s(t) \\
&= s^{\mathrm{T}}(t)\Big\{G_i(A_i\hat{x}(t) + B_i\hat{u}(t) + L_iC_i(e(t) + \hat{e}(t))) \\
&\qquad - G_i(A_i + B_iK_i)\hat{x}(t)\Big\} \\
&= s^{\mathrm{T}}(t)\Big\{G_iB_iK_i(\hat{x}(t_k) - \hat{x}(t)) - \eta\,\mathrm{sgn}(s(t_k))\Big\} \\
&\leq \|s(t)\|\|G_iB_iK_i\|\|\hat{e}(t)\| - \eta s^{\mathrm{T}}(t)\mathrm{sgn}(s(t_k)).
\end{aligned}
$$

Considering Lemma 8.1, one has

$$
{}^{C}_{t_0}D^{\alpha}_{t}\hat{V}(t) \leq -(\eta - \|G_iB_iK_i\|\|\hat{e}(t)\|)\|s(t)\| < -\iota\|s(t)\|,
$$

which implies that the system trajectory will move towards a small vicinity of the sliding mode surface $s(t) = 0$. However, when it arrives at the sliding surface, $\mathrm{sgn}((s)_j(t)) = \mathrm{sgn}((s)_j(t_k))$ will not hold, that is to say, we cannot guarantee that ${}^{C}_{t_0}D^{\alpha}_{t}V_2(t) \leq 0$ holds. Therefore, the system trajectory will move away from the sliding mode surface. On the other hand, when it stays outside the region $\mathfrak{B}$, the trajectory will move towards a smaller vicinity of the sliding mode surface again, therefore, the system trajectory will be restricted to a certain region, namely

$$\mathfrak{B}_0 = \{\hat{\boldsymbol{x}}(t)|\ |(s)_j(t)|^2 + |(s)_j(t_k)|^2 \le \frac{\rho}{m}\|\hat{\boldsymbol{x}}(t)\|^2\}.$$

It is obvious that the sliding surface function tends to zero when the system trajectory tends to zero. Therefore, based on the above analysis, the sliding surface will occur and the boundary is given as $\mathfrak{B} = \{\hat{\boldsymbol{x}}(t)|\|s(t)\| < \sqrt{\rho}\|\hat{\boldsymbol{x}}(t)\|\}$. $\qquad\square$

8.3.4 Discussion on Zeno behavior

Different from the results presented in [12,32], under the proposed event-triggered control strategy, the positive lower bound on the event-triggered inter-execution interval needs to be derived to avoid infinite triggering. Therefore, an ETC depending on the last triggered state is presented to avoid the Zeno behavior.

Corollary 8.1. Define a self-triggered condition (STC) as follows:

$$\begin{cases} t_{k+1} = \min\{t_k + \Delta, t_{k+1}\}, \\ t_{k+1} = \inf\{t > t_k|\ \|\hat{\boldsymbol{e}}(t)\|^2 \ge \varrho\|\hat{\boldsymbol{x}}(t_k)\|^2 \vee |(s_e)_j(t)| \ge \sqrt{\frac{\varrho}{m}}\|\hat{\boldsymbol{x}}(t_k)\|\}, \end{cases} \tag{8.29}$$

where $\varrho = \frac{\rho}{2+2\rho}$. Then the finite-time stability of sliding mode dynamic (8.8) can be guaranteed under STC (8.29).

Proof. First, it is known from ETC (8.6) that for any $t \in [t_k, t_{k+1})$, one has

$$\hat{\boldsymbol{e}}^{\mathrm{T}}(t)\hat{\boldsymbol{e}}(t) < \varrho\hat{\boldsymbol{x}}^{\mathrm{T}}(t_k)\hat{\boldsymbol{x}}(t_k) = \rho(\hat{\boldsymbol{x}}(t) - \hat{\boldsymbol{e}}(t))^{\mathrm{T}}(\hat{\boldsymbol{x}}(t) - \hat{\boldsymbol{e}}(t))$$

$$\le 2\varrho(\hat{\boldsymbol{x}}^{\mathrm{T}}(t)\hat{\boldsymbol{x}}(t) + \hat{\boldsymbol{e}}^{\mathrm{T}}(t)\hat{\boldsymbol{e}}(t))$$

The latter be rewritten as

$$\|\hat{\boldsymbol{e}}(t)\|^2 < \frac{2\varrho}{1 - 2\varrho}\|\hat{\boldsymbol{x}}(t)\|^2 = \rho\|\hat{\boldsymbol{x}}(t)\|^2.$$

Second, combining ETC (8.6) with Lemma 5 in [7], one has

$$|(s_e)_j(t)| < \frac{\sqrt{\varrho}}{\sqrt{m}}\|\hat{\boldsymbol{x}}(t_k)\| = \frac{\sqrt{\varrho}}{\sqrt{m}}\|\hat{\boldsymbol{x}}(t) - \hat{\boldsymbol{e}}(t)\|$$

$$\le \frac{\sqrt{\varrho}}{\sqrt{m}}(\|\hat{\boldsymbol{x}}(t)\| + \|\hat{\boldsymbol{e}}(t)\|) \le \frac{\sqrt{\varrho}}{\sqrt{m}}(1 + \sqrt{\rho})\|\hat{\boldsymbol{x}}(t)\|$$

$$= \frac{\sqrt{\rho}}{\sqrt{m}}\|\boldsymbol{x}(t)\|.$$

Therefore, according to Theorem 8.2, the finite-time stability of the sliding mode dynamics (8.8) can be guaranteed under the proposed STC (8.29). $\qquad\square$

Remark 8.6. Corollary 8.1 indicates that the presented STC (8.29) is a necessary condition for the ETC (8.6). Hence, Theorem 8.2 provides another method to guarantee the finite-time stability of the sliding mode dynamics (8.8).

Theorem 8.4. *Under the presented event-triggered control strategy, a positive lower bound of inter-event interval is obtained as*

$$T_{\min} = \min_{k}\{t_{k+1} - t_k\} = \min_{k}\{\hat{T}_k, \tilde{T}_k, \Delta\} > 0, \quad k = 0, 1, 2, \ldots, \tag{8.30}$$

where

$$\hat{T}_k \geq \left\{ \frac{\Gamma(1+\alpha)}{\|\hat{A}_i\|} \ln\left(\frac{\|\hat{A}_i\|}{\Upsilon_2(\Delta)} \sqrt{\varrho}\|\hat{x}(t_k)\| + 1 \right) \right\}^{\frac{1}{\alpha}}$$

and

$$\tilde{T}_k > \left\{ \frac{\Gamma(1+\alpha)\sqrt{\varrho}}{\Upsilon_2(\Delta)\sqrt{\rho}} \|\hat{x}(t_k)\| \exp\left(-\frac{\|\hat{A}_i\|}{\Gamma(1+\alpha)}\Delta^{\alpha} \right) \right\}^{\frac{1}{\alpha}}.$$

Therefore, Zeno behavior will not exist.

Proof. First, if $T_{\min} = \Delta$, Zeno behavior does not occur, for sure. Assume that $T_{\min} < \Delta$, then according to Theorem 8.1, we have that $\|\hat{x}(t)\| \to 0$ and $\|e(t)\| \to 0$ hold when $t \to +\infty$, that is to say, there exists a positive scalar δ_3 such that $\|\hat{x}(t)\| \leq \delta_3$ and $\|e(t)\| \leq \delta_3$ hold. Then, from $\hat{e}(t) = \hat{x}(t) - \hat{x}(t_k)$, we have

$$\left| {}_{t_0}^{C}D_t^{\alpha}(\hat{e})_j(t) \right| = \left| {}_{t_0}^{C}D_t^{\alpha}(\hat{x})_j(t) \right|$$

$$\leq \|(A_i + \tilde{B}_1 L_i C_i)_j\|\|\hat{x}(t)\| + \|(\tilde{B}_i L_i C_i)_j\|\|e(t)\|$$

$$+ \|(B_i K_i - \tilde{B}_1 L_i C_i)_j\|\|\hat{x}(t_k)\| + \eta\|(B_i(G_i B_i)^{-1})_j\|$$

$$\leq \|(B_i K_i - \tilde{B}_1 L_i C_i)_j\|\|\hat{x}(t_k)\| + (\|(A_i + \tilde{B}_1 L_i C_i)_j\|$$

$$+ \|(\tilde{B}_i L_i C_i)_j\|)\delta_3 + \eta\|(B_i(G_i B_i)^{-1})_j\|$$

$$\triangleq (\tilde{\eta})_j(t_k).$$

Denote $y(t) = \hat{e}'(t)$, then there exists a nonnegative continuous function $z(t)$ such that

$$ {}_{t_0}D_t^{\alpha-1}(y)_j(t) = \frac{\sqrt{(\tilde{\eta})_j(t_k)}}{\sqrt{n}} - (z)_j(t) \tag{8.31}$$

holds. Then, taking the Laplace transform on both sides of (8.31), we have

$$s(Y)_j(s) = \frac{\sqrt{(\tilde{\eta})_j(t_k)}}{\sqrt{n}} s^{1-\alpha} - s^{2-\alpha}(Z)_j(s), \tag{8.32}$$

where $Y(s) = \mathcal{L}[y(t)]$ and $Z(s) = \mathcal{L}[z(t)]$.

From (8.32), it is easily obtained that $\lim\limits_{s \to 0} s(Y)_j(s) = 0$, which is equivalent to $\lim\limits_{t \to +\infty} (y)_j(t) = 0$ via the final value theorem. Namely, we can also deduce that $\lim\limits_{t \to +\infty} y(t) = 0$ holds. Therefore, there exists a positive scalar $\hat{\delta}$ such that $\|\hat{e}'(t)\| \le \hat{\delta}$ holds for any $t \in [t_0, +\infty)$.

Thereupon, taking $\hat{e}(t) = \hat{x}(t) - \hat{x}(t_k)$, we have

$$
\begin{aligned}
{}^{C}_{t_0}D^{\alpha}_{t}\hat{e}(t) = {}^{C}_{t_0}D^{\alpha}_{t}\hat{x}(t) &= A_i\hat{x}(t) + B_i\hat{u}(t) + L_iC_i(e(t) + \hat{e}(t)) \\
&= A_i\hat{x}(t) + B_iK_i\hat{x}(t_k) - B_i(G_iB_i)^{-1}G_iL_i(y(t) - \hat{y}(t) + C_i\hat{e}(t)) \\
&\quad - \eta B_i(G_iB_i)^{-1}G_i\mathrm{sgn}(s(t_k)) + L_iC_i(e(t) + \hat{e}(t)) \\
&= (A_i + \hat{B}_1L_iC_i)\hat{e}(t) + (A_i + B_iK_i)\hat{x}(t_k) + \hat{B}_1L_i(y(t) - \hat{y}(t)) \\
&\quad - \eta B_i(G_iB_i)^{-1}\mathrm{sgn}(s(t_k)).
\end{aligned}
$$

For any $t \in [t_k, t_{k+1})$, according to Definition 2.6, one has

$$
\begin{aligned}
{}^{C}_{t_k}D^{\alpha}_{t}\|\hat{e}(t)\| \le \|{}^{C}_{t_k}D^{\alpha}_{t}\hat{e}(t)\| &= \left\|{}^{C}_{t_0}D^{\alpha}_{t}\hat{e}(t) - \frac{1}{\Gamma(1-\alpha)}\int_{t_0}^{t_k}\frac{\hat{e}'(\tau)}{(t-\tau)^{\alpha}}d\tau\right\| \\
&\le \|A_i + \hat{B}_1L_iC_i\|\|\hat{e}(t)\| + \|(A_i + B_iK_i)\|\|\hat{x}(t_k)\| \\
&\quad + \|\hat{B}_1L_i\|\|C_i\|\|e(t)\| + \eta\|B_i(G_iB_i)^{-1}\| \\
&\quad + \frac{\hat{\delta}}{\Gamma(1-\alpha)}\int_{t_0}^{t_k}\frac{1}{(t-\tau)^{\alpha}}d\tau \\
&\le \|A_i + \hat{B}_1L_iC_i\|\|\hat{e}(t)\| + \|(A_i + B_iK_i)\|\|\hat{x}(t_k)\| \\
&\quad + \|\hat{B}_1L_i\|\|C_i\|\delta + \eta\|B_i(G_iB_i)^{-1}\| + \frac{\hat{\delta}}{\Gamma(2-\alpha)}\Delta^{1-\alpha} \\
&\triangleq \|\hat{A}_i\|\|\hat{e}(t)\| + \Upsilon_2(\Delta),
\end{aligned}
\tag{8.33}
$$

where $\|\hat{A}_i\| = \|A_i + \hat{B}_1L_iC_i\|$ and

$$
\begin{aligned}
\Upsilon_2(\Delta) = \|A_i + B_iK_i\|\|\hat{x}(t_k)\| + \|\tilde{B}_iL_i\|\|C_i\|\delta + \eta\|B_i(G_iB_i)^{-1}\| \\
+ \frac{\hat{\delta}}{\Gamma(2-\alpha)}\Delta^{\alpha}.
\end{aligned}
$$

Then, considering the fractional-order integral of (8.33) from t_k to t, one has

$$
\begin{aligned}
{}_{t_k}D^{-\alpha}_{t}({}^{C}_{t_k}D^{\alpha}_{t}\|\hat{e}(t)\|) &\le {}_{t_k}D^{-\alpha}_{t}(\|\hat{A}_i\|\|\hat{e}(t)\| + \Upsilon_2(\Delta)) \\
&= \frac{1}{\Gamma(\alpha)}\int_{t_k}^{t}\frac{\|\hat{A}_i\|\|\hat{e}(\tau)\| + \Upsilon_2(\Delta)}{(t-\tau)^{1-\alpha}}d\tau.
\end{aligned}
\tag{8.34}
$$

In addition, from Lemma 2.4 and since $\|\hat{e}(t)\| = 0$ at $t = t_k$, (8.34) can be rewritten as

$$\|\hat{e}(t)\| \leq \frac{1}{\Gamma(\alpha)} \int_{t_k}^{t} \frac{\|\hat{A}_i\| \|\hat{e}(\tau)\| + \Upsilon_2(\Delta)}{(t - \tau)^{1-\alpha}} d\tau.$$

Denote $\varpi(t) = \|\hat{A}_i\| \|\hat{e}(t)\| + \Upsilon_2(\Delta)$, then we have that

$$\varpi(t) \leq \frac{\|\hat{A}_i\|}{\Gamma(\alpha)} \int_{t_k}^{t} \frac{\varpi(\tau)}{(t - \tau)^{1-\alpha}} d\tau + \Upsilon_2(\Delta).$$

Hence, based on Lemma 2.9, one can easily obtain

$$\varpi(t) \leq \Upsilon_2(\Delta) \exp\left(\frac{\|\hat{A}_i\|}{\Gamma(1+\alpha)}(t - t_k)^\alpha\right). \tag{8.35}$$

It should be noted that we will deduce two crucial results from (8.35), that is,

(I) for any $t \in [t_k, t_{k+1})$, one has

$$\begin{aligned}
\|\hat{e}(t)\| &\leq \frac{\Upsilon_2(\Delta)}{\|\hat{A}_i\|} \left\{ \exp\left(\frac{\|\hat{A}_i\|}{\Gamma(1+\alpha)}(t - t_k)^\alpha\right) - 1 \right\} \\
&\leq \frac{\Upsilon_2(\Delta)}{\|\hat{A}_i\|} \left\{ \exp\left(\frac{\|\hat{A}_i\|}{\Gamma(1+\alpha)}\Delta^\alpha\right) - 1 \right\};
\end{aligned} \tag{8.36}$$

(II) for any $t \in [t_k, t_{k+1})$, one has

$$t - t_k \geq \left\{ \frac{\Gamma(1+\alpha)}{\|\hat{A}_i\|} \ln\left(\frac{\|\hat{A}_i\|}{\Upsilon_2(\Delta)}\|\hat{e}(t)\| + 1\right) \right\}^{\frac{1}{\alpha}}. \tag{8.37}$$

Furthermore, from STC (8.29), we have $\|\hat{e}(t_{k+1})\| = \sqrt{\varrho}\|\hat{x}(t_k)\|$, which implies that a positive inter-event time $\hat{T}_k$ satisfies

$$\hat{T}_k \geq \left\{ \frac{\Gamma(1+\alpha)}{\|\hat{A}_i\|} \ln\left(\frac{\|\hat{A}_i\|}{\Upsilon_2(\Delta)}\sqrt{\varrho}\|\hat{x}(t_k)\| + 1\right) \right\}^{\frac{1}{\alpha}} > 0. \tag{8.38}$$

From $|(s_e)_j(t)| < \frac{\sqrt{\rho}}{\sqrt{m}}\|\hat{x}(t)\|$ and $\hat{x}(t) = \hat{e}(t) + \hat{x}(t_k)$, we have

$$|(s_e)_j(t)| < \frac{\sqrt{\rho}}{\sqrt{m}}(\|\hat{e}(t)\| + \|\hat{x}(t_k)\|).$$

Then, for any $t \in [t_k, t_{k+1})$, taking the fractional-order derivative of $|(s_e)_j(t)|$, one has

$$\, _{t_k}^{C} D_t^{\alpha} |(s_e)_j(t)| < \frac{\sqrt{\rho}}{\sqrt{m}} \, _{t_k}^{C} D_t^{\alpha} \|\hat{e}(t)\|.$$

Combining (8.33) with (8.37), one has

$$\, _{t_k}^{C} D_t^{\alpha} |(s_e)_j(t)| < \frac{\sqrt{\rho}}{\sqrt{m}} \Upsilon_2(\Delta) \exp\left(\frac{\|\hat{A}_i\|}{\Gamma(1+\alpha)} \Delta^{\alpha} \right). \tag{8.39}$$

Similarly, taking the fractional-order integral of (8.39) from t_k to t, since $|(s_e)_j(t)(t_k)| = 0$, one gets

$$|(s_e)_j(t)| < \frac{\sqrt{\rho}}{\sqrt{m}} \frac{\Upsilon_2(\Delta)}{\Gamma(1+\alpha)} \exp\left(\frac{\|\hat{A}_i\|}{\Gamma(1+\alpha)} \Delta^{\alpha} \right)(t-t_k)^{\alpha}.$$

That is to say,

$$t - t_k > \left\{ \frac{\sqrt{m}}{\sqrt{\rho}} |(s_e)_j(t)| \frac{\Gamma(1+\alpha)}{\Upsilon_2(\Delta)} \exp\left(-\frac{\|\hat{A}_i\|}{\Gamma(1+\alpha)} \Delta^{\alpha} \right) \right\}^{\frac{1}{\alpha}}.$$

Furthermore, from STC (8.29), we have $|(s_e)_j(t)(t_{k+1})| = \sqrt{\frac{\varrho}{m}} \|\hat{x}(t_k)\|$, which implies that a positive inter-event time $\tilde{T}_k$ satisfies

$$\tilde{T}_k > \left\{ \frac{\Gamma(1+\alpha)\sqrt{\varrho}}{\Upsilon_2(\Delta)\sqrt{\rho}} \|\hat{x}(t_k)\| \exp\left(-\frac{\|\hat{A}_i\|}{\Gamma(1+\alpha)} \Delta^{\alpha} \right) \right\}^{\frac{1}{\alpha}} \geq 0. \tag{8.40}$$

Therefore, it can be seen that a positive lower bound $T_{\min}$ satisfies

$$T_{\min} = \min_k \{t_{k+1} - t_k\} = \min_k \{\hat{T}_k, \tilde{T}_k, \Delta\} > 0, \quad k = 0, 1, 2, \ldots$$

That is to say, Zeno behavior can be avoided under the proposed event-triggered control strategy. $\square$

Remark 8.7. In a few works, for instance, [12], the following positive inter-event time is presented to prove the nonexistence of Zeno behavior, namely

$$T_{\min} = t_{k+1} - t_k > \left(\frac{\Gamma(1+\alpha)\ln(\sqrt{\rho}\|\hat{x}(t_k)\|)}{\|A\| + \beta(\|\hat{x}(t_k)\|)} \right)^{\frac{1}{\alpha}}, \quad k = 0, 1, 2, \ldots,$$

where $\beta(\|\hat{x}_{t_k}\|) \geq 0$. However, it is worth pointing out that the above $T_{\min} > 0$ may be not guaranteed in some cases. While, from (8.30) in Theorem 8.4, it is easily obtained that $T_{\min} > 0$ holds for any $t \in [t_0, +\infty)$.

Remark 8.8. As we know, the control system under the event-triggered mechanism can be seen as a hybrid dynamical system. As a special phenomenon of hybrid systems, Zeno behavior will happen if the existence of the minimum positive lower bound of the inter-event time cannot be guaranteed. Zeno behavior may cause the nonexistence of equilibrium solutions for the controlled system after some finite instant, which makes the theoretical analysis invalid, and also easily leads to excessive waste of communication resources. This is also an important reason why Zeno behavior is unacceptable in the design of event-triggered mechanisms. Recently, some excellent research results on how to exclude Zeno behavior have been reported in [33,34]. Note that Theorem 8.4, which theoretically gives a positive lower bound of the inter-event time by combining with the theory of fractional calculus, verifies that the event-triggered mechanism proposed in this chapter can avoid the occurrence of Zeno behavior.

Remark 8.9. Recently, with the rapid development and improvement of the event-triggered mechanism, the dynamic event-triggered mechanism, as an effective tool, has been proposed to reduce the number of triggers and make the system have better dynamic properties. Some new results for switched systems have been reported in [35,36], to name a few references. While focusing on the fractional-order switched systems, to the best of our knowledge, there are few works on the dynamic event-triggered sliding mode control problem of fractional-order switched systems so far. Therefore, in the future research, the dynamic event-triggered SMC strategy will be fully considered for the fractional-order uncertain switched systems.

8.4 Numerical examples

In this section, three numerical examples are provided to illustrate the effectiveness and feasibility of the proposed event-triggered SMC method.

Example 8.1. Consider the following fractional-order uncertain switched systems with order $\alpha = 0.95$. The switching function $\sigma(t)$ is shown in Fig. 8.1, in which $\sigma(t) = 1$ for State 1 and $\sigma(t) = 0$ for State 2:

State 1:

$$A_1 = \begin{bmatrix} -9.077 & 6.469 \\ -8.057 & 3.897 \end{bmatrix}, \quad B_1 = \begin{bmatrix} 0.200 \\ -0.400 \end{bmatrix}, \quad C_1 = \begin{bmatrix} -0.366 & 0.900 \end{bmatrix},$$

$$E_{11} = \begin{bmatrix} -0.931 \\ -0.123 \end{bmatrix}, \quad H_{11} = \begin{bmatrix} -0.237 & 0.531 \end{bmatrix},$$

$$E_{21} = E_{11}, \quad H_{21} = 0.590, \quad F_{11}(t) = \sin(t), \quad F_{21}(t) = \cos(t);$$

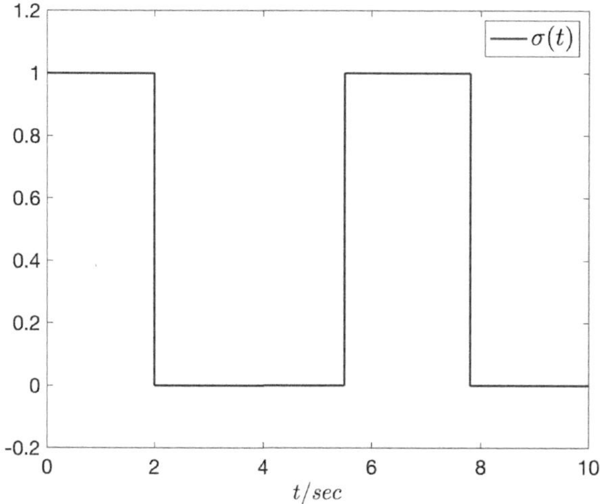

FIGURE 8.1   Switching signal.

State 2:

$$A_2 = \begin{bmatrix} -14.100 & 33.860 \\ -34.680 & -35.651 \end{bmatrix}, \quad B_2 = \begin{bmatrix} 0.200 \\ -0.400 \end{bmatrix}, \quad C_2 = \begin{bmatrix} -3.456 & 0.588 \end{bmatrix},$$

$$E_{12} = \begin{bmatrix} -0.486 \\ -3.781 \end{bmatrix}, \quad H_{12} = \begin{bmatrix} -3.602 & -2.092 \end{bmatrix},$$

$$E_{22} = E_{21}, \quad H_{22} = -2.982, \quad F_{12}(t) = \sin(t), \quad F_{22}(t) = \cos(t).$$

The initial conditions of system (8.2) and observer system (8.3) are given as $x(0) = [0.1, -0.1]^{\mathrm{T}}$ and $\hat{x}(0) = [-0.1, 0.5]^{\mathrm{T}}$, respectively. Select $G_1 = B_1^{\mathrm{T}}$ and $G_2 = B_2^{\mathrm{T}}$, then, by solving (8.26) in Theorem 8.2 via Matlab LMI Toolbox, we can obtain that

$$X_1 = \begin{bmatrix} 10.9316 & 8.6655 \\ 8.6655 & 10.6315 \end{bmatrix}, \quad Y_1 = \begin{bmatrix} 4.5268 & 5.8181 \end{bmatrix},$$

$$Z_1 = \begin{bmatrix} 1.5315 & -7.0360 \\ 0.0976 & -3.0182 \end{bmatrix},$$

$$\varepsilon_{11} = 5.6731, \quad \varepsilon_{21} = 5.2960, \quad \varepsilon_{31} = 4.8059,$$

$$\varepsilon_{41} = 4.1881, \quad \varepsilon_{51} = 3.9470, \quad \varepsilon_{61} = 3.9078;$$

and

$$X_2 = \begin{bmatrix} 6.3208 & 0.6042 \\ 0.6042 & 5.5378 \end{bmatrix}, \quad Y_2 = \begin{bmatrix} -0.0049 & -0.0168 \end{bmatrix},$$

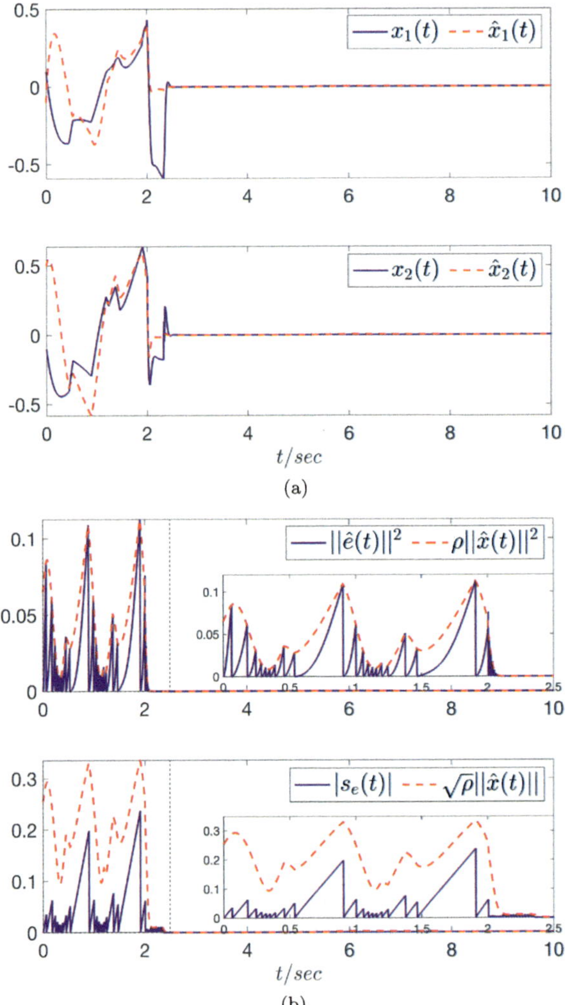

FIGURE 8.2   Simulation results under ETC (8.6): (a) system state and observer state trajectories; (b) $\|\hat{e}(t)\|^2$ along the bound $\rho\|\hat{x}(t)\|^2$ and $|s_e(t)|$ along the bound $\sqrt{\rho}\|\hat{x}(t)\|$; (c) response curves of state estimation errors and sample observer errors; (d) sliding surface function and controller curves.

$$\mathbf{Z}_2 = \begin{bmatrix} 3.3692 & 7.1984 \\ 1.6902 & 3.6061 \end{bmatrix},$$

$$\varepsilon_{12} = 7.0018, \quad \varepsilon_{22} = 2.7917, \quad \varepsilon_{32} = 9.1257,$$

$$\varepsilon_{42} = 2.7919, \quad \varepsilon_{52} = 2.7920, \quad \varepsilon_{62} = 2.7918.$$

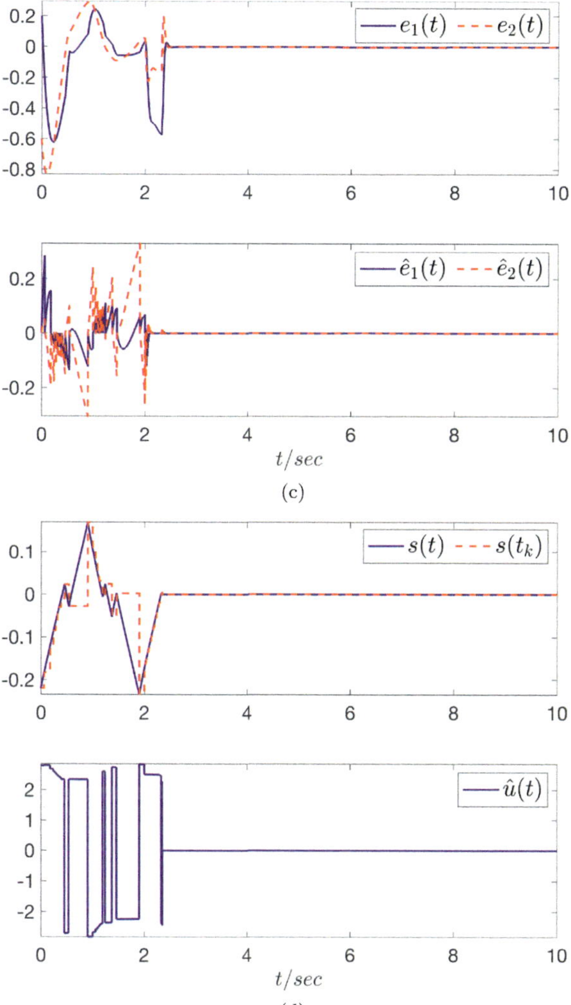

FIGURE 8.2   (*continued*)

Hence, the feedback matrix $\boldsymbol{K}_i$ and observer gain matrix $\boldsymbol{L}_i$ can be computed as follows:

$$\boldsymbol{K}_1 = \begin{bmatrix} -0.0557 & 0.5926 \end{bmatrix}, \quad \boldsymbol{L}_1 = \begin{bmatrix} -0.7081 \\ -0.3422 \end{bmatrix};$$

$$\boldsymbol{K}_2 = \begin{bmatrix} -0.0005 & -0.0030 \end{bmatrix}, \quad \boldsymbol{L}_2 = \begin{bmatrix} -0.1382 \\ -0.0693 \end{bmatrix}.$$

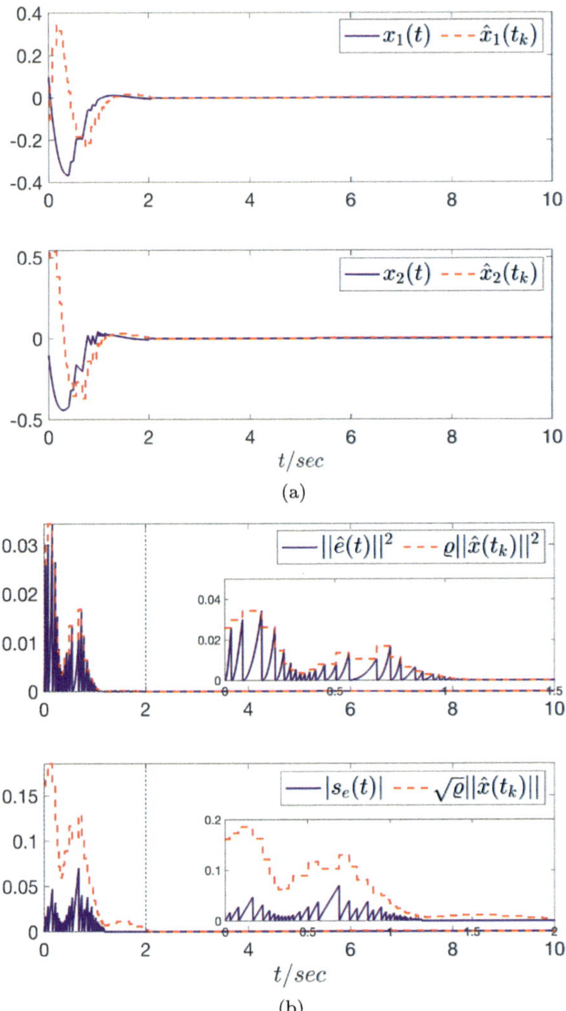

FIGURE 8.3   Simulation results under STC (8.29): (a) system state and observer state trajectories; (b) $\|\hat{e}(t)\|^2$ along the bound $\varrho\|\hat{x}(t_k)\|^2$ and $|s_e(t)|$ along the bound $\sqrt{\varrho}\|\hat{x}(t_k)\|$; (c) response curves of state estimation errors and sample observer errors; (d) sliding surface function and controller curves.

Let $\alpha = 0.95$, $\varepsilon = 0.08$, $\mu = 3.734$, $\beta_1 = 0.0017$, and $\beta_2 = 8.420$. Then, according to Theorem 8.1, there exist scalars $\hat{\mu} = 3.8$, $\gamma_1 = 2.2361$, and $\gamma_2 = 0.0045$ such that (8.10)–(8.12) hold. By computing, we can obtain that $T_\sigma^* = 2.4639$. The curves of the fractional-order switched system (8.2) and observer system (8.3) under ETC (8.6) and STC (8.29) are shown in Figs. 8.2 and 8.3, respectively. It can be seen that the switched systems state trajectory $x(t)$ and observer state $\hat{x}(t)$ are both converging to zero in finite-time

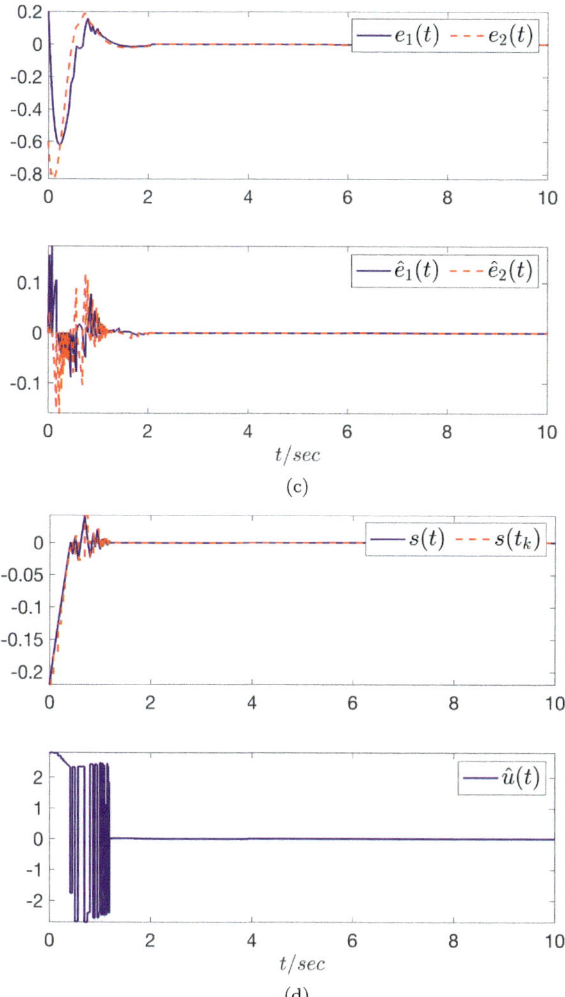

FIGURE 8.3 (*continued*)

under the proposed event-triggered SMC law (8.28). Meanwhile, Fig. 8.4 depicts the triggered instants and periods under different event-triggered schemes with respect to $\Delta = 2$. Finally, some quantitative comparisons with respect to the influence on the number of control updates are shown in Table 8.1, from which one sees that the sampling time is largely reduced compared to other control strategies.

Example 8.2. To further illustrate the advantages of the proposed event-triggered control strategy, a comparison with [12] is given for a non-switched system. The parameters are shown as follows:

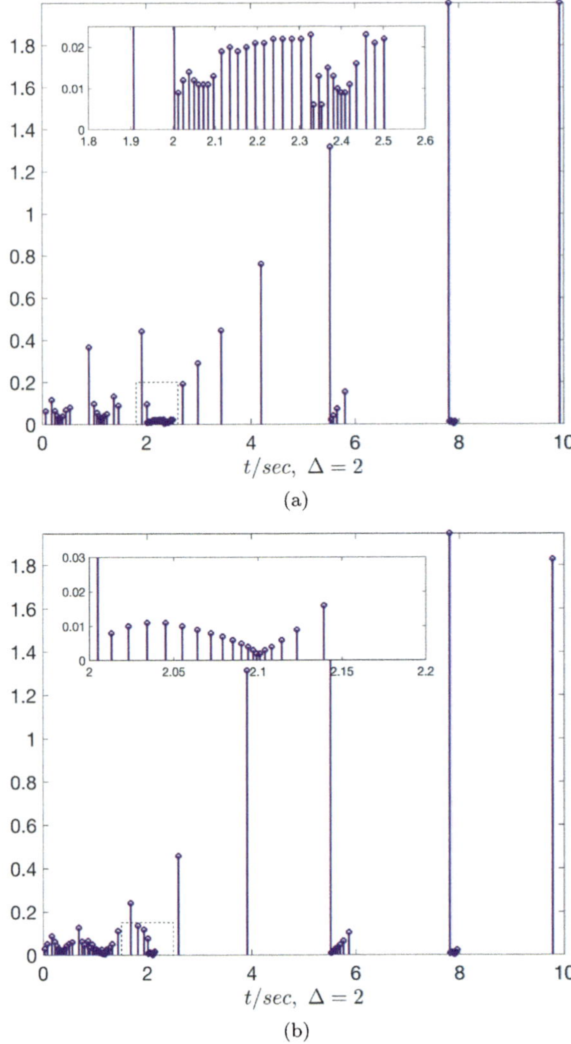

FIGURE 8.4   Triggered instants and inter-execution periods under (a) Condition (8.6) and (b) Condition (8.29).

TABLE 8.1   Simulation results under different event-triggered mechanisms.

	ETC (8.6)	STC (8.29)	Time-trigger
Sample numbers	77	98	10000
Average period (s)	0.130	0.102	0.001

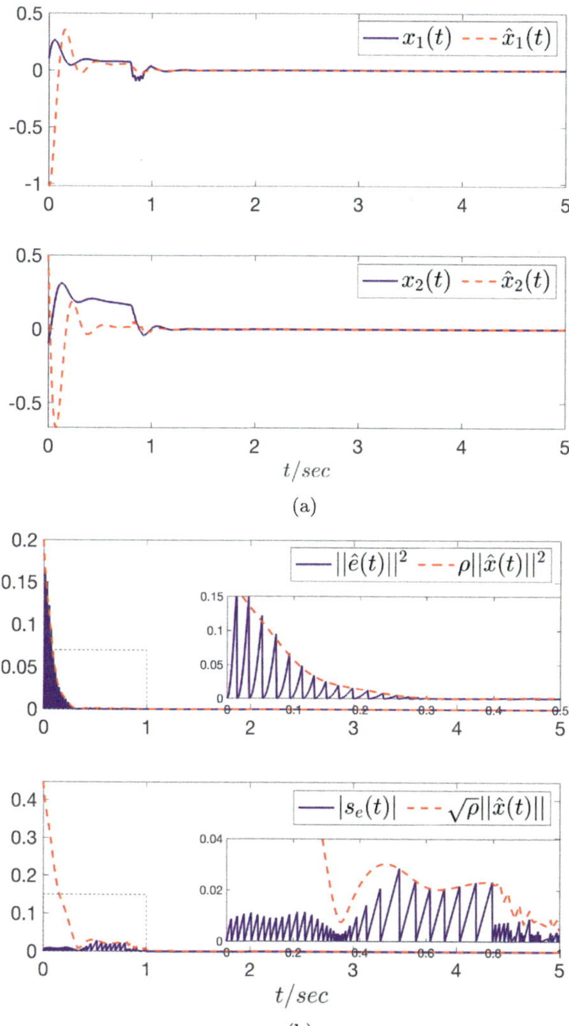

FIGURE 8.5 Simulation results under ETC (8.6): (a) system state and observer state trajectories; (b) $\|\hat{e}(t)\|^2$ along the bound $\rho\|\hat{x}(t)\|^2$ and $|s_e(t)|$ along the bound $\sqrt{\rho}\|\hat{x}(t)\|$; (c) response curves of state estimation errors and sample observer errors; (d) triggered periods.

$$A = \begin{bmatrix} -3.846 & -16.400 \\ 19.537 & -7.162 \end{bmatrix}, \quad B = \begin{bmatrix} 0.200 \\ -0.400 \end{bmatrix}, \quad C = \begin{bmatrix} 0.045 & -1.757 \end{bmatrix},$$

$$E_1 = \begin{bmatrix} 0.903 \\ 0.226 \end{bmatrix}, \quad H_1 = \begin{bmatrix} 0.117 & 1.320 \end{bmatrix},$$

$$E_2 = E_1, \quad H_2 = 1.435, \quad F_1(t) = \sin(t), \quad F_2(t) = \cos(t).$$

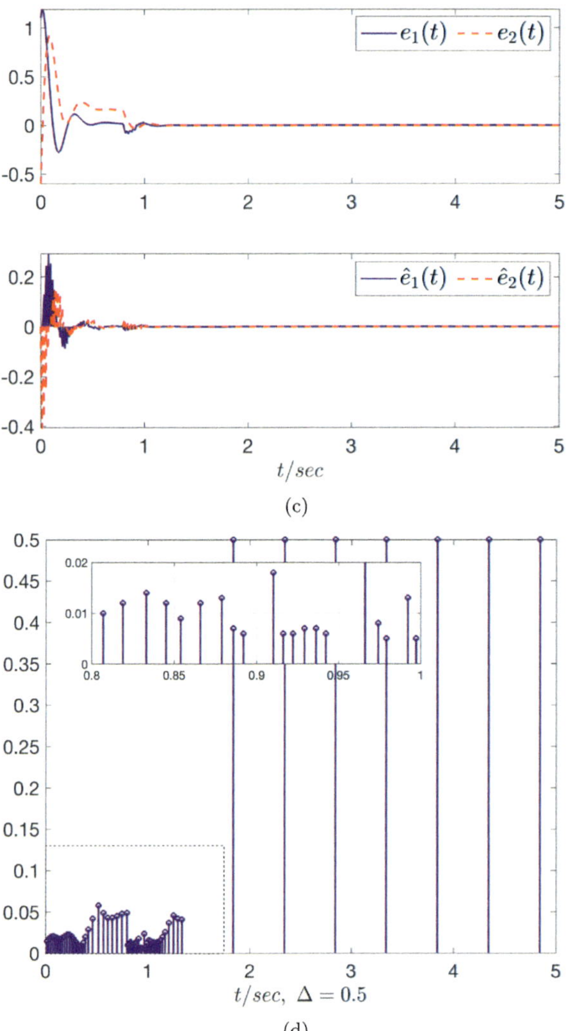

(c)

(d)

FIGURE 8.5   *(continued)*

Selecting $G = B^{\mathrm{T}}$, the feedback matrix K and observer gain matrix L can be easily obtained as

$$K = \left[\; 0.0322 \quad -0.0254 \;\right], \quad L = \left[\begin{array}{c} -0.0751 \\ -0.0286 \end{array}\right].$$

Under the event-triggered SMC law, the simulation results are depicted in Figs. 8.5 to 8.7. Fig. 8.5 shows the response curves of nonswitched sys-

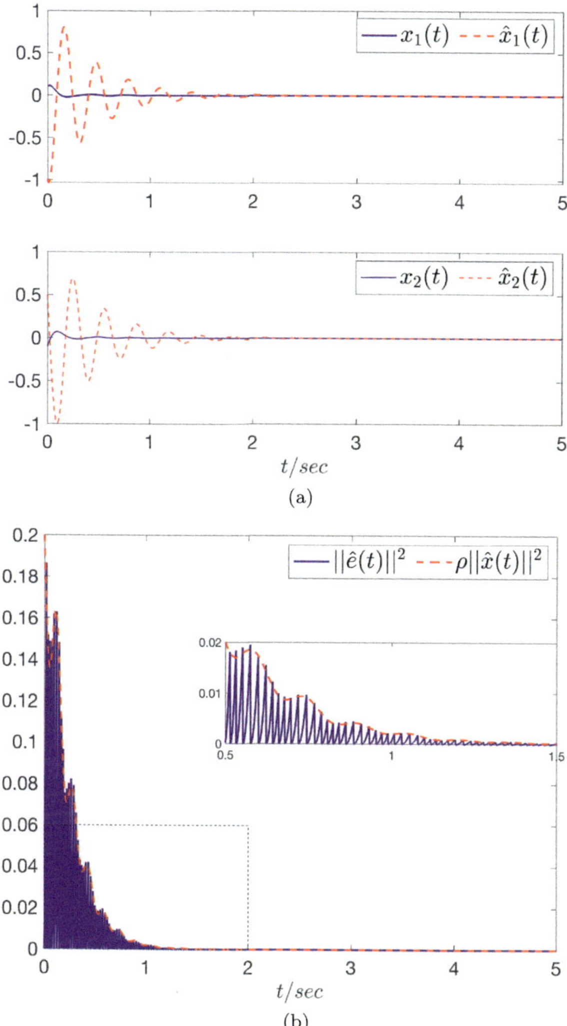

FIGURE 8.6 Simulation results under ETC-1 in [12]: (a) system state and observer state trajectories; (b) $\|\hat{e}(t)\|^2$ along with the bound $\rho\|\hat{x}(t)\|^2$; (c) response curves of state estimation errors and sample observer errors; (d) triggered periods.

tem and observer under ETC (8.6). In addition, to illustrate the superiority of the proposed control strategy, two event-triggered control strategies adopted in [12] are presented in Figs. 8.6 and 8.7, respectively. It is also clearly seen that the proposed event-triggered SMC schemes in this chapter are superior in reducing the sampling times compared to the control methods in [12].

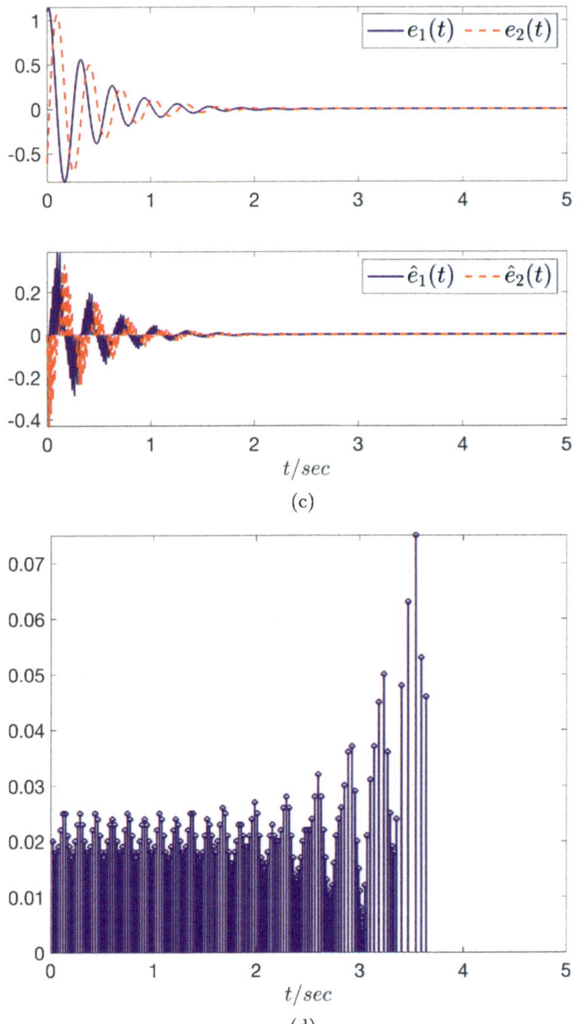

FIGURE 8.6   (*continued*)

Example 8.3. In this part, consider the following fractional-order switched system borrowed from [18] to further illustrate the advantages of the proposed event-triggered control strategy:

$$\begin{cases} {}^{C}_{t_0}D_t^{\alpha}\boldsymbol{x}(t) = \boldsymbol{A}_{\sigma(t)}\boldsymbol{x}(t) + \boldsymbol{B}_{\sigma(t)}\boldsymbol{u}(t), \\ \boldsymbol{y}(t) = \boldsymbol{C}_{\sigma(t)}\boldsymbol{x}(t), \end{cases} \tag{8.41}$$

where $\sigma(t) = 1$ for State 1 and $\sigma(t) = 0$ for State 2.

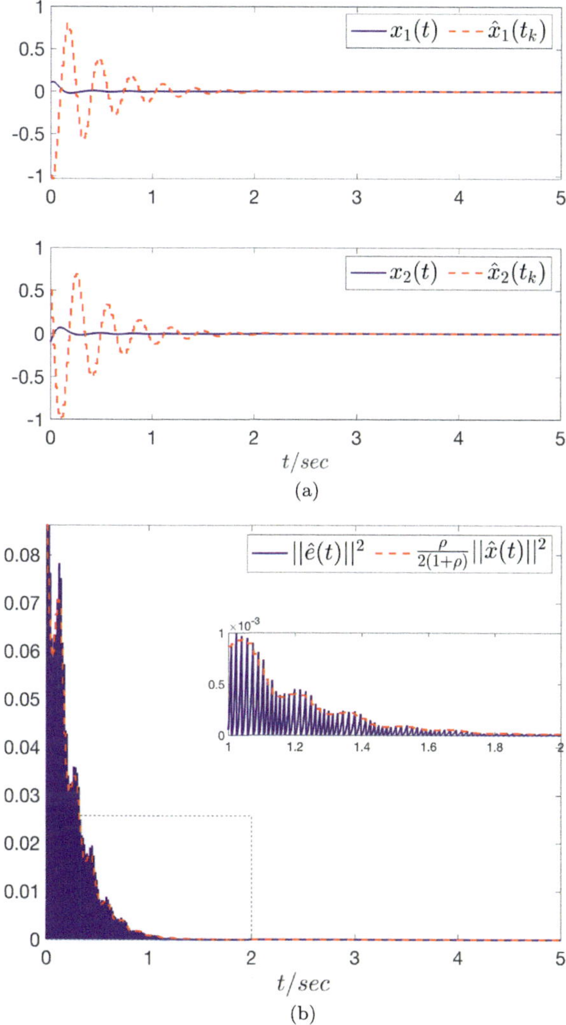

FIGURE 8.7 Simulation results under ETC-2 in [12]: (a) system state and observer state trajectories; (b) $\|\hat{e}(t)\|^2$ along the bound $\frac{\rho}{2(1+\rho)}\|\hat{\boldsymbol{x}}(t_k)\|^2$; (c) response curves of state estimation errors and sample observer errors; (d) triggered periods.

State 1:

$$\boldsymbol{A}_1 = \begin{bmatrix} -10 & 0 \\ 0 & -0.5 \end{bmatrix}, \quad \boldsymbol{B}_1 = \begin{bmatrix} 10 \\ 01 \end{bmatrix}, \quad \boldsymbol{C}_1 = \begin{bmatrix} 0 & 1 \end{bmatrix};$$

State 2:

$$\boldsymbol{A}_2 = \begin{bmatrix} -4 & 0 \\ 0 & -2 \end{bmatrix}, \quad \boldsymbol{B}_2 = \begin{bmatrix} 0.20 \\ 00.25 \end{bmatrix}, \quad \boldsymbol{C}_2 = \begin{bmatrix} 0 & 1 \end{bmatrix}.$$

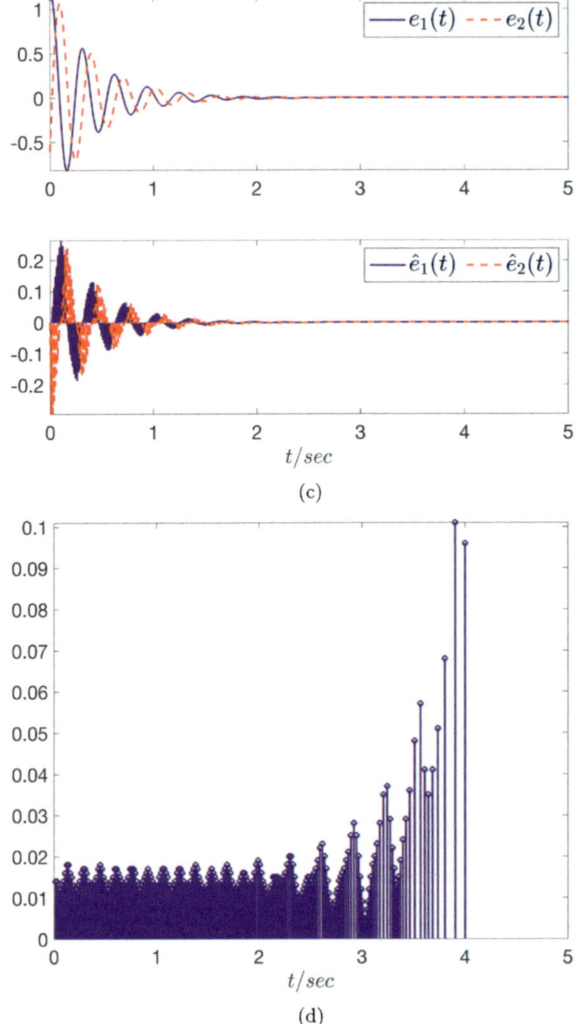

(c)

(d)

FIGURE 8.7 *(continued)*

The initial conditions of system (8.41) and observer system are given as $\boldsymbol{x}(0) = [10, -2]^{\mathrm{T}}$ and $\hat{\boldsymbol{x}}(0) = [1, -1]^{\mathrm{T}}$, respectively. Select $\boldsymbol{G}_1 = \boldsymbol{B}_1^{\mathrm{T}}$ and $\boldsymbol{G}_2 = \boldsymbol{B}_2^{\mathrm{T}}$, then, by solving (8.26) in Theorem 8.2 via Matlab LMI Toolbox, we can obtain

$$\boldsymbol{X}_1 = \begin{bmatrix} 1.7216 & 0 \\ 0 & 4.1341 \end{bmatrix}, \quad \boldsymbol{Y}_1 = \begin{bmatrix} 2.4377 & 0 \\ 0 & -3.8361 \end{bmatrix},$$

$$\boldsymbol{Z}_1 = \begin{bmatrix} -2.4463 & 0 \\ 0 & 3.3868 \end{bmatrix}, \quad \varepsilon_{k1} = 16.3526, \quad k = 1, 2, \ldots, 6;$$

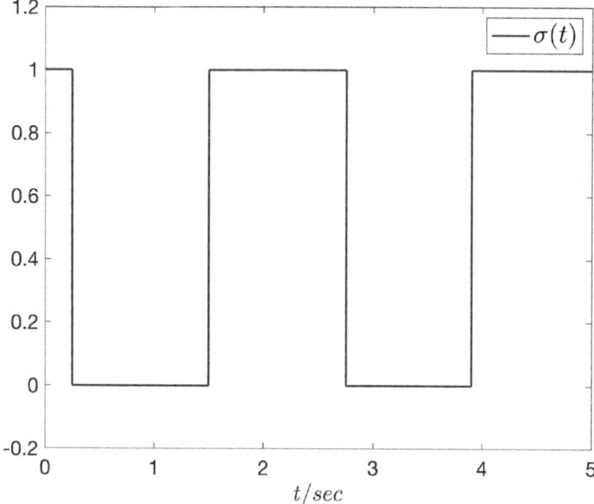

FIGURE 8.8 Switching signal.

and

$$X_2 = \begin{bmatrix} 2.6453 & 0 \\ 0 & 4.2835 \end{bmatrix}, \quad Y_2 = \begin{bmatrix} 10.1790 & 0 \\ 0 & 2.7717 \end{bmatrix},$$

$$Z_2 = \begin{bmatrix} -2.0358 & 0 \\ 0 & -0.6929 \end{bmatrix}, \quad \varepsilon_{k2} = 15.0551, \quad k = 1, 2, \dots, 6;$$

Hence, the feedback matrix K_i and observer gain matrix L_i are:

$$K_1 = \begin{bmatrix} 1.4159 & 0 \\ 0 & -0.9279 \end{bmatrix}, \quad L_1 = \begin{bmatrix} 0 \\ 0.8193 \end{bmatrix};$$

$$K_2 = \begin{bmatrix} 3.8479 & 0 \\ 0 & 0.6471 \end{bmatrix}, \quad L_2 = \begin{bmatrix} 0 \\ -0.1618 \end{bmatrix}.$$

Under the presented event-triggered SMC law, the simulation results are depicted in Figs. 8.8 to 8.10 and Table 8.2. Fig. 8.8 shows the switching signal while Fig. 8.9 depicts the response curves of state trajectory $x(t)$ of switched system (8.41) under ETC (8.6) and [18], respectively, from which it is observed that our method can achieve better system stability. Lastly, the triggered instants and inter-execution periods under different event-trigger mechanisms are presented in Fig. 8.10.

8.5 Conclusion

The problem of event-triggered SMC for fractional-order uncertain switched systems has been examined in this chapter. First, a fractional-

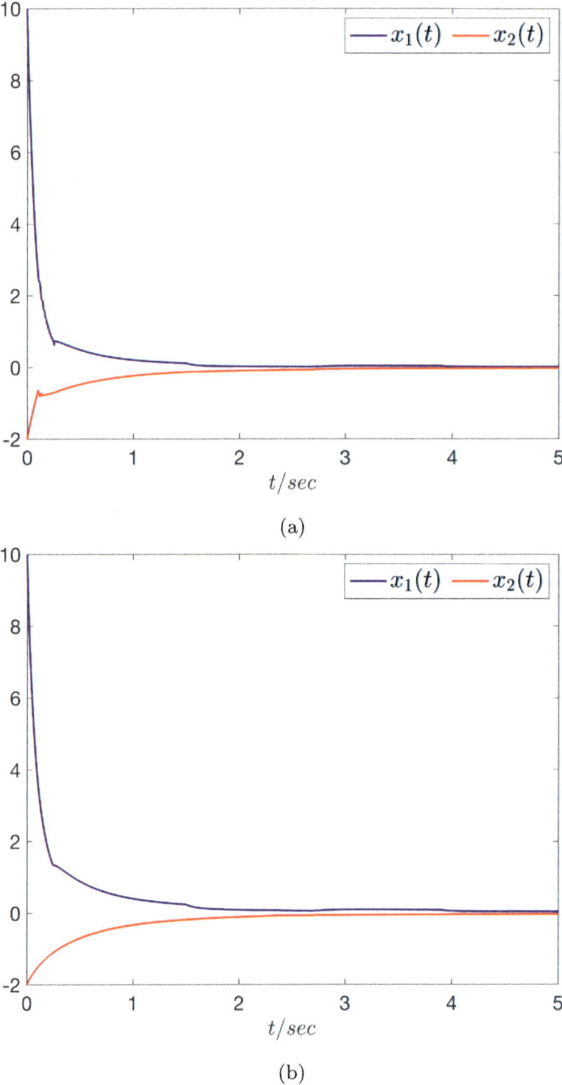

FIGURE 8.9   State trajectories $x_i(t)$: (a) under the ETC (8.6); (b) under the presented ETC in [18].

order observer has been designed based on the proposed event-triggered mechanism. Second, an event-triggered sliding mode controller has been designed to guarantee the finite-time stability of the sliding mode dynamic. Compared with the time-triggered control strategy and other related results, switching times in a finite-time interval could be further reduced. Third, the Zeno behavior has been discussed through the

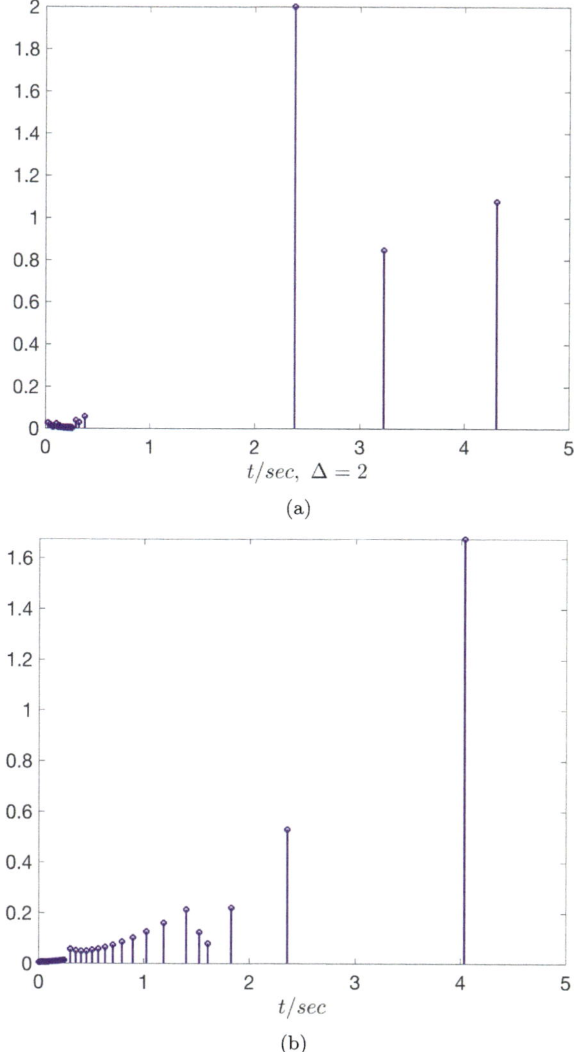

FIGURE 8.10 Comparative results under different trigger strategies: (a) ETC (8.6); (b) ETC with $\psi_1 = 1$ and $\psi_2 = 1$ in [18]; (c) ETC with $\psi_1 = 1$ and $\psi_2 = 0$ in [18]; and (d) ETC with $\psi_1 = 0$ and $\psi_2 = 1$ in [18].

Gronwall–Bellman inequality. Lastly, three examples have been given to illustrate the effectiveness and superiority of the proposed event-triggered control strategy. In our future research, the dynamic event-triggered SMC strategy will be fully considered for the fractional-order uncertain switched systems.

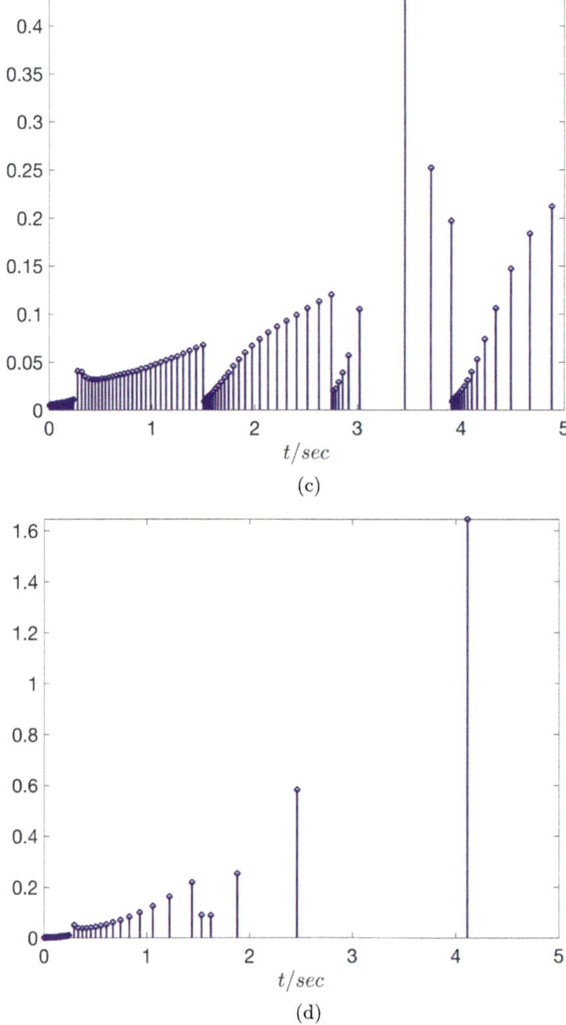

FIGURE 8.10   *(continued)*

TABLE 8.2   Simulation results under different event-triggered mechanisms.

Triggered mechanism		Event-triggered function	No. of samples	Figure
ETC (8.6)		$\|\hat{e}(t)\|^2 < \rho \|\hat{x}(t)\|^2 \wedge |(s_e)_j(t)| < \frac{\sqrt{\rho}}{\sqrt{m}} \|\hat{x}(t)\|$	43	8.10(a)
	$(\psi_1, \psi_2) = (1,1)$	$e^{\mathrm{T}}(t)\tilde{M}_{\sigma_t}(t)e(t) < \psi_1 \rho_{\sigma_t} x^{\mathrm{T}}(t)\tilde{N}_{\sigma_t}(t)x(t) + \psi_2 e^{-\gamma t}$	48	8.10(b)
ETC in [18]	$(\psi_1, \psi_2) = (1,0)$	$e^{\mathrm{T}}(t)\tilde{M}_{\sigma_t}(t)e(t) < \psi_1 \rho_{\sigma_t} x^{\mathrm{T}}(t)\tilde{N}_{\sigma_t}(t)x(t)$	111	8.10(c)
	$(\psi_1, \psi_2) = (0,1)$	$e^{\mathrm{T}}(t)\tilde{M}_{\sigma_t}(t)e(t) < \psi_2 e^{-\gamma t}$	79	8.10(d)

References

[1] D. Liberzon, A.S. Morse, Basic problems in stability and design of switched systems, IEEE Control Systems Magazine 19 (5) (1999) 59–70.

[2] F. Blanchini, S. Miani, Switching and switched systems, in: Set-Theoretic Methods in Control, Springer, 2015, pp. 405–466.

[3] L. Chen, Y. Zhu, C.K. Ahn, Adaptive neural network-based observer design for switched systems with quantized measurements, IEEE Transactions on Neural Networks and Learning Systems (2021) 1–14.

[4] Y. Liu, Y. Niu, Y. Zou, Non-fragile observer-based sliding mode control for a class of uncertain switched systems, Journal of the Franklin Institute 351 (2) (2014) 952–963.

[5] I. Podlubny, An introduction to fractional derivatives, fractional differential equations, to methods of their solution and some of their applications, Mathematics in Science and Engineering 198 (1999) 340.

[6] X. Chen, Y. Chen, B. Zhang, D. Qiu, A modeling and analysis method for fractional-order DC–DC converters, IEEE Transactions on Power Electronics 32 (9) (2016) 7034–7044.

[7] L. Liu, X. Cao, Z. Fu, S. Song, H. Xing, Finite-time control of uncertain fractional-order positive impulsive switched systems with mode-dependent average dwell time, Circuits, Systems, and Signal Processing 37 (9) (2018) 3739–3755.

[8] C. Zhang, B.-S. Han, Stability analysis of stochastic delayed complex networks with multi-weights based on Razumikhin technique and graph theory, Physica A. Statistical Mechanics and Its Applications 538 (2020) 122827.

[9] S. Sui, C.L.P. Chen, S. Tong, Neural-network-based adaptive DSC design for switched fractional-order nonlinear systems, IEEE Transactions on Neural Networks and Learning Systems 32 (10) (2020) 4703–4712.

[10] T. Zhan, S. Ma, W. Li, W. Pedrycz, Exponential stability of fractional-order switched systems with mode-dependent impulses and its application, IEEE Transactions on Cybernetics 52 (11) (2021) 11516–11525.

[11] J.-E. Zhang, Stabilization of uncertain fractional-order complex switched networks via impulsive control and its application to blind source separation, IEEE Access 6 (2018) 32780–32789.

[12] T. Feng, Y.-E. Wang, L. Liu, B. Wu, Observer-based event-triggered control for uncertain fractional-order systems, Journal of the Franklin Institute 357 (14) (2020) 9423–9441.

[13] A.K. Behera, B. Bandyopadhyay, M. Cucuzzella, A. Ferrara, X. Yu, A survey on event-triggered sliding mode control, IEEE Journal of Emerging and Selected Topics in Industrial Electronics 2 (3) (2021) 206–217.

[14] H. Komurcugil, S. Biricik, S. Bayhan, Z. Zhang, Sliding mode control: overview of its applications in power converters, IEEE Industrial Electronics Magazine 15 (1) (2020) 40–49.

[15] X. Li, Z. Sun, Y. Tang, H.R. Karimi, Adaptive event-triggered consensus of multiagent systems on directed graphs, IEEE Transactions on Automatic Control 66 (4) (2020) 1670–1685.

[16] B. Jiang, H.R. Karimi, Y. Kao, C. Gao, Takagi–Sugeno model based event-triggered fuzzy sliding-mode control of networked control systems with semi-Markovian switchings, IEEE Transactions on Fuzzy Systems 28 (4) (2019) 673–683.

[17] T. Hu, Z. He, X. Zhang, S. Zhong, Event-triggered consensus strategy for uncertain topological fractional-order multiagent systems based on Takagi–Sugeno fuzzy models, Information Sciences 551 (2021) 304–323.

[18] Y. Shang, L. Liu, Y. Di, Z. Fu, B. Fan, Guaranteed cost and finite-time event-triggered control of fractional-order switched systems, Transactions of the Institute of Measurement and Control 43 (12) (2021) 2724–2733.

[19] H. Zhang, J. Wang, Active steering actuator fault detection for an automatically-steered electric ground vehicle, IEEE Transactions on Vehicular Technology 66 (5) (2016) 3685–3702.

[20] Y. Liu, R. Chen, X. Na, Y. Luo, H. Zhang, Robust fault estimation of vehicular yaw rate sensor using a type-2 fuzzy approach, IEEE Transactions on Industrial Electronics 68 (10) (2020) 10029–10039.

[21] R. Yang, W.X. Zheng, Y. Yu, Event-triggered sliding mode control of discrete-time two-dimensional systems in Roesser model, Automatica 114 (2020) 108813.

[22] J. Feng, F. Hao, Event-triggered sliding mode control for time-delay uncertain systems, Asian Journal of Control 23 (3) (2021) 1407–1418.

[23] X. Fan, Z. Wang, Event-triggered sliding mode control for singular systems with disturbance, Nonlinear Analysis: Hybrid Systems 40 (2021) 101011.

[24] L. Wang, J. Dong, Event-based distributed adaptive fuzzy consensus for nonlinear fractional-order multiagent systems, IEEE Transactions on Systems, Man, and Cybernetics: Systems 52 (9) (2021) 5901–5912.

[25] Y. Yang, Y. Niu, Z. Zhang, Dynamic event-triggered sliding mode control for interval type-2 fuzzy systems with fading channels, ISA Transactions 110 (2021) 53–62.

[26] F. Amato, M. Ariola, P. Dorato, Finite-time control of linear systems subject to parametric uncertainties and disturbances, Automatica 37 (9) (2001) 1459–1463.

[27] A. Polyakov, L. Fridman, Stability notions and Lyapunov functions for sliding mode control systems, Journal of the Franklin Institute 351 (4) (2014) 1831–1865.

[28] L. Liu, Z. Fu, X. Cai, X. Song, Non-fragile sliding mode control of discrete singular systems, Communications in Nonlinear Science and Numerical Simulation 18 (3) (2013) 735–743.

[29] N. Yu, W. Zhu, Event-triggered impulsive chaotic synchronization of fractional-order differential systems, Applied Mathematics and Computation 388 (2021) 125554.

[30] Z. Liu, H.R. Karimi, J. Yu, Passivity-based robust sliding mode synthesis for uncertain delayed stochastic systems via state observer, Automatica 111 (2020) 108596.

[31] B. Jiang, C. Gao, Decentralized adaptive sliding mode control of large-scale semi-Markovian jump interconnected systems with dead-zone input, IEEE Transactions on Automatic Control 67 (3) (2021) 1521–1528.

[32] H. Ye, J. Gao, Y. Ding, A generalized Gronwall inequality and its application to a fractional differential equation, Journal of Mathematical Analysis and Applications 328 (2) (2007) 1075–1081.

[33] H. Yu, F. Hao, The existence of Zeno behavior and its application to finite-time event-triggered control, Science China. Information Sciences 63 (1) (2020) 1–3.

[34] H. Yu, T. Chen, On Zeno behavior in event-triggered finite-time consensus of multiagent systems, IEEE Transactions on Automatic Control 66 (10) (2020) 4700–4714.

[35] Y. Ma, Z. Li, J. Zhao, H_∞ control for switched systems based on dynamic event-triggered strategy and quantization under state-dependent switching, IEEE Transactions on Circuits and Systems. I, Regular Papers 67 (9) (2020) 3175–3186.

[36] G. Zong, X. Sun, D. Yang, M. Chadli, K. Shi, Robust finite-time H_∞ control of switched systems and its applications: a dynamic event-triggered method, International Journal of General Systems 51 (1) (2022) 71–93.

Nonfragile sliding mode control of fractional-order complex networked systems via combination event-triggered approach

9.1 Introduction

Recently, complex networked systems (CNSs) have gained extensive attention from academia, and have been widely used in various fields of engineering problems due to their capacity of characterizing the topologies including connection diversity, node diversity, and dynamic complexity; see [1–7] and the references therein. For instance, combining with the complex network theory, the authors of [8] proposed a novel broad learning system to implement EEG-based driver fatigue detection to ensure traffic safety and reduce economic and social losses. Nonfragility, as one of the important factors for evaluating system performance indicators, has also been studied by many scholars. For instance, the authors of [9] used sampled-data feedback control for the first time to study the nonfragile exponential synchronous control problem of CNSs with time-varying coupled delays. Besides, in actual complex network, it has been noticed that the transmission of data is often subject to cyber-attacks, such as eavesdropping attacks [10], deception attacks [11], etc., causing changes of some behaviors of CNSs. Such changes may lead the system to enter an unsafe state, thereby damaging the system. To reduce the influence of cyber-attacks, a novel multi-channel transmission structure was constructed in [12] to explore the problem of distributed fusion estimation of multi-sensor nonlinear CNSs under random access protocol.

Sliding Mode Control of Fractional-order Systems
https://doi.org/10.1016/B978-0-44-331696-8.00013-6

189

It should be noted that all the above-mentioned control methods were concentrated on integer-order CNSs rather than fractional-order CNSs (FO-CNSs). It is well known that fractional-order systems can better describe complex physical phenomena and physical systems than integer-order systems, and fractional calculus has been applied to physics, automatic control, and biochemistry due to its memory and genetic properties [13–15]. Focusing on the FO-CNSs, the authors of [16] discussed the projective cluster synchronization problem of systems with coupled delay by presenting a novel adaptive pinning control strategy. In addition, several sufficient conditions were obtained to ensure the projective cluster synchronization for the considered FO-CNSs based on the Lyapunov stability theory and Barbalat lemma. In parallel with this, the authors considered the finite-time synchronization problem of FO-CNSs by using the intermittent control method in [17]. In [18], the authors designed a nonfragile state estimation for the fractional-order memristive BAM neural networks with and without time delays, respectively. Meanwhile, scholars also discussed the robust finite-time synchronization control problem of FO-CNSs with multiweights and uncertain couplings by designing a nonfragile observer frame shown in [19].

On the one hand, event-triggered control, due to its advantages such as fewer execution times and less energy consumption, which has attracted scholars' attention, has become an important topic; see [20–28] and the references therein. Parallel to these developments, there do exist excellent results about the nonfragile state estimation extending CNSs to FO-CNSs by using the event-triggered control; see [29,30]. For instance, by designing an adaptive event-triggered strategy, the authors of [29] considered the nonfragile state estimation problem of FO-CNSs subject to adversarial cyber-attacks. Focusing on the event-triggered mechanism, it generally involves two important problems, i.e., (i) designing an appropriate control strategy to obtain the desired goal; (ii) effectively avoiding Zeno behavior in the system, which is an important issue in fractional-order control systems. Different from the classical time-triggered control method, to avoid Zeno behavior, we just need to make sure that the minimal inter-event time is positive, i.e., $\Delta T = \min\{t_{k+1} - t_k | k = 0, 1, \ldots\} > 0$ holds. However, as far as we know, problem (i) has not been well solved for FO-CNSs in the existing literature. Particularly, there was no reasonable method given of how to get the minimal positive inter-event time. Therefore, how to obtain such a positive lower bound for this considered control strategy is presented here.

On the other hand, considering the existing literature, such as [31,29, 32–36], there exist two main forms of event-triggered conditions, that is, case (i) $[\xi(t) - \xi(t_k)]^T \Omega [\xi(t) - \xi(t_k)] \leq \varrho \xi(t)^T \Omega \xi(t)$, $t \in [t_k, t_{k+1})$, where $\xi(t)$ denotes the state of the physical model which can be obtained, $\Omega > 0$ is a given matrix, and $\varrho > 0$ is a constant; this case is called the threshold function type; case (ii) $\|\boldsymbol{\xi}(t) - \boldsymbol{\xi}(t_k)\|^2 \leq \varrho_0 e^{-\varepsilon t}$, $t \in [t_k, t_{k+1})$, where $\varrho_0 > 0$ and $\varepsilon > 0$; this case is called the exponential function type. In [37], the combination event-triggered condition (CETC) was first proposed by Li et al., which can effectively combine the advantages of above two types, thereby decreasing the number of updates. Since then, the authors of [38] also presented an intermittent CETC to consider the exponential boundedness problem of fractional-order multi-agent systems. However, there are few results reported on the nonfragile SMC for FO-CNSs based on the combination triggering mechanism, which motivates us to explore this issue.

Based on the above discussions, this chapter will consider the SMC problem for FO-CNSs with the combination event-triggered mechanism. The main contributions of this work can be summarized as follows:

(1) A combination event-triggered mechanism-based SMC of FO-CNSs is presented for the first time. This CETC contains the merits of traditional event-triggered mechanisms, which can save more communication resources, compared with other control strategies;

(2) According to the presented CETC, the stability problem for the considered FO-CNSs is discussed by using a nonfragile SMC strategy, which is the first attempt in this problem;

(3) By designing a suitable sliding surface function and event-triggered SMC law, several sufficient conditions involving strict LMIs are derived to ensure the stability of the closed-loop system;

(4) Compared with the previous literature, a corrected event-triggered mechanism is given for the considered FO-CNSs. Meanwhile, based on the Gronwall–Bellman inequality, a positive lower bound $\Delta T > 0$ is given, meaning that the presented control method can effectively avoid Zeno behavior in FO-CNSs.

9.2 Problem statement

In this part, consider the FO-CNSs with $\mathcal{N}$ nodes as follows:

$$\begin{cases} {}^{C}_{t_0}D^{\alpha}_t \boldsymbol{x}_i(t) = \boldsymbol{A}\boldsymbol{x}_i(t) + \sum_{j=1}^{\mathcal{N}} h_{ij}\boldsymbol{\Gamma}\boldsymbol{x}_j(t) + \beta_i(t)\boldsymbol{f}(\boldsymbol{x}_i(t)) + \boldsymbol{B}\boldsymbol{u}_i(t), \\ \boldsymbol{y}_i(t) = \boldsymbol{C}_i\boldsymbol{x}_i(t), \\ \tilde{\boldsymbol{y}}_i(t) = \boldsymbol{y}_i(t) + \gamma_i(t)\boldsymbol{g}(\boldsymbol{y}_i(t)), \quad i = 1, 2, \ldots, \mathcal{N}, \end{cases} \tag{9.1}$$

where ${}^{C}_{t_0}D^{\alpha}_t$ is the Caputo fractional-order derivative operator of order $\alpha \in (0, 1)$, and for node i, $\boldsymbol{x}_i(t) \in \mathbb{R}^n$ denotes the pseudo-state of the ith system in the Euclidean space; $\boldsymbol{u}_i(t) \in \mathbb{R}^m$ represents the control input signal; $\boldsymbol{y}_i(t)$ and $\tilde{\boldsymbol{y}}_i(t) \in \mathbb{R}^l$ demonstrate the measurement output signal and actual output of the estimator, respectively; A, B, C_i are real matrices, and B is full column rank, C_i is row full rank; $H = [h_{ij}]_{\mathcal{N} \times \mathcal{N}}$ denotes the coupled configuration matrix, satisfying $h_{ij} \geq 0$ ($i \neq j$) and $h_{ii} = -\sum\limits_{j=1}^{\mathcal{N}} h_{ij}$; $f(\cdot)$ and $g(\cdot)$ denote two nonlinear matrix functions with $f(0) = 0$ and $g(0) = 0$; moreover, $\beta_i(t)$, $\gamma_i(t) \in \{0, 1\}$ represent stochastic variables which have Bernoulli distribution, i.e., $\mathbb{E}\{\beta_i(t)\} = \beta_i$ and $\mathbb{E}\{\gamma_i(t)\} = \gamma_i$, with β_i, $\gamma_i \in [0, 1]$ being known scalars.

In order to derive the main results, some assumptions for the FO-CNSs (9.1) are presented.

Assumption 9.1. For any $x_i, x_j \in \mathbb{R}^n$, $y_i, y_j \in \mathbb{R}^m$, functions $f(\cdot)$ and $g(\cdot)$ satisfy the following Lipschitz conditions:

$$\|f(x_i) - f(x_j)\|^2 \leq \|F_1(x_i - x_j)\|^2, \quad \|g(y_i) - g(y_j)\|^2 \leq \|F_2(y_i - y_j)\|^2,$$

where F_1, F_2 are given matrices.

Assumption 9.2. Function $g(\cdot)$ is norm bounded, i.e., there exists a constant $\upsilon > 0$ such that $\|g(\cdot)\| \leq \upsilon$ holds.

Remark 9.1. As we know, fractional-order complex networked systems have been widely used in daily life, especially in the field of neural network systems. For example, in [39], the authors explored the global synchronization problem for a class of fractional complex-valued neural networks, subject to pantograph delays and inhibitory factors, by using the feedback control methods, and the expression of this fractional mathematical model is given by

$$\begin{cases} {}^{C}_{t_0}D^{\gamma}_t x_i(t) = -A_i x_i(t) + \sum\limits_{j=1}^{n} \alpha_{ij} f_j(x_j(t)) - \sum\limits_{j=1}^{n} \beta_{ij} g_j(x_j(pt)) + w_i(t), \\ {}^{C}_{t_0}D^{\gamma}_t y_i(t) = -A_i y_i(t) + \sum\limits_{j=1}^{n} \alpha_{ij} f_j(y_j(t)) - \sum\limits_{j=1}^{n} \beta_{ij} g_j(y_j(pt)) + u_i(t) + w_i(t), \\ i = 1, 2, \ldots, \mathcal{N}. \end{cases}$$

Meanwhile, in [40], the quasi-projective synchronization control problem of fractional-order neural network systems with time-delay has also been explored. Its mathematical model is described as

$$\begin{cases} {}^{C}_{t_0}D^{\gamma}_t x_i(t) = -A_i x_i(t) + \sum\limits_{j=1}^{n} \alpha_{ij}(t) f_j(x_j(t)) - \sum\limits_{j=1}^{n} \beta_{ij} g_j(x_j(t - \tau_j)) + u_i(t), \\ {}^{C}_{t_0}D^{\gamma}_t y_i(t) = -A_i y_i(t) + \sum\limits_{j=1}^{n} \alpha_{ij}(t) f_j(y_j(t)) - \sum\limits_{j=1}^{n} \beta_{ij} g_j(y_j(t - \tau_j)) \\ \qquad\qquad + u_i(t) + w_i(t), \\ i = 1, 2, \ldots, \mathcal{N}. \end{cases}$$

The above research results show the effect of fractional-order derivatives on the state synchronization of networked systems.

9.3 Main results

9.3.1 Nonfragile fractional-order observer design

In this section, we design the following nonfragile fractional-order state observer:

$$\begin{cases} {}^{C}_{t_0}D_t^{\alpha}\hat{x}_i(t) = A\hat{x}_i(t) + \sum_{j=1}^{\mathcal{N}} h_{ij}\boldsymbol{\Gamma}\hat{x}_j(t) + Bu_i(t) - (L_i + \Delta L_i(t))(\tilde{y}_i(t) - \hat{y}_i(t)), \\ \hat{y}_i(t) = C_i\hat{x}_i(t_k^i), \quad i = 1, 2, \ldots, \mathcal{N}, \end{cases}$$

(9.2)

where $t \in [t_k^i, t_{k+1}^i)$ is specified in (9.5), $\hat{x}_i(t) \in \mathbb{R}^n$ and $\hat{y}_i(t) \in \mathbb{R}^l$ denote the observer state and output of the estimator, respectively; L_i and $\Delta L_i(t)$ are estimator gains and uncertainties of the estimator, such that $\Delta L_i(t) = M_i F(t) N_i$, with $F^{\mathrm{T}}(t)F(t) \leq I$.

Let $e_i(t) = x_i(t) - \hat{x}_i(t)$ denote the state estimation error vector, then the ith observer error system is

$$\begin{aligned} &{}^{C}_{t_0}D_t^{\alpha}e_i(t) \\ &= Ae_i(t) + \sum_{j=1}^{\mathcal{N}} h_{ij}\boldsymbol{\Gamma}e_j(t) + \beta_i(t)f(x_i(t)) + (L_i + \Delta L_i(t))(\tilde{y}_i(t) - \hat{y}_i(t)). \end{aligned}$$

(9.3)

Meanwhile, according to the SMC theory, the sliding surface function $s_i(t)$ is designed as follows:

$$\begin{cases} s_i(t) = G\big(\hat{x}_i(t) - \hat{x}_i(t_0)\big) - G_{t_0}D_t^{-\alpha}z_i(t), \\ z_i(t) = (A + BK_i)\hat{x}_i(t) + \sum_{j=1}^{\mathcal{N}} h_{ij}\boldsymbol{\Gamma}\hat{x}_j(t), \end{cases}$$

(9.4)

where $G \in \mathbb{R}^{m \times n}$ is such that GB is nonsingular, and $K_i \in \mathbb{R}^{m \times n}$ guarantees that $A + BK_i$ is Hurwitz.

Remark 9.2. It is well known that the traditional sliding mode reachability condition only guarantees that the system arrives at the switching surface from any point in the state space in a finite time, and does not make any specification as to what its trajectory is. So it cannot guarantee that the system always meets the desired dynamic performance index in the whole reach phase. In order to solve this problem, various convergence laws can be designed to ensure the dynamic quality of the convergence process, including isochronous convergence law, exponential convergence law, power convergence law, and so on. The most direct method is to adopt the method of integral sliding mode, which eliminates the arrival phase by setting the initial state of the integrator reasonably, so that the initial state of the system is on the sliding mode surface at the beginning, so as to improve the robustness of the control system. Therefore, a fractional-order integral sliding mode surface shaped as in Eq. (9.4) is designed by combining the theory of fractional-order calculus, and then the sliding mode controller is designed by using the Caputo fractional-order calculus property.

9.3.2 Event-triggered SMC strategy

In this section, we will design a novel combination triggering mechanism to guarantee the stability of the error system (9.3):

$$t^i_{k+1} = \min\{t^i_k + \nabla, \, t^i_{k+1}\}, \tag{9.5}$$

where

$$t^i_{k+1} \triangleq \inf\left\{t > t^i_k \,\middle|\, \|\hat{e}_i(t)\|^2 \geq \xi_{1i}(t) \vee |(\hat{s}_i)_r(t)| \geq \xi_{2i}(t)\right\},$$

$$\xi_{1i}(t) = \theta_{1i}(\kappa)\|\hat{x}_i(t)\|^2 + \theta_{2i}(\kappa)(e^{-\varepsilon_i t} + \rho_i),$$

$$\xi_{2i}(t) = \frac{\sqrt{\theta_{1i}(\kappa)}}{\sqrt{m}}\|\hat{x}_i(t)\| + \frac{\sqrt{\theta_{2i}(\kappa)}}{\sqrt{m}}(e^{-\varepsilon_i t} + \rho_i)^{\frac{1}{2}},$$

with $\nabla > 0$ being a given constant, $\hat{e}_i(t) = \hat{x}_i(t) - \hat{x}_i(t^i_k)$ denoting the sampled error vector, and $\hat{s}_i(t) = s_i(t) - s_i(t^i_k)$ standing for the sampled error of sliding surface function; $\theta_{1i}(\kappa)$, $\theta_{2i}(\kappa)$, ε_i, and ρ_i are all nonnegative scalars.

Particularly, CETC (9.5) includes other two classical cases:

(a) If $\mathrm{sgn}(\theta_{1i}(\kappa)) = 1$ and $\mathrm{sgn}(\theta_{2i}(\kappa)) = 0$, then (9.5) reduces to the norm event-triggered condition (NETC);

(b) If $\mathrm{sgn}(\theta_{1i}(\kappa)) = 0$ and $\mathrm{sgn}(\theta_{2i}(\kappa)) = 1$, then (9.5) reduces to the exponential event-triggered condition (EETC).

Meanwhile, since B is of full column rank, there exists a matrix $G = (B^T B)^{-1} B^T$ such that $GB = I$ holds. Due to $^C_{t_0}D^\alpha_t s_i(t) = 0$, the equivalent controller can be obtained as

$$u_{i\mathrm{eq}}(t) = K_i\hat{x}_i(t) + G(L_i + \Delta L_i)(\tilde{y}_i(t) - \hat{y}_i(t)). \tag{9.6}$$

Therefore, substituting (9.6) into (9.2), we have

$$
{}_{t_0}^{C}D_t^\alpha \hat{x}_i(t)
$$

$$
= (A + BK_i)\hat{x}_i(t) + \sum_{j=1}^{\mathcal{N}} h_{ij}\,\Gamma \hat{x}_j(t) - (I - BG)(L_i + \Delta L_i)(\tilde{y}_i(t) - \hat{y}_i(t)).
$$

$$(9.7)$$

Considering (9.7) and (9.3), by using the Kronecker product, one has

$$
\begin{cases}
{}_{t_0}^{C}D_t^\alpha \hat{x}(t) = (\hat{A} + \hat{B}K)\hat{x}(t) - \hat{G}_0(L + \Delta L)(Ce(t) + C\hat{e}(t) + \hat{y}(t)g(y(t))), \\
{}_{t_0}^{C}D_t^\alpha e(t) = (\hat{A} + (L + \Delta L)C)e(t) + (L + \Delta L)C\hat{e}(t) + \hat{\beta}(t)f(x(t)) \\
\qquad\qquad + (L + \Delta L)\hat{y}(t)g(y(t)),
\end{cases}
$$

$$(9.8)$$

where $x(t) = \mathrm{col}_{\mathcal{N}}\{x_i(t)\}$, $\hat{x}(t) = \mathrm{col}_{\mathcal{N}}\{\hat{x}_i(t)\}$, $e(t) = \mathrm{col}_{\mathcal{N}}\{e_i(t)\}$, $\hat{e}(t) = \mathrm{col}_{\mathcal{N}}\{\hat{e}_i(t)\}$, $f(x(t)) = \mathrm{col}_{\mathcal{N}}\{f(x_i(t))\}$, $g(y(t)) = \mathrm{col}_{\mathcal{N}}\{g(y_i(t))\}$, $\hat{A} = I_{\mathcal{N}} \otimes A + H \otimes \Gamma$, $\hat{B} = I_{\mathcal{N}} \otimes B$, $K = \mathrm{diag}\{K_i\}$, $L + \Delta L = \mathrm{diag}\{L_i + \Delta L_i\}$, $\hat{G}_0 = I_{\mathcal{N}} \otimes (I - BG)$, $C = \mathrm{diag}\{C_i\}$, $\hat{\beta}(t) = \beta(t) \otimes I_{\mathcal{N}}$, and $\hat{y}(t) = y(t) \otimes I_{\mathcal{N}}$, with $\beta(t) = \mathrm{diag}\{\beta_i(t)\}$ and $y(t) = \mathrm{diag}\{y_i(t)\}$.

Remark 9.3. For the sliding mode dynamics (9.8), since $\beta_i(t)$, $\gamma_i(t) \in \{0, 1\}$ for any $t \in [t_0, +\infty)$, we know that

$$
\lambda_{\max}\{\hat{\beta}^{\mathrm{T}}(t)\hat{\beta}(t)\} \le 1, \quad \lambda_{\max}\{\hat{y}^{\mathrm{T}}(t)\hat{y}(t)\} \le 1.
$$

It should be pointed that the method proposed in [29] brought stochastic variables β_i and γ_i into LMI. Hence, for different stochastic variables, it will obtain different estimator gains L_i by calculating strict LMI. Compared with [29], the proposed method for dealing with stochastic variables is more general.

Meanwhile, as $x(t) = e(t) + \hat{x}(t)$, Assumption 9.1 can be rewritten as follows:

$$
\|f(x(t))\|^2 \le \|\hat{F}_1 x(t)\|^2 \le 2(\|\hat{F}_1 \hat{x}(t)\|^2 + \|\hat{F}_1 e(t)\|^2),
$$

$$
\|g(y(t))\|^2 \le \|\hat{F}_2 C x(t)\|^2 \le 2(\|\hat{F}_2 C \hat{x}(t)\|^2 + \|\hat{F}_2 C e(t)\|^2),
$$

where $\hat{F}_1 = I_{\mathcal{N}} \otimes F_1$ and $\hat{F}_2 = I_{\mathcal{N}} \otimes F_2$.

Theorem 9.1. *For given scalars $\eta_i > 0$, if there exist symmetric positive definite matrices P_1, $P_2 \in \mathbb{R}^{n\mathcal{N} \times n\mathcal{N}}$ and positive scalars ω_k ($k = 1, 2, \ldots, 7$) such that the following inequality holds:*

$$
\begin{bmatrix}
\Sigma_{11} & \Sigma_{12} & \Sigma_{13} \\
* & \Sigma_{22} & 0 \\
* & * & \Sigma_{33}
\end{bmatrix} < 0,
$$

$$(9.9)$$

where

$$\Sigma_{11} = \begin{bmatrix} \tilde{\Sigma}_{11} & -P_1\hat{G}_0 LC & -P_1\hat{G}_0 LC \\ * & \tilde{\Sigma}_{22} & P_2 LC \\ * & * & -I \end{bmatrix},$$

$$\Sigma_{13} = \begin{bmatrix} C^{\mathrm{T}}\hat{F}_2^{\mathrm{T}} & 0 & C^{\mathrm{T}}\hat{F}_2^{\mathrm{T}} & 0 & P_1\hat{G}_0 LC & 0 \\ 0 & C^{\mathrm{T}}\hat{F}_2^{\mathrm{T}} & 0 & C^{\mathrm{T}}\hat{F}_2^{\mathrm{T}} & 0 & P_2 LC \\ 0 & 0 & 0 & 0 & 0 & 0 \end{bmatrix},$$

$$\Sigma_{12} = \begin{bmatrix} 0 & 0 & C^{\mathrm{T}}\hat{F}_2^{\mathrm{T}} & 0 & 0 & 0 & C^{\mathrm{T}}\hat{F}_2^{\mathrm{T}} & 0 & \hat{F}_1^{\mathrm{T}} & 0 \\ C^{\mathrm{T}}N^{\mathrm{T}} & 0 & 0 & C^{\mathrm{T}}\hat{F}_2^{\mathrm{T}} & C^{\mathrm{T}}N^{\mathrm{T}} & 0 & 0 & C^{\mathrm{T}}\hat{F}_2^{\mathrm{T}} & 0 & \hat{F}_1^{\mathrm{T}} \\ 0 & C^{\mathrm{T}}N^{\mathrm{T}} & 0 & 0 & 0 & C^{\mathrm{T}}N^{\mathrm{T}} & 0 & 0 & 0 & 0 \end{bmatrix},$$

$$\Sigma_{22} = \mathrm{diag}\left\{ -\omega_1 I, -\omega_2 I, \tfrac{-\omega_3 I}{2p_1}, \tfrac{-\omega_3 I}{2p_1}, -\omega_4 I, -\omega_5 I, \tfrac{-\omega_6 I}{2p_1}, \tfrac{-\omega_6 I}{2p_1}, \tfrac{-\omega_7 I}{2}, \tfrac{-\omega_7 I}{2} \right\},$$

$$\Sigma_{33} = \mathrm{diag}\left\{ \tfrac{-\eta_1 I}{2}, \tfrac{-\eta_1 I}{2}, \tfrac{-\eta_2 I}{2}, \tfrac{-\eta_2 I}{2}, \tfrac{-I}{\eta_1 p_2}, \tfrac{-I}{\eta_2 p_2} \right\},$$

with $\tilde{\Sigma}_{11} = \mathrm{sym}\{P_1(\hat{A} + \hat{B}K)\} + (\omega_1 + \omega_2 + \omega_3)P_1\hat{G}_0 MM^{\mathrm{T}}\hat{G}_0^{\mathrm{T}}P_1 + \theta_1(\kappa)I$, $\tilde{\Sigma}_{22} = \mathrm{sym}\{P_2(\hat{A} + LC)\} + (\omega_4 + \omega_5 + \omega_6)P_2 MM^{\mathrm{T}}P_2 + \omega_7\|P_2\|^2$, $\tilde{G} = C^{\mathrm{T}}(CC^{\mathrm{T}})^{-1}$, $p_1 = \lambda_{\max}\{N^{\mathrm{T}}N\}$, $p_2 = \lambda_{\max}\{\tilde{G}\tilde{G}^{\mathrm{T}}\}$, *with* $M = \mathrm{diag}\{M_i\}$ *and* $N = \mathrm{diag}\{N_i\}$. *Then, the sliding mode dynamics (9.8) is stable under the presented CETC (9.5).*

Proof. Construct the following Lyapunov function:

$$V(t) = \hat{x}^{\mathrm{T}}(t)P_1\hat{x}(t) + e^{\mathrm{T}}(t)P_2 e(t).$$

Then, combining with Lemma 2.2, one has

$$\begin{aligned}
{}_{t_0}^{C}D_t^{\alpha}V(t) \leq\; & 2\hat{x}^{\mathrm{T}}(t)P_1{}_{t_0}^{C}D_t^{\alpha}\hat{x}(t) + 2e^{\mathrm{T}}(t)P_2{}_{t_0}^{C}D_t^{\alpha}e(t) \\
=\; & 2\hat{x}^{\mathrm{T}}(t)P_1\Big[(\hat{A} + \hat{B}K)\hat{x}(t) \\
& \qquad - \hat{G}_0(L + \Delta L)(Ce(t) + C\hat{e}(t) + \hat{\gamma}(t)g(y(t)))\Big] \\
& + 2e^{\mathrm{T}}(t)P_2\Big[(\hat{A} + (L + \Delta L)C)e(t) + (L + \Delta L)C\hat{e}(t) \\
& \qquad + \hat{\beta}(t)f(x(t)) + (L + \Delta L)\hat{\gamma}(t)g(y(t))\Big].
\end{aligned}$$

Moreover, according to Lemma 2.13, there exist some positive constants ω_k and η_i such that

$$-2\hat{x}^{\mathrm{T}}(t)P_1\hat{G}_0\Delta L Ce(t) \leq \omega_1\|\hat{x}^{\mathrm{T}}(t)P_1\hat{G}_0 M\|^2 + \tfrac{1}{\omega_1}\|NCe(t)\|^2,$$

$$-2\hat{x}^{\mathrm{T}}(t)P_1\hat{G}_0\Delta L C\hat{e}(t) \leq \omega_2\|\hat{x}^{\mathrm{T}}(t)P_1\hat{G}_0 M\|^2 + \tfrac{1}{\omega_2}\|NC\hat{e}(t)\|^2,$$

$$-2\hat{\boldsymbol{x}}^{\mathrm{T}}(t)\boldsymbol{P}_1\hat{\boldsymbol{G}}_0\Delta\boldsymbol{L}\hat{\boldsymbol{\gamma}}(t)\boldsymbol{g}(\boldsymbol{y}(t)) \le \omega_3\|\hat{\boldsymbol{x}}^{\mathrm{T}}(t)\boldsymbol{P}_1\hat{\boldsymbol{G}}_0\boldsymbol{M}\|^2 + \tfrac{1}{\omega_3}\|\boldsymbol{N}\hat{\boldsymbol{\gamma}}(t)\boldsymbol{g}(\boldsymbol{y}(t))\|^2$$

$$\le \omega_3\|\hat{\boldsymbol{x}}^{\mathrm{T}}(t)\boldsymbol{P}_1\hat{\boldsymbol{G}}_0\boldsymbol{M}\|^2 + \tfrac{p_1}{\omega_3}\|\boldsymbol{g}(\boldsymbol{y}(t))\|^2$$

$$\le \omega_3\|\hat{\boldsymbol{x}}^{\mathrm{T}}(t)\boldsymbol{P}_1\hat{\boldsymbol{G}}_0\boldsymbol{M}\|^2$$

$$+ \tfrac{2p_1}{\omega_3}\left(\|\hat{\boldsymbol{F}}_2\boldsymbol{C}\hat{\boldsymbol{x}}(t)\|^2 + \|\hat{\boldsymbol{F}}_2\boldsymbol{C}\boldsymbol{e}(t)\|^2\right),$$

$$2\boldsymbol{e}^{\mathrm{T}}(t)\boldsymbol{P}_2\Delta\boldsymbol{L}\boldsymbol{C}\boldsymbol{e}(t) \le \omega_4\|\boldsymbol{e}^{\mathrm{T}}(t)\boldsymbol{P}_2\boldsymbol{M}\|^2 + \tfrac{1}{\omega_4}\|\boldsymbol{N}\boldsymbol{C}\boldsymbol{e}(t)\|^2,$$

$$2\boldsymbol{e}^{\mathrm{T}}(t)\boldsymbol{P}_2\Delta\boldsymbol{L}\boldsymbol{C}\hat{\boldsymbol{e}}(t) \le \omega_5\|\boldsymbol{e}^{\mathrm{T}}(t)\boldsymbol{P}_2\boldsymbol{M}\|^2 + \tfrac{1}{\omega_5}\|\boldsymbol{N}\boldsymbol{C}\hat{\boldsymbol{e}}(t)\|^2,$$

$$2\boldsymbol{e}^{\mathrm{T}}(t)\boldsymbol{P}_2\Delta\boldsymbol{L}\hat{\boldsymbol{\gamma}}(t)\boldsymbol{g}(\boldsymbol{y}(t)) \le \omega_6\|\boldsymbol{e}^{\mathrm{T}}(t)\boldsymbol{P}_2\boldsymbol{M}\|^2 + \tfrac{1}{\omega_6}\|\boldsymbol{N}\hat{\boldsymbol{\gamma}}(t)\boldsymbol{g}(\boldsymbol{y}(t))\|^2$$

$$\le \omega_6\|\boldsymbol{e}^{\mathrm{T}}(t)\boldsymbol{P}_2\boldsymbol{M}\|^2$$

$$+ \tfrac{2p_1}{\omega_6}\left(\|\hat{\boldsymbol{F}}_2\boldsymbol{C}\hat{\boldsymbol{x}}(t)\|^2 + \|\hat{\boldsymbol{F}}_2\boldsymbol{C}\boldsymbol{e}(t)\|^2\right),$$

$$2\boldsymbol{e}^{\mathrm{T}}(t)\boldsymbol{P}_2\hat{\boldsymbol{\beta}}(t)\boldsymbol{f}(\boldsymbol{x}(t)) \le \omega_7\|\boldsymbol{e}^{\mathrm{T}}(t)\boldsymbol{P}_2\|^2 + \tfrac{1}{\omega_7}\|\hat{\boldsymbol{\beta}}(t)\boldsymbol{f}(\boldsymbol{x}(t))\|^2$$

$$\le \omega_7\|\boldsymbol{e}^{\mathrm{T}}(t)\boldsymbol{P}_2\|^2$$

$$+ \tfrac{2}{\omega_7}\left(\|\hat{\boldsymbol{F}}_1\hat{\boldsymbol{x}}(t)\|^2 + \|\hat{\boldsymbol{F}}_1\boldsymbol{e}(t)\|^2\right),$$

and

$$-2\hat{\boldsymbol{x}}^{\mathrm{T}}(t)\boldsymbol{P}_1\hat{\boldsymbol{G}}_0\boldsymbol{L}\hat{\boldsymbol{\gamma}}(t)\boldsymbol{g}(\boldsymbol{y}(t)) \le \eta_1\|\hat{\boldsymbol{x}}^{\mathrm{T}}(t)\boldsymbol{P}_1\hat{\boldsymbol{G}}_0\boldsymbol{L}\boldsymbol{C}\tilde{\boldsymbol{G}}\|^2 + \tfrac{1}{\eta_1}\|\hat{\boldsymbol{\gamma}}(t)\boldsymbol{g}(\boldsymbol{y}(t))\|^2$$

$$\le \eta_1 p_2\|\hat{\boldsymbol{x}}^{\mathrm{T}}(t)\boldsymbol{P}_1\hat{\boldsymbol{G}}_0\boldsymbol{L}\boldsymbol{C}\|^2$$

$$+ \tfrac{2}{\eta_1}(\|\hat{\boldsymbol{F}}_2\boldsymbol{C}\hat{\boldsymbol{x}}(t)\|^2 + \|\hat{\boldsymbol{F}}_2\boldsymbol{C}\boldsymbol{e}(t)\|^2),$$

$$2\boldsymbol{e}^{\mathrm{T}}(t)\boldsymbol{P}_2\boldsymbol{L}\hat{\boldsymbol{\gamma}}(t)\boldsymbol{g}(\boldsymbol{y}(t)) \le \eta_2\|\boldsymbol{e}^{\mathrm{T}}(t)\boldsymbol{P}_2\boldsymbol{L}\boldsymbol{C}\tilde{\boldsymbol{G}}\|^2 + \tfrac{1}{\eta_2}\|\hat{\boldsymbol{\gamma}}(t)\boldsymbol{g}(\boldsymbol{y}(t))\|^2$$

$$\le \eta_2 p_2\|\boldsymbol{e}^{\mathrm{T}}(t)\boldsymbol{P}_2\boldsymbol{L}\boldsymbol{C}\|^2$$

$$+ \tfrac{2}{\eta_2}(\|\hat{\boldsymbol{F}}_2\boldsymbol{C}\hat{\boldsymbol{x}}(t)\|^2 + \|\hat{\boldsymbol{F}}_2\boldsymbol{C}\boldsymbol{e}(t)\|^2).$$

Furthermore, considering the CETC (9.5), for any $t \in [t_k^i, t_{k+1}^i)$, one has

$$\|\hat{\boldsymbol{e}}(t)\|^2 \le \theta_1(\kappa)\|\hat{\boldsymbol{x}}(t)\|^2 + \mathcal{N}\theta_2(\kappa)(e^{-\varepsilon t} + \rho),$$

where $\theta_1(\kappa) = \max\limits_{i}\{\theta_{1i}(\kappa)\}$, $\theta_2(\kappa) = \max\limits_{i}\{\theta_{2i}(\kappa)\}$, $\varepsilon = \min\limits_{i}\{\varepsilon_i\}$, and $\rho = \max\limits_{i}\{\rho_i\}$ are all nonnegative scalars.

Therefore, combining the above inequalities, one gets

$$_{t_0}^{C}D_t^{\alpha}V(t) \le \boldsymbol{\xi}^{\mathrm{T}}(t)\hat{\boldsymbol{\Sigma}}\boldsymbol{\xi}(t) + \mathcal{N}\theta_2(\kappa)(e^{-\varepsilon t} + \rho),$$

where $\boldsymbol{\xi}^{\mathrm{T}}(t) = [\hat{\boldsymbol{x}}^{\mathrm{T}}(t), \boldsymbol{e}^{\mathrm{T}}(t), \hat{\boldsymbol{e}}^{\mathrm{T}}(t)]$, and

$$\hat{\boldsymbol{\Sigma}} = \begin{bmatrix} \hat{\boldsymbol{\Sigma}}_{11} & -\boldsymbol{P}_1\hat{\boldsymbol{G}}_0\boldsymbol{L}\boldsymbol{C} & -\boldsymbol{P}_1\hat{\boldsymbol{G}}_0\boldsymbol{L}\boldsymbol{C} \\ * & \hat{\boldsymbol{\Sigma}}_{22} & \boldsymbol{P}_2\boldsymbol{L}\boldsymbol{C} \\ * & * & \hat{\boldsymbol{\Sigma}}_{33} \end{bmatrix},$$

with

$$\hat{\Sigma}_{11} = \mathrm{sym}\{P_1(\hat{A} + \hat{B}K)\} + (\omega_1 + \omega_2 + \omega_3)P_1\hat{G}_0 M M^\mathrm{T} \hat{G}_0^\mathrm{T} P_1$$
$$+ \left(\tfrac{2}{\omega_3} + \tfrac{2}{\omega_6}\right) p_1 C^\mathrm{T} \hat{F}_2^\mathrm{T} \hat{F}_2 C + \tfrac{2}{\omega_7} \hat{F}_1^\mathrm{T} \hat{F}_1 + \eta_1 p_2 P_1 \hat{G}_0 L C C^\mathrm{T} L^\mathrm{T} \hat{G}_0 P_1$$
$$+ \left(\tfrac{2}{\eta_1} + \tfrac{2}{\eta_2}\right) C^\mathrm{T} \hat{F}_2^\mathrm{T} \hat{F}_2 C + \theta_1(\kappa) I,$$

$$\hat{\Sigma}_{22} = \mathrm{sym}\{P_2(\hat{A} + LC)\} + (\omega_4 + \omega_5 + \omega_6)P_2 M M^\mathrm{T} P_2 + \omega_7 P_2 P_2$$
$$+ \left(\tfrac{1}{\omega_1} + \tfrac{1}{\omega_4}\right) C^\mathrm{T} N^\mathrm{T} N C + \tfrac{2}{\omega_7} \hat{F}_1^\mathrm{T} \hat{F}_1$$
$$+ \left[\left(\tfrac{2}{\omega_3} + \tfrac{2}{\omega_6}\right) p_1 + \tfrac{2}{\eta_1} + \tfrac{2}{\eta_2}\right] C^\mathrm{T} \hat{F}_2^\mathrm{T} \hat{F}_2 C + \eta_2 p_2 P_2 L C C^\mathrm{T} L^\mathrm{T} P_2,$$

$$\hat{\Sigma}_{33} = -I + \left(\tfrac{1}{\omega_2} + \tfrac{1}{\omega_5}\right) C^\mathrm{T} N^\mathrm{T} N C.$$

Furthermore, one can show that (9.9) is equivalent to $\hat{\Sigma} < 0$ based on the Schur complement lemma. Then, there exists a constant $\sigma > 0$ such that

$$^{C}_{t_0}D_t^\alpha V(t) \le -\sigma V(t) + \mathcal{N}\theta_2(\kappa)(e^{-\varepsilon t} + \rho) \le -\sigma V(t) + \mathcal{N}\theta_2(\kappa)(1 + \rho). \quad (9.10)$$

By applying Lemma 2.5 to the inequality (9.10), we have

$$V(t) \le V(t_0) E_{\alpha,1}[-\sigma(t - t_0)^\alpha] + \mathcal{N}\theta_2(\kappa)(1 + \rho)(t - t_0)^\alpha E_{\alpha,\alpha+1}[-\sigma(t - t_0)^\alpha].$$

On the one hand, according to Lemma 2.11, since $\arg(-\sigma(t - t_0)^\alpha) = -\pi$, there exists a scalar $\lambda_1 > 0$ satisfying

$$\left| E_{\alpha,1}(-\sigma(t - t_0)^\alpha) \right| \le \frac{\lambda_1}{1 + |-\sigma(t - t_0)^\alpha|} \to 0,$$

when $t \to +\infty$, which implies that

$$\lim_{t \to +\infty} V(t_0) E_{\alpha,1}(-\sigma(t - t_0)^\alpha) = 0.$$

On the other hand, according to Lemma 2.12, there also exists a scalar $\lambda_2 > 0$ satisfying

$$\left| (t - t_0)^\alpha E_{\alpha,\alpha+1}(-\sigma(t - t_0)^\alpha) \right| \le \lambda_2(t - t_0)^\alpha e^{-\sigma(t-t_0)} \to 0,$$

when $t \to +\infty$, which implies that

$$\lim_{t \to +\infty} \mathcal{N}\theta_2(\kappa)(e^{-\varepsilon t} + \rho)(t - t_0)^\alpha E_{\alpha,\alpha+1}(-\sigma(t - t_0)^\alpha) = 0.$$

Therefore, based on the above analysis, the final result is

$$\lim_{t \to +\infty} V(t) = 0,$$

which means that system (9.8) is stable under the presented CETC (9.5).
□

Remark 9.4. Particularly, taking notice of the CETC (9.5), if one chooses $\theta_2(\kappa) = 0$, it will reduce to the NETC. According to (9.10), one has

$$
{}^{C}_{t_0}D^{\alpha}_t V(t) \leq -\sigma V(t).
$$

Therefore, it can be concluded that the sliding mode dynamics (9.8) is also asymptotically stable from Theorem 9.2.

Remark 9.5. Noting that matrices P_i, K_i, and L_i in inequality (9.9) are unknown, and ω_k are all unknown scalars, inequality (9.9) cannot be solved. Therefore, to solve this inequality by using Matlab$^{\circledR}$ Toolbox, the following theorem is proposed.

Theorem 9.2. *For given positive scalars* η_i, λ_j, *if there exists positive scalars* ω_k $(k = 1, 2, \ldots, 7)$, *and matrices* X_1, X_2, Y, Z *satisfying the following inequality:*

$$
\begin{bmatrix}
\Pi_{11} & \Pi_{12} & \Pi_{13} & \Pi_{14} \\
* & \Pi_{22} & 0 & 0 \\
* & * & \Pi_{33} & 0 \\
* & * & * & -\theta_1(\kappa)I
\end{bmatrix} < 0,
\tag{9.11}
$$

where

$$
\Pi_{11} =
\begin{bmatrix}
\tilde{\Pi}_{11} & -\hat{G}_0 Z & -\hat{G}_0 Z \\
* & \tilde{\Pi}_{22} & Z \\
* & * & \lambda_1^2 I - 2\lambda_1 X_1
\end{bmatrix},
$$

$$
\Pi_{13} =
\begin{bmatrix}
X_1 C^{\mathrm{T}} \hat{F}_2^{\mathrm{T}} & 0 & X_1 C^{\mathrm{T}} \hat{F}_2^{\mathrm{T}} & 0 & \hat{G}_0 Z & 0 \\
0 & X_2 C^{\mathrm{T}} \hat{F}_2^{\mathrm{T}} & 0 & X_2 C^{\mathrm{T}} \hat{F}_2^{\mathrm{T}} & 0 & Z \\
0 & 0 & 0 & 0 & 0 & 0
\end{bmatrix},
$$

$$
\Pi_{12} =
\begin{bmatrix}
0 & 0 & X_1 C^{\mathrm{T}} \hat{F}_2^{\mathrm{T}} & 0 & 0 & 0 & X_1 C^{\mathrm{T}} \hat{F}_2^{\mathrm{T}} & 0 & X_1 \hat{F}_1^{\mathrm{T}} & 0 \\
X_2 C^{\mathrm{T}} N^{\mathrm{T}} & 0 & 0 & X_2 C^{\mathrm{T}} \hat{F}_2^{\mathrm{T}} & X_2 C^{\mathrm{T}} N^{\mathrm{T}} & 0 & 0 & X_2 C^{\mathrm{T}} \hat{F}_2^{\mathrm{T}} & 0 & X_2 \hat{F}_1^{\mathrm{T}} \\
0 & X_2 C^{\mathrm{T}} N^{\mathrm{T}} & 0 & 0 & 0 & X_2 C^{\mathrm{T}} N^{\mathrm{T}} & 0 & 0 & 0 & 0
\end{bmatrix},
$$

$$
\Pi_{14} = \begin{bmatrix} \theta_1(\kappa)X_1 & 0 & 0 \end{bmatrix}^{\mathrm{T}},
$$

$$
\Pi_{22} = \mathrm{diag}\left\{-\omega_1 I, -\omega_2 I, \frac{-\omega_3 I}{2p_1}, \frac{-\omega_3 I}{2p_1}, -\omega_4 I, -\omega_5 I, \frac{-\omega_6 I}{2p_1}, \frac{-\omega_6 I}{2p_1}, \frac{-\omega_7 I}{2}, -\frac{\omega_7 I}{2} I\right\},
$$

$$
\Pi_{33} = \mathrm{diag}\left\{\frac{-\eta_1 I}{2}, \frac{-\eta_1 I}{2}, \frac{-\eta_2 I}{2}, \frac{-\eta_2 I}{2}, \frac{\lambda_2^2 I - 2\lambda_2 X_2}{\eta_1 p_2}, \frac{\lambda_3^2 I - 2\lambda_3 X_2}{\eta_2 p_2}\right\},
$$

with $\tilde{\Pi}_{11} = \mathrm{sym}\{\hat{A}X_1 + \hat{B}Y\} + (\omega_1 + \omega_2 + \omega_3)\hat{G}_0 M M^{\mathrm{T}} \hat{G}_0^{\mathrm{T}}$ *and* $\tilde{\Pi}_{22} = \mathrm{sym}\{\hat{A}X_2 + Z\} + (\omega_4 + \omega_5 + \omega_6)M M^{\mathrm{T}} + \omega_7 I$. *In addition, the gain matrices can be obtained as* $K = Y X_1^{-1}$ *and* $L = Z(C X_2)^{\dagger}$, *respectively. Therefore, the sliding mode dynamics (9.8) is stable under the presented CETC (9.5).*

Proof. Multiplying both sides of (9.9) with matrix

$$\mathrm{diag}\left\{ P_1^{-1}, P_2^{-1}, P_2^{-1}, \underbrace{I, \ldots, I}_{14\text{ times}}, P_2^{-1}, P_2^{-1} \right\},$$

and letting $X_1 = P_1^{-1}$, $X_2 = P_2^{-1}$, $Y = KX_1$, as well as $Z = LCX_2$, one obtains

$$\begin{bmatrix} \boldsymbol{\Phi}_{11} & \boldsymbol{\Pi}_{12} & \boldsymbol{\Phi}_{13} \\ * & \boldsymbol{\Pi}_{22} & 0 \\ * & * & \boldsymbol{\Phi}_{33} \end{bmatrix} < 0, \tag{9.12}$$

where

$$\boldsymbol{\Phi}_{11} = \begin{bmatrix} \tilde{\boldsymbol{\Phi}}_{11} & -\hat{G}_0 LCX_2 & -\hat{G}_0 Z \\ * & \tilde{\boldsymbol{\Phi}}_{22} & Z \\ * & * & -X_1 X_1 \end{bmatrix},$$

$$\boldsymbol{\Phi}_{13} = \begin{bmatrix} X_1 C^{\mathrm{T}} \hat{F}_2^{\mathrm{T}} & 0 & X_1 C^{\mathrm{T}} \hat{F}_2^{\mathrm{T}} & 0 & \hat{G}_0 Z & 0 \\ 0 & X_2 C^{\mathrm{T}} \hat{F}_2^{\mathrm{T}} & 0 & X_2 C^{\mathrm{T}} \hat{F}_2^{\mathrm{T}} & 0 & Z \\ 0 & 0 & 0 & 0 & 0 & 0 \end{bmatrix},$$

$$\boldsymbol{\Phi}_{33} = \mathrm{diag}\left\{ \frac{-\eta_1 I}{2}, \frac{-\eta_1 I}{2}, \frac{-\eta_2 I}{2}, \frac{-\eta_2 I}{2}, \frac{-X_2 X_2}{\eta_1 p_2}, \frac{-X_2 X_2}{\eta_2 p_2} \right\},$$

with $\tilde{\boldsymbol{\Phi}}_{11} = \mathrm{sym}\{\hat{A}X_1 + \hat{B}Y\} + (\omega_1 + \omega_2 + \omega_3)\hat{G}_0 MM^{\mathrm{T}}\hat{G}_0^{\mathrm{T}} + \theta_1(\kappa)X_1 X_1$ and $\tilde{\boldsymbol{\Phi}}_{22} = \mathrm{sym}\{\hat{A}X_2 + Z\} + (\omega_4 + \omega_5 + \omega_6)MM^{\mathrm{T}} + \omega_7 I$. Meanwhile, note that there exist positive scalars λ_j such that $-X_1 X_1 \leq \lambda_1^2 I - 2\lambda_1 X_1$, $-X_2 X_2 \leq \lambda_2^2 I - 2\lambda_2 X_2$, and $-X_2 X_2 \leq \lambda_3^2 I - 2\lambda_3 X_2$ hold. Therefore, by using the famous Schur complement lemma, one can obtain that (9.12) is equivalent to (9.11). Therefore, the sliding mode dynamics (9.8) is stable under the presented CETC (9.5). $\qquad\square$

9.3.3 Reachability analysis

For the ith system, the event-triggered SMC law is designed as follows:

$$u_i(t) = K_i \hat{x}_i(t_k^i) - \zeta\,\mathrm{sgn}(s_i(t_k^i)), \tag{9.13}$$

where $\zeta > 0$, $t \in [t_k^i, t_{k+1}^i)$, and $k = 0, 1, 2, \ldots$

Remark 9.6. Unlike traditional SMC strategies in [41,42], the biggest problem is how to deal with $s_i^{\mathrm{T}}(t)\mathrm{sgn}(s_i(t_k^i))$. So far, some significant contributions have been published in [43–46]. Based on the analysis and results in the above literature, one crucial lemma in the following can be established if the trigger mechanism is given.

Lemma 9.1. *If* $\operatorname{sgn}((s_i)_r(t)) = \operatorname{sgn}((s_i)_r(t_k^i))$ $(r = 1, 2, \ldots, m)$ *holds for any* $t \in [t_k^i, t_{k+1}^i)$, *then the observer system trajectory* $\hat{x}_i(t)$ *will stay out of the following region:*

$$\mathfrak{B} = \{\hat{x}_i(t) \mid \|s(t)\| < \sqrt{\theta_{1i}(\kappa)}\|\hat{x}_i(t)\| + \sqrt{\theta_{2i}(\kappa)}(e^{-\varepsilon_i t} + \rho_i)^{\frac{1}{2}}\}.$$

Theorem 9.3. *For the given system* (9.1), *the nonfragile observer* (9.2), *the sliding surface function* (9.4), *and the event-triggered controller* (9.13) *is updated to the* i*th system at the triggering sequence* $\{t_k^i\}_{k=0}^{+\infty}$. *If there exist positive scalars* ϵ_{1i}, ϵ_{2i}, *and* ζ *such that*

$$\begin{cases} (\|GL_iC_i\| + \|GM_i\|\|N_iC_i\|)\|e_i(t)\| \leq \epsilon_{1i}, \\ (\|K_i\| + \|GL_iC_i\| + \|GM_i\|\|N_iC_i\|)\|\hat{e}_i(t)\| \leq \epsilon_{2i}, \\ \zeta > (\|GL_i\| + \|GM_i\|\|N_i\|)\upsilon + \epsilon_{1i} + \epsilon_{2i} \end{cases} \tag{9.14}$$

hold, then the practical sliding mode occurs within the region $\mathfrak{B}$.

Proof. Construct the following Lyapunov function:

$$\tilde{V}_i(t) = \frac{1}{2}s_i^{\mathrm{T}}(t)s_i(t). \tag{9.15}$$

Then, taking the derivative of (9.15) of order α and combining with controller (9.13), we have

$$\begin{aligned} {}_{t_0}^{C}D_t^\alpha \tilde{V}_i(t) &\leq s_i^{\mathrm{T}}(t){}_{t_0}^{C}D_t^\alpha s_i(t) \\ &= s_i^{\mathrm{T}}(t)\Big\{ -G(L_i + \Delta L_i)C_i e_i(t) - [K_i + G(L_i + \Delta L_i)C_i]\hat{e}_i(t) \\ &\quad - G(L_i + \Delta L_i)\gamma_i(t)g(y_i(t)) - \zeta\operatorname{sgn}(s_i(t_k^i))\Big\} \\ &\leq \|s_i(t)\|\Big\{(\|GL_iC_i\| + \|GM_i\|\|N_iC_i\|)\|e_i(t)\| \\ &\quad + (\|K_i\| + \|GL_iC_i\| + \|GM_i\|\|N_iC_i\|)\|\hat{e}_i(t)\| \\ &\quad + (\|GL_i\| + \|GM_i\|\|N_i\|)\upsilon\Big\} \\ &\quad - \zeta s_i^{\mathrm{T}}(t)\operatorname{sgn}(s_i(t_k^i)) \\ &\leq \|s_i(t)\|\Big\{\epsilon_{1i} + \epsilon_{2i} + (\|GL_i\| + \|GM_i\|\|N_i\|)\upsilon\Big\} \\ &\quad - \zeta s_i^{\mathrm{T}}(t)\operatorname{sgn}(s_i(t_k^i)). \end{aligned}$$

Considering Lemma 9.1, for any $t \in [t_k^i, t_{k+1}^i)$, when the observer state $\hat{x}_i(t)$ stays outside the region $\mathfrak{B}$, one has $\operatorname{sgn}((s_i)_r(t)) = \operatorname{sgn}((s_i)_r(t_k^i))$, therefore,

$$ {}_{t_0}^{C}D_t^\alpha \tilde{V}_i(t) \leq -\Big\{\zeta - \epsilon_{1i} - \epsilon_{2i} - (\|GL_i\| + \|GM_i\|\|N_i\|)\upsilon\Big\}\|s_i(t)\|. $$

In addition, $_{t_0}^{C}D_t^{\alpha}\tilde{V}_i(t) \leq 0$ means that the state trajectory will move towards the neighborhood of $s_i(t) = 0$. Therefore, if $\hat{x}_i(t)$ reaches the sliding surface, $\text{sgn}((s_i)_r(t)) \neq \text{sgn}((s_i)_r(t_k^i))$, i.e., $_{t_0}^{C}D_t^{\alpha}\tilde{V}_i(t) \leq 0$ will not hold. Hence, the trajectories of system will move away from the sliding surface and stay inside a certain region, that is,

$$\|s_i(t) - s_i(t_k^i)\| < \sqrt{\theta_{1i}(\kappa)}\|\hat{x}_i(t)\| + \sqrt{\theta_{2i}(\kappa)}(e^{-\varepsilon_i t} + \rho_i)^{\frac{1}{2}}.$$

By setting $s_i(t_k^i) = \mathbf{0}$, the maximum value of the system trajectory can be obtained. In conclusion, the sliding surface will occur and the boundary will be given by $\mathfrak{B} = \{\hat{x}_i(t)|\|s(t)\| < \sqrt{\theta_{1i}(\kappa)}\|\hat{x}_i(t)\| + \sqrt{\theta_{2i}(\kappa)}(e^{-\varepsilon_i t} + \rho_i)^{\frac{1}{2}}\}$.
$\square$

Remark 9.7. Lemma 9.1 provides a bound of sliding surface to ensure the existence of practical sliding mode function. Meanwhile, Theorem 9.3 indicates that the boundedness of FO-CNSs (9.1) can be obtained by adjusting threshold parameters. Particularly, if one chooses $\theta_{1i}(\kappa) = 0$ and $\varepsilon_i = 0$, then the CETC (9.5) will reduce to the constant triggering condition in [44,46]. Therefore, the presented combination triggering mechanism in this chapter includes those in [44,46].

9.3.4 Discussion of Zeno behavior

In order to successfully discuss the Zeno behavior of FO-CNSs, a combination self-triggered condition (CSTC) is first given in this section.

Corollary 9.1. *Define a new CSTC as follows:*

$$t_{k+1}^i = \min\{t_k^i + \nabla, \hat{\mathfrak{t}}_{k+1}^i\}, \tag{9.16}$$

where

$$\hat{\mathfrak{t}}_{k+1}^i \triangleq \inf\left\{t > t_k^i \,\Big|\, \|\hat{e}_i(t)\|^2 \geq \tilde{\xi}_{1i}(t_k^i) \vee |(\hat{s}_i)_r(t)| \geq \tilde{\xi}_{2i}(t_k^i)\right\},$$

$$\tilde{\xi}_{1i}(t_k^i) = \tilde{\theta}_{1i}(\kappa)\|\hat{x}_i(t_k^i)\|^2 + \tilde{\theta}_{2i}(\kappa)(e^{-\varepsilon_i(t_k^i + \nabla)} + \rho_i),$$

$$\tilde{\xi}_{2i}(t_k^i) = \frac{\sqrt{\tilde{\theta}_{1i}(\kappa)}}{\sqrt{m}}\|\hat{x}_i(t_k^i)\| + \frac{\sqrt{\tilde{\theta}_{2i}(\kappa)}}{\sqrt{m}}(e^{-\varepsilon_i(t_k^i + \nabla)} + \rho_i)^{\frac{1}{2}},$$

with $\tilde{\theta}_{1i}(\kappa) = \frac{\theta_{1i}(\kappa)}{2 + 2\theta_{1i}(\kappa)}$ and $\tilde{\theta}_{2i}(\kappa) = \frac{\theta_{2i}(\kappa)}{1 + \theta_{1i}(\kappa)}$, respectively. Then the stability of system (9.8) can be also guaranteed under the presented CSTC (9.16).

Proof. First, considering system (9.8), for any $t \in [t_k^i, t_{k+1}^i)$, from CSTC (9.16), one has that

$$\|\hat{e}_i(t)\|^2 < \tilde{\xi}_{1i}(t_k^i) \leq 2\tilde{\theta}_{1i}(\kappa)(\|\hat{x}_i(t)\|^2 + \|\hat{e}_i(t)\|^2) + \tilde{\theta}_{2i}(\kappa)(e^{-\varepsilon_i t} + \rho_i), \tag{9.17}$$

therefore, combining with the form of $\tilde{\theta}_{1i}(\kappa)$ and $\tilde{\theta}_{2i}(\kappa)$, the latter can be rewritten as

$$\|\hat{\boldsymbol{e}}_i(t)\|^2 < \frac{2\tilde{\theta}_{1i}(\kappa)}{1-2\tilde{\theta}_{1i}(\kappa)}\|\hat{\boldsymbol{x}}_i(t)\|^2 + \frac{\tilde{\theta}_{2i}(\kappa)}{1-2\tilde{\theta}_{1i}(\kappa)}(e^{-\varepsilon_i t}+\rho_i)$$
$$= \theta_{1i}(\kappa)\|\hat{\boldsymbol{x}}_i(t)\|^2 + \theta_{2i}(\kappa)(e^{-\varepsilon_i t}+\rho_i). \tag{9.18}$$

Second, combining with the CSTC (9.16) and inequality (9.18), by using Lemma 2.10, one has

$$|(\hat{\boldsymbol{s}}_i)_r(t)| < \tilde{\xi}_{2i}(t_k^i)$$
$$\leq \frac{\sqrt{\tilde{\theta}_{1i}(\kappa)}}{\sqrt{m}}\left(2\|\hat{\boldsymbol{x}}_i\|^2 + 2\|\hat{\boldsymbol{e}}_i(t)\|\right)^{\frac{1}{2}} + \frac{\sqrt{\tilde{\theta}_{2i}(\kappa)}}{\sqrt{m}}(e^{-\varepsilon_i t}+\rho_i)^{\frac{1}{2}}$$
$$< \frac{\sqrt{\tilde{\theta}_{1i}(\kappa)}}{\sqrt{m}}\left(\frac{2}{1-2\tilde{\theta}_{1i}(\kappa)}\|\hat{\boldsymbol{x}}_i(t)\|^2 + \frac{2\tilde{\theta}_{2i}(\kappa)}{1-2\tilde{\theta}_{1i}(\kappa)}(e^{-\varepsilon_i t}+\rho_i)\right)^{\frac{1}{2}}$$
$$\quad + \frac{\sqrt{\tilde{\theta}_{2i}(\kappa)}}{\sqrt{m}}(e^{-\varepsilon_i t}+\rho_i)^{\frac{1}{2}}$$
$$\leq \frac{\sqrt{\tilde{\theta}_{1i}(\kappa)}}{\sqrt{m}}\left(\frac{\sqrt{2}}{\sqrt{1-2\tilde{\theta}_{1i}(\kappa)}}\|\hat{\boldsymbol{x}}_i(t)\| + \frac{\sqrt{2\tilde{\theta}_{2i}(\kappa)}}{\sqrt{1-2\tilde{\theta}_{1i}(\kappa)}}(e^{-\varepsilon_i t}+\rho_i)^{\frac{1}{2}}\right)$$
$$\quad + \frac{\sqrt{\tilde{\theta}_{2i}(\kappa)}}{\sqrt{m}}(e^{-\varepsilon_i t}+\rho_i)^{\frac{1}{2}}$$
$$= \frac{1}{\sqrt{m}}\frac{\sqrt{2\tilde{\theta}_{1i}(\kappa)}}{\sqrt{1-2\tilde{\theta}_{1i}(\kappa)}}\|\hat{\boldsymbol{x}}_i(t)\|$$
$$\quad + \frac{1}{\sqrt{m}}\left(\frac{\sqrt{2\tilde{\theta}_{1i}(\kappa)\tilde{\theta}_{2i}(\kappa)}}{\sqrt{1-2\tilde{\theta}_{1i}(\kappa)}} + \sqrt{\tilde{\theta}_{2i}(\kappa)}\right)(e^{-\varepsilon_i t}+\rho_i)^{\frac{1}{2}}$$
$$\leq \frac{1}{\sqrt{m}}\frac{\sqrt{2\tilde{\theta}_{1i}(\kappa)}}{\sqrt{1-2\tilde{\theta}_{1i}(\kappa)}}\|\hat{\boldsymbol{x}}_i(t)\| + \frac{1}{\sqrt{m}}\frac{\sqrt{\tilde{\theta}_{2i}(\kappa)}}{\sqrt{1-2\tilde{\theta}_{1i}(\kappa)}}(e^{-\varepsilon_i t}+\rho_i)^{\frac{1}{2}}.$$

Therefore, combining with the form of $\tilde{\theta}_{1i}(\kappa)$ and $\tilde{\theta}_{2i}(\kappa)$, we can also obtain

$$|(\hat{\boldsymbol{s}}_i)_r(t)| \leq \frac{1}{\sqrt{m}}\frac{\sqrt{2\tilde{\theta}_{1i}(\kappa)}}{\sqrt{1-2\tilde{\theta}_{1i}(\kappa)}}\|\hat{\boldsymbol{x}}_i(t)\| + \frac{1}{\sqrt{m}}\frac{\sqrt{\tilde{\theta}_{2i}(\kappa)}}{\sqrt{1-2\tilde{\theta}_{1i}(\kappa)}}(e^{-\varepsilon_i t}+\rho_i)^{\frac{1}{2}}$$
$$= \frac{\sqrt{\theta_{1i}(\kappa)}}{\sqrt{m}}\|\hat{\boldsymbol{x}}_i(t)\| + \frac{\sqrt{\theta_{2i}}}{\sqrt{m}}(e^{-\varepsilon_i t}+\rho_i)^{\frac{1}{2}}. \tag{9.19}$$

In summary, under the presented CSTC (9.16), the stability of system (9.8) can be also guaranteed through Theorem 9.2. $\qquad\square$

Remark 9.8. It is easy to see from the above analysis that Corollary 9.1 provides a necessary condition of the presented CETC (9.5). Therefore, by utilizing CSTC (9.16), it can also guarantee the stability of FO-CNSs (9.1).

Remark 9.9. Just as mentioned in Remark 2.1, in some related literature, the Zeno behavior of fractional-order control system has not been adequately considered. For example, the authors of [31] gave the following minimum inter-event time $T_{\min}$ to illustrate that Zeno behavior can be avoided:

$$T_{\min} = t_{k+1} - t_k > \left(\frac{\Gamma(1+\alpha)\ln(\sqrt{\rho}\|\hat{x}_{t_k}\|)}{\|A\| + \beta(\|\hat{x}_{t_k}\|)} \right)^{\frac{1}{\alpha}}, \tag{9.20}$$

where $\beta(\|\hat{x}_{t_k}\|) \geq 0$, $\rho > 0$, $k = 0, 1, 2, \ldots$ However, from (9.20), it is easily obtained that $T_{\min} > 0$ is not guaranteed when $\|\hat{x}_{t_k}\| \to 0$. Therefore, to solve this problem, Theorem 9.4 is proposed.

Theorem 9.4. *For the given control strategy, the positive lower bound of inter-event interval satisfies*

$$T_{\min}^i = \min_k\{t_{k+1}^i - t_k^i\} = \min_k\{\hat{T}_k^i, \tilde{T}_k^i, \nabla\} > 0, \quad k = 0, 1, \ldots, \tag{9.21}$$

where $\hat{T}_k^i$ and $\tilde{T}_k^i$ are given in (9.28) and (9.32), respectively.

Proof. First, if $T_{\min}^i = \nabla$, then the system clearly does not experience Zeno behavior.

Second, assuming that $T_{\min}^i < \nabla$, by Theorem 9.1, we have $\xi(t) \to 0$ when $t \to +\infty$, i.e., there exists a positive constant δ_1 such that $\|\hat{x}(t)\| < \delta_1$ and $\|e(t)\| < \delta_1$. Consider the following:

$$\begin{aligned}
{}_{t_0}^{C}D_t^\alpha \hat{e}_i(t) &= {}_{t_0}^{C}D_t^\alpha \hat{x}_i(t) \\
&= A\hat{x}_i(t) + \sum_{j=1}^{\mathcal{N}} h_{ij}\Gamma\hat{x}_j(t) + B(K_i\hat{x}_i(t_k^i) - \zeta\,\mathrm{sgn}(s_i(t_k^i))) \\
&\quad - (L_i + \Delta L_i)(\tilde{y}_i(t) - \hat{y}_i(t)) \\
&= (A - (L_i + \Delta L_i)C_i)\hat{x}_i(t) + \sum_{j=1}^{\mathcal{N}} h_{ij}\Gamma\hat{x}_j(t) \\
&\quad + (BK_i - (L_i + \Delta L_i)C_i)\hat{x}_i(t_k^i) - (L_i + \Delta L_i)C_i e_i(t) \\
&\quad - \zeta B\,\mathrm{sgn}(s_i(t_k^i)) - (L_i + \Delta L_i)\gamma_i(t)g(y_i(t)).
\end{aligned} \tag{9.22}$$

By using the Kronecker product, Eq. (9.22) becomes

$$\begin{aligned}
{}_{t_0}^{C}D_t^\alpha \hat{e}(t) &= (\hat{A} - (L + \Delta L)C)\hat{x}(t) + (\hat{B}K - (L + \Delta L)C)\hat{x}(t_k^i) \\
&\quad - (L + \Delta L)Ce(t) - \zeta\hat{B}\,\mathrm{sgn}(s(t_k)) - (L + \Delta L)\hat{\gamma}(t)g(y(t)).
\end{aligned}$$

Therefore, for the (r')th equation, one has

$$|{}^{\mathcal{C}}_{t_0}D_t^\alpha(\hat{e})_{r'}(t)| \le \|(\hat{A} - (L + \Delta L)C)_{r'}\|\|\hat{x}(t)\|$$
$$+ \|(\hat{B}K - (L + \Delta L)C)_{r'}\|\|\hat{x}(t_k)\|$$
$$+ \|((L + \Delta L)C)_{r'}\|\|e(t)\| + \zeta\|(\hat{B})_{r'}\| + \|(L + \Delta L)_{r'}\|\upsilon$$
$$\triangleq (\varphi)_{r'}(t_k), \quad r' = 1, 2, \ldots, n\mathcal{N}.$$

Meanwhile, by using the Laplace transform and final value theorem, it is easily obtained that there exists a positive constant δ_2 such that $\|\hat{e}_i'(t)\| \le \delta_2$. And from $\hat{e}_i(t) = \hat{x}_i(t) - \hat{x}_i(t_k^i)$ one has

$$^{\mathcal{C}}_{t_k^i}D_t^\alpha\|\hat{e}_i(t)\| \le \|{}^{\mathcal{C}}_{t_k^i}D_t^\alpha\hat{e}_i(t)\| = \left\|{}^{\mathcal{C}}_{t_0}D_t^\alpha\hat{e}_i(t) - \frac{1}{\Gamma(1-\alpha)}\int_{t_0}^{t_k^i}\frac{\hat{e}_i'(\tau)}{(t-\tau)^\alpha}d\tau\right\|$$

$$\le \|{}^{\mathcal{C}}_{t_0}D_t^\alpha\hat{e}_i(t)\| + \frac{1}{\Gamma(1-\alpha)}\int_{t_0}^{t_k^i}\frac{\|\hat{e}_i'(\tau)\|}{(t-\tau)^\alpha}d\tau$$

$$\le \left\|(A - (L_i + \Delta L_i)C_i)\hat{e}_i(t) + (A + BK_i)\hat{x}_i(t_k^i)\right.$$

$$+ \sum_{j=1}^{\mathcal{N}}h_{ij}\Gamma x_j(t) - \zeta B\mathrm{sgn}(s_i(t_k^i))$$

$$\left. - (L_i + \Delta L_i)(C_ie_i(t) + \gamma(t)g(y_i(t)))\right\| + \frac{1}{\Gamma(1-\alpha)}\int_{t_0}^{t_k^i}\frac{\delta_2}{(t-\tau)^\alpha}d\tau$$

$$\le \|A - (L_i + \Delta L_i)C_i\|\|\hat{e}_i(t)\| + \|A + BK_i\|\|\hat{x}_i(t_k^i)\|$$

$$+ \sum_{j=1}^{\mathcal{N}}|h_{ij}|\|\Gamma\|\delta_1 + \zeta\|B\| + \|L_i + \Delta L_i\|(\|C_i\|\delta_1 + \upsilon_i)$$

$$+ \frac{\delta_2}{\Gamma(2-\alpha)}\nabla^{1-\alpha}$$

$$\triangleq \|\tilde{L}_i\|\|\hat{e}_i(t)\| + \Upsilon(\nabla),$$

$$(9.23)$$

where $\|\tilde{L}_i\| = \|A - (L_i + \Delta L_i)C_i\|$ and $\Upsilon(\nabla) = \|A + BK_i\|\|\hat{x}_i(t_k^i)\| + \sum_{j=1}^{\mathcal{N}}|h_{ij}|\|\Gamma\|\delta_1 + \zeta\|B\| + \|L_i + \Delta L_i\|(\|C_i\|\delta_1 + \upsilon_i) + \frac{\delta_2}{\Gamma(2-\alpha)}\nabla^{1-\alpha}$. Then, taking the integral of (9.23) of order α from t_k^i to t, we have

$$_{t_k^i}D_t^{-\alpha}({}^{\mathcal{C}}_{t_k^i}D_t^\alpha\|\hat{e}_i(t)\|) \le {}_{t_k^i}D_t^{-\alpha}(\|\tilde{L}_i\|\|\hat{e}_i(t)\| + \Upsilon(\nabla)).$$

In addition, from Lemma 2.4 and since $\|\hat{e}_i(t)\| = 0$ at $t = t_k^i$, one has

$$\|\hat{e}_i(t)\| - \|\hat{e}_i(t_k^i)\| \le \frac{1}{\Gamma(\alpha)}\int_{t_k^i}^{t}\frac{\|\tilde{L}_i\|\|\hat{e}_i(\tau)\| + \Upsilon(\nabla)}{(t - \tau)^{1-\alpha}}d\tau.$$

Denoting $\varrho(t) = \|\tilde{L}_i\|\|\hat{e}_i(t)\| + \Upsilon(\nabla)$, then we have

$$\varrho(t) \leq \frac{\|\tilde{L}_i\|}{\Gamma(\alpha)} \int_{t_k^i}^{t} \frac{\varrho(\tau)}{(t-\tau)^{1-\alpha}} d\tau + \Upsilon(\nabla). \tag{9.24}$$

According to Lemma 2.9, we have

$$\varrho(t) \leq \Upsilon(\nabla) \exp\left(\frac{\|\tilde{L}_i\|}{\Gamma(1+\alpha)}(t-t_k^i)^\alpha\right). \tag{9.25}$$

Two important results can be obtained from (9.25), namely

(I) for given $t \in [t_k^i, t_{k+1}^i)$, one has

$$\begin{aligned}
\|\hat{e}_i(t)\| &\leq \frac{\Upsilon(\nabla)}{\|\tilde{L}_i\|} \exp\left(\frac{\|\tilde{L}_i\|}{\Gamma(1+\alpha)}(t-t_k^i)^\alpha\right) - \frac{\Upsilon(\nabla)}{\|\tilde{L}_i\|} \\
&\leq \frac{\Upsilon(\nabla)}{\|\tilde{L}_i\|}\left[\exp\left(\frac{\|\tilde{L}_i\|}{\Gamma(1+\alpha)}\nabla^\alpha\right) - 1\right];
\end{aligned} \tag{9.26}$$

(II) for given $t \in [t_k^i, t_{k+1}^i)$, one has

$$t - t_k^i \geq \left\{\frac{\Gamma(1+\alpha)}{\|\tilde{L}_i\|} \ln\left(\frac{\|\tilde{L}_i\|}{\Upsilon(\nabla)}\|\hat{e}_i(t)\| + 1\right)\right\}^{\frac{1}{\alpha}}. \tag{9.27}$$

Furthermore, according to CSTC (9.16), we have $\|\hat{e}_i(t_{k+1}^i)\|^2 = \tilde{\theta}_{1i}(\kappa)\|\hat{x}_i(t_k^i)\|^2 + \tilde{\theta}_{2i}(\kappa)(e^{-\varepsilon_i(t_k^i+\nabla)} + \rho)$, which means that the minimum inter-event time $\hat{T}_k^i$ satisfies

$$\begin{aligned}
\hat{T}_k^i &\geq \left\{\frac{\Gamma(1+\alpha)}{\|\tilde{L}_i\|} \ln\left(\frac{\|\tilde{L}_i\|}{\Upsilon(\nabla)}(\tilde{\theta}_{1i}(\kappa)\|\hat{x}_i(t_k^i)\|^2 + \tilde{\theta}_{2i}(\kappa)(e^{-\varepsilon_i(t_k^i+\nabla)} + \rho))^{\frac{1}{2}} + 1\right)\right\}^{\frac{1}{\alpha}} \\
&\geq \left\{\frac{\Gamma(1+\alpha)}{\|\tilde{L}_i\|} \ln\left(\frac{\sqrt{2}}{2}\frac{\|\tilde{L}_i\|}{\Upsilon(\nabla)}(\sqrt{\tilde{\theta}_{1i}(\kappa)}\|\hat{x}_i(t_k^i)\| + \sqrt{\tilde{\theta}_{2i}(\kappa)}(e^{-\varepsilon_i(t_k^i+\nabla)} + \rho)^{\frac{1}{2}}) + 1\right)\right\}^{\frac{1}{\alpha}} \\
&> 0.
\end{aligned} \tag{9.28}$$

Third, since $\hat{x}_i(t) = \hat{e}_i(t) + \hat{x}_i(t_k^i)$, we have

$$\begin{aligned}
|(\hat{s}_i)_r(t)| &< \xi_{2i}(t) \\
&< \frac{\sqrt{\theta_{1i}(\kappa)}}{\sqrt{m}}(\|\hat{e}_i(t)\| + \|\hat{x}_i(t_k^i)\|) + \frac{\sqrt{\theta_{2i}(\kappa)}}{\sqrt{m}}(e^{-\varepsilon_i t} + \rho)^{\frac{1}{2}} \\
&\leq \frac{\sqrt{\theta_{1i}(\kappa)}}{\sqrt{m}}(\|\hat{e}_i(t)\| + \|\hat{x}_i(t_k^i)\|) + \frac{\sqrt{\theta_{2i}(\kappa)}}{\sqrt{m}}(e^{-\frac{\varepsilon_i}{2}t} + \sqrt{\rho}).
\end{aligned}$$

Then, for given $t \in [t_k^i, t_{k+1}^i)$, considering the fractional-order integral of $|(\hat{s}_i)_r(t)|$ and noting that $\,^C_{t_k^i}D_t^\alpha e^{-\frac{\varepsilon_i}{2}t} \leq 0$, we have

$$
\begin{aligned}
\,^C_{t_k^i}D_t^\alpha |(\hat{s}_i)_r(t)| &< \frac{\sqrt{\theta_{1i}(\kappa)}}{\sqrt{m}} \,^C_{t_k^i}D_t^\alpha \|\hat{e}_i(t)\| \\
&\leq \frac{\sqrt{\theta_{1i}(\kappa)}}{\sqrt{m}} (\|\tilde{L}_i\| \|\hat{e}_i(t)\| + \Upsilon(\nabla)) \\
&\leq \frac{\sqrt{\theta_{1i}(\kappa)}}{\sqrt{m}} \left\{ \|\tilde{L}_i\| \frac{\Upsilon(\nabla)}{\|\tilde{L}_i\|} \left[\exp\left(\frac{\|\tilde{L}_i\|}{\Gamma(1+\alpha)} \nabla^\alpha \right) - 1 \right] + \Upsilon(\nabla) \right\} \\
&\leq \frac{\sqrt{\theta_{1i}(\kappa)}}{\sqrt{m}} \Upsilon(\nabla) \exp\left(\frac{\|\tilde{L}_i\|}{\Gamma(1+\alpha)} \nabla^\alpha \right).
\end{aligned}
\tag{9.29}
$$

Taking the integral of (9.29) from t_k^i to t, since $|(\hat{s}_i)_r(t_k^i)| = 0$, one has

$$
|(\hat{s}_i)_r(t)| < \frac{\sqrt{\theta_{1i}(\kappa)}}{\sqrt{m}} \frac{\Upsilon(\nabla)}{\Gamma(1+\alpha)} \exp\left(\frac{\|\tilde{L}_i\|}{\Gamma(1+\alpha)} \nabla^\alpha \right) (t - t_k^i)^\alpha,
\tag{9.30}
$$

that is to say,

$$
t - t_k^i > \left\{ \frac{\sqrt{m}}{\sqrt{\theta_{1i}(\kappa)}} \frac{\Gamma(1+\alpha)}{\Upsilon(\nabla)} |(\hat{s}_i)_r(t)| \exp\left(-\frac{\|\tilde{L}_i\|}{\Gamma(1+\alpha)} \nabla^\alpha \right) \right\}^{\frac{1}{\alpha}}.
\tag{9.31}
$$

Furthermore, according to CSTC (9.16), we have

$$
|(\hat{s}_i)_r(t_{k+1}^i)| = \frac{\sqrt{\tilde{\theta}_{1i}(\kappa)}}{\sqrt{m}} \|\hat{x}_i(t_k^i)\| + \frac{\sqrt{\tilde{\theta}_{2i}(\kappa)}}{\sqrt{m}} (e^{-\varepsilon_i(t_k^i + \nabla)} + \rho)^{\frac{1}{2}} \},
$$

which implies that the minimum inter-event time $\tilde{T}_k^i$ satisfies

$$
\begin{aligned}
\tilde{T}_k^i &> \left\{ \frac{\sqrt{m}}{\sqrt{\theta_{1i}(\kappa)}} \frac{\Gamma(1+\alpha)}{\Upsilon(\nabla)} \left[\frac{\sqrt{\tilde{\theta}_{1i}(\kappa)}}{\sqrt{m}} \|\hat{x}_i(t_k^i)\| + \frac{\sqrt{\tilde{\theta}_{2i}(\kappa)}}{\sqrt{m}} (e^{-\varepsilon_i(t_k^i + \nabla)} + \rho)^{\frac{1}{2}} \right] \right. \\
&\quad \left. \times \exp\left(-\frac{\|\tilde{L}_i\|}{\Gamma(1+\alpha)} \nabla^\alpha \right) \right\}^{\frac{1}{\alpha}} \\
&= \left\{ \frac{\Gamma(1+\alpha)}{\Upsilon(\nabla)} \left[\frac{\sqrt{\tilde{\theta}_{1i}(\kappa)}}{\sqrt{\theta_{1i}(\kappa)}} \|\hat{x}_i(t_k^i)\| + \frac{\sqrt{\tilde{\theta}_{2i}(\kappa)}}{\sqrt{\theta_{1i}(\kappa)}} (e^{-\varepsilon_i(t_k^i + \nabla)} + \rho)^{\frac{1}{2}} \right] \right. \\
&\quad \left. \times \exp\left(-\frac{\|\tilde{L}_i\|}{\Gamma(1+\alpha)} \nabla^\alpha \right) \right\}^{\frac{1}{\alpha}} \\
&\geq 0.
\end{aligned}
\tag{9.32}
$$

There exists a minimum lower bound $T_{\min}^i$ which satisfies

$$
T_{\min}^i = \min_k \{t_{k+1}^i - t_k^i\} = \min_k \{\hat{T}_k^i, \tilde{T}_k^i, \nabla\} > 0, \quad k = 0, 1, \dots
\tag{9.33}
$$

In conclusion, Zeno behavior can be effectively avoided. $\qquad\square$

Remark 9.10. Compared with the result in [31], there exist two advantages in our work. One has been illustrated in Remark 2.1; another is the obtained inequality (9.25) from (9.24) by utilizing the Gronwall–Bellman inequality. In [31], Feng et al. gained the result about $\|\hat{e}_i(t)\|$ directly from (9.24). Hence, they derived (9.20). As mentioned in Remark 9.7, this approach may be questionable. Therefore, we design an auxiliary variable $\varrho(t)$ first and then obtain the result about $\|\hat{e}_i(t)\|$ in this chapter. This way, we can guarantee that there exists a positive lower bound $T_{\min}^i$, i.e., $T_{\min}^i > 0$ holds.

Remark 9.11. It is well known that for integer-order complex network control systems, the theoretical research framework of the event-triggered control problem has been well studied, but for fractional-order complex network systems, due to the nonlocal characteristics of fractional-order calculus operators, it often gives wrong conclusions in the study of Zeno behavior, as shown in Remark 2.1. Based on the above analysis, this chapter designed a combination event-triggering mechanism containing a threshold parameter to effectively avoid Zeno behavior of the closed-loop system, which is one of the most important innovations of this chapter.

9.4 Numerical example

Example 9.1. Consider the following FO-CNSs consisting of three nodes:

$$
\begin{cases}
{}^{C}_{t_0}D_t^{\alpha}x_i(t) = Ax_i(t) + \sum_{j=1}^{3} h_{ij}\Gamma x_j(t) + \beta_i(t)f(x_i(t)) + Bu_i(t), \\
y_i(t) = C_i x_i(t), \\
\tilde{y}_i(t) = y_i(t) + \gamma_i(t)g(y_i(t)), \quad i = 1, 2, 3,
\end{cases}
\tag{9.34}
$$

where $x_i(t) = \begin{bmatrix} x_{i1}(t) \\ x_{i2}(t) \\ x_{i3}(t) \end{bmatrix}$, $A = \begin{bmatrix} -0.872 & 0.676 & 1.773 \\ -1.548 & -1.195 & 1.686 \\ 0.360 & 1.788 & -0.893 \end{bmatrix}$, $H = [h_{ij}]_{3\times3} = \begin{bmatrix} -1 & 0 & 1 \\ 0 & -1 & 1 \\ 1 & 1 & -2 \end{bmatrix}$, $\Gamma = \begin{bmatrix} 1 & 0 & 0 \\ 0 & 1 & 0 \\ 0 & 0 & 1 \end{bmatrix}$, $B = \begin{bmatrix} 1 \\ 1 \\ 1 \end{bmatrix}$, $C_1 = \begin{bmatrix} 0 \\ 1 \\ 1 \end{bmatrix}$, $C_2 = \begin{bmatrix} 1 \\ 0 \\ 1 \end{bmatrix}$, $C_3 = \begin{bmatrix} 1 \\ 1 \\ 0 \end{bmatrix}$, $f(x_i) = \begin{bmatrix} f_1(x_{i1}) \\ 0 \\ 0 \end{bmatrix}$, with $f_1(x_{i1}(t)) = 0.1x_{i1}(t) + 0.1(|x_{i1}(t) + 1| - |x_{i1}(t) - 1|)$. And consider the initial states chosen as follows:

$$
x_1 = \begin{bmatrix} -0.6, & 0.5, & 0.7 \end{bmatrix}^{\mathrm{T}}, \quad x_2 = \begin{bmatrix} 0.4, & -0.3, & 0.2 \end{bmatrix}^{\mathrm{T}}, \quad x_3 = \begin{bmatrix} 0.2, & 0.1, & -0.3 \end{bmatrix}^{\mathrm{T}},
$$
$$
\hat{x}_1 = \begin{bmatrix} 0.2, & -0.4, & -0.1 \end{bmatrix}^{\mathrm{T}}, \quad \hat{x}_2 = \begin{bmatrix} -0.8, & 0.3, & -0.5 \end{bmatrix}^{\mathrm{T}},
$$

$$\hat{x}_3 = \big[-0.4,\ 0.5,\ -0.8\big]^{\mathrm{T}}.$$

Also select the nonlinear function $g(y_i(t)) = 0.1\sin(y_i(t))$, and the mathematical expectations of networked attacks as $\gamma_1 = 0.30$, $\gamma_2 = 0.35$, and $\gamma_3 = 0.40$.

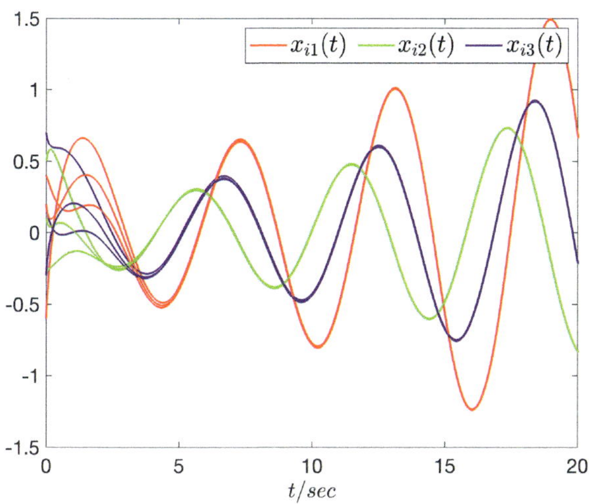

FIGURE 9.1 State trajectories of system (9.34) without controllers.

First, from Fig. 9.1, it is clear that the states $x_i(t)$ of the open-loop system are unstable.

Second, according to Assumption 9.1, the Lipschitz matrices of $f(\cdot)$ and $g(\cdot)$ can be selected as $F_1 = \mathrm{diag}\{0.3, 0, 0\}$ and $F_2 = 0.1$, respectively. The parameters of the nonfragile observer are given as follows: $M_1 = \mathrm{diag}\{0.1, 0.2, 0.4\}$, $M_2 = \mathrm{diag}\{0.2, 0.4, 0.2\}$, $M_3 = \mathrm{diag}\{0.4, 0.2, 0.1\}$, $N_1 = \big[1,\ 1,\ 0\big]^{\mathrm{T}}$, $N_2 = \big[1,\ 0,\ 1\big]^{\mathrm{T}}$, $N_3 = \big[0,\ 1,\ 1\big]^{\mathrm{T}}$. Meanwhile, choosing $\eta_1 = 0.0935$, $\eta_2 = 0.0818$, $\lambda_1 = 0.3641$, $\lambda_2 = 0.0879$, and $\lambda_3 = 0.1801$, and by solving the LMI (9.11) in Theorem 9.2, the corresponding gain matrices K_i and L_i are obtained as follows:

$$K_1 = \big[-0.4675,\ -0.2244,\ -3.3538\big],$$
$$K_2 = \big[-1.5696,\ 0.9222,\ -3.6586\big],$$
$$K_3 = \big[-2.4107,\ -0.9323,\ -0.9102\big],$$
$$L_1 = \big[-2.3995,\ -2.1201,\ -2.0963\big]^{\mathrm{T}},$$
$$L_2 = \big[-1.1688,\ 0.2870,\ -0.8733\big]^{\mathrm{T}},$$
$$L_3 = \big[-0.8047,\ -0.8142,\ -0.5024\big]^{\mathrm{T}}.$$

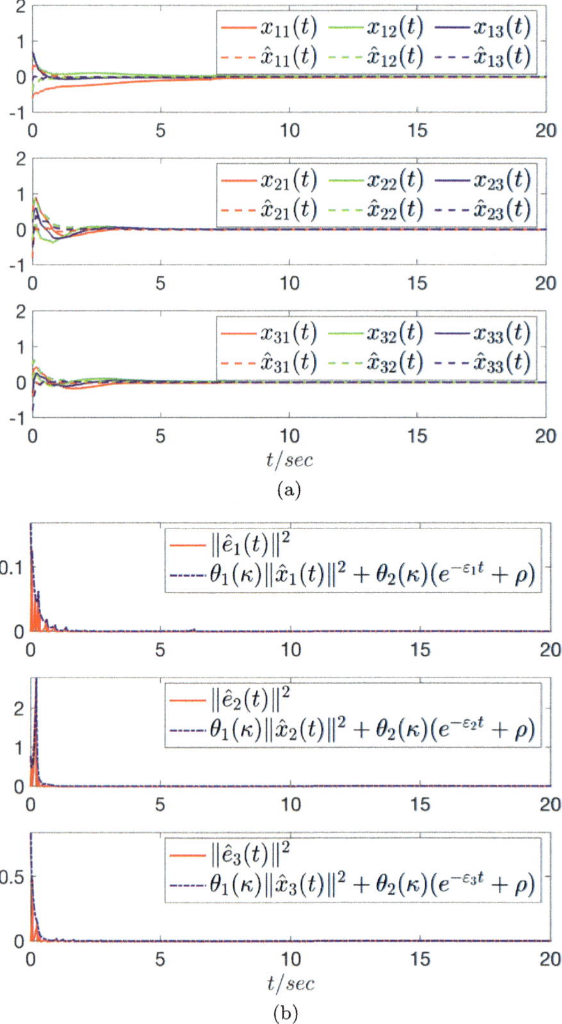

FIGURE 9.2   Simulation results under CETC (9.5): (a) state trajectories of (9.5) and observer trajectories of (9.34); (b) $\|\hat{e}_i(t)\|^2$ along with its bound; (c) $|s_{ie}(t)|$ along with its bound.

Third, considering the presented CETC (9.5), the chosen parameters are $\nabla = 2$, $\theta_{1i}(\kappa) = 0.8$, $\theta_{2i}(\kappa) = 0.0001$, $\rho_i = 0.0001$, $\varepsilon_{1i} = 0.725$, $\varepsilon_{2i} = 0.8$, $\varepsilon_{3i} = 0.95$, and $\xi = 0.25$ in the controllers (9.13). The simulation results are demonstrated in Figs. 9.2 to 9.4. Obviously, from Fig. 9.2(a), it is clear that both system state $x_i(t)$ and observer state $\hat{x}_i(t)$ will tend to zero in a finite-time. Meanwhile, from Figs. 9.2(b) and 9.2(c), we can see that once an event involving $\|\hat{e}_i(t)\|^2$ or $|s_{ie}(t)|$ is triggered, which indicates sampling errors reached their prescribed bound, these sampling errors will have to

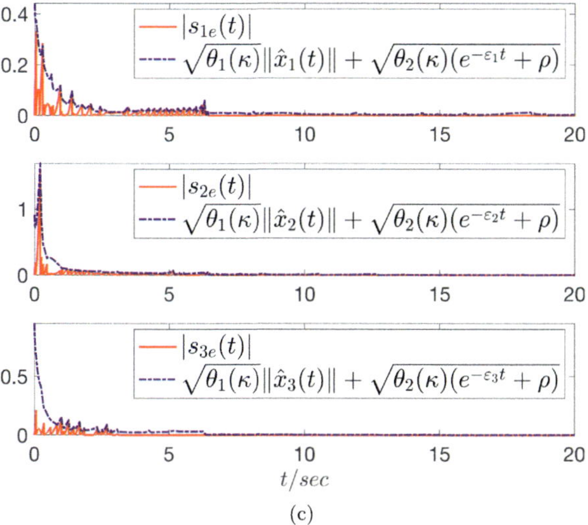

(c)

FIGURE 9.2 (*continued*)

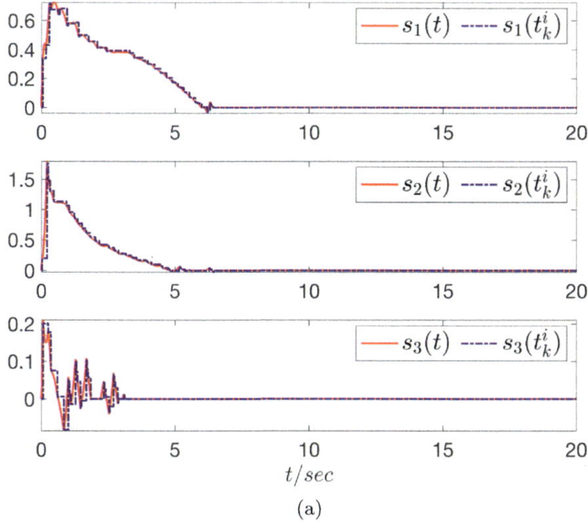

(a)

FIGURE 9.3 Simulation results under CETC (9.5): (a) sliding surfaces $s_i(t)$ and sampling $s_i(t_k^i)$; (b) event-triggered SMC law curves (9.13); (c) actual output curves $\bar{y}_i(t)$ and observer output $\hat{y}_i(t)$.

be cleared immediately and a new round should start until they reach the new event-triggered bound again. Under the implementation of ZOH, the sliding surface, the corresponding SMC law, and output curves are shown in Fig. 9.3. In addition, Fig. 9.4 depicts release instants and error curves.

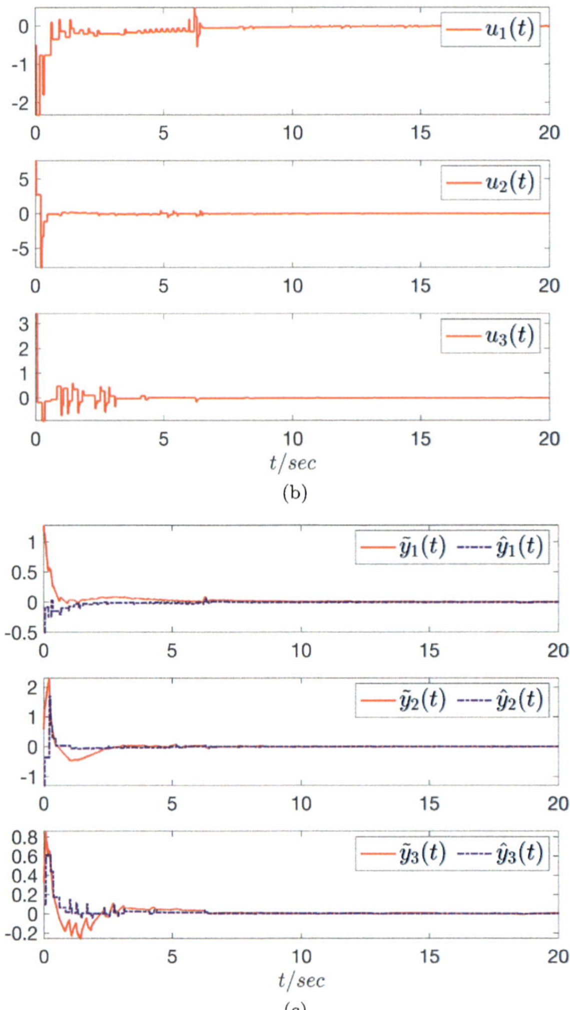

FIGURE 9.3   (*continued*)

Similarly, consider the presented CSTC (9.16), several simulation results are also presented in Figs. 9.5 and 9.6 under the same parameters.

Finally, a bar graph in Fig. 9.7 indicates the numbers of employed sampling under CETC (9.5) and CSTC (9.16), from which we know that the number of employed sampling for the combination trigger condition is less than that for the norm and exponential conditions.

Example 9.2. In this example, consider the following fractional-order power system which was introduced in [47,48]:

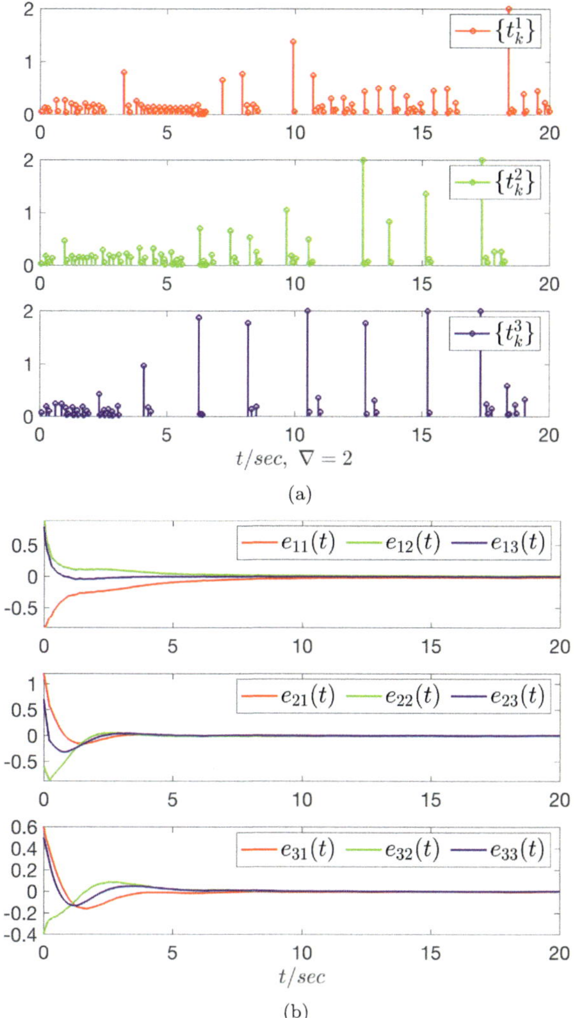

FIGURE 9.4   Simulation results under CETC (9.16): (a) release instants and release intervals; (b) state estimation error curves; (c) sampled error curves.

$$\begin{cases} {}^{C}_{t_0}D^{\alpha}_t v_1(t) = v_2(t), \\ {}^{C}_{t_0}D^{\alpha}_t v_2(t) = -\psi v_2(t) - \eta \sin(v_1(t)) + \rho + \mu \cos(\phi t), \end{cases} \tag{9.35}$$

where $v_1(t)$ and $v_2(t)$ denote the rotor angle and rotating speed, respectively. The parameters are $\eta = \frac{p_s}{H}$, $\psi = \frac{D}{H}$, $\rho = \frac{p_m}{H}$, and $\mu = \frac{p_e}{H}$, with p_s, p_m, p_e, D, and H represents the electrical power, mechanical power, disturbance power amplitude, damping coefficient, and the inertia time constant, respectively; parameter ϕ stands for the power frequency. Mean-

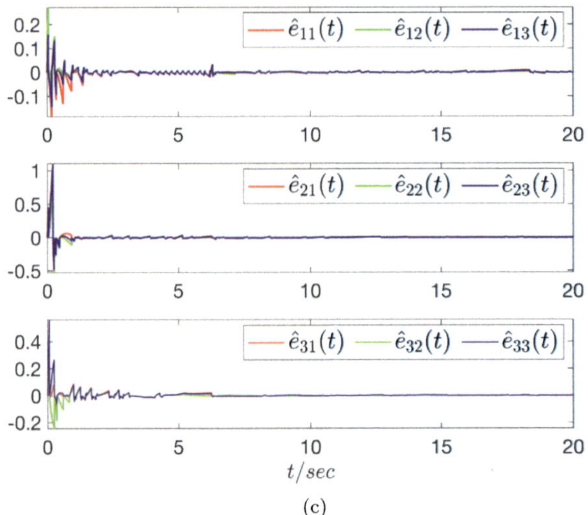

(c)

FIGURE 9.4   *(continued)*

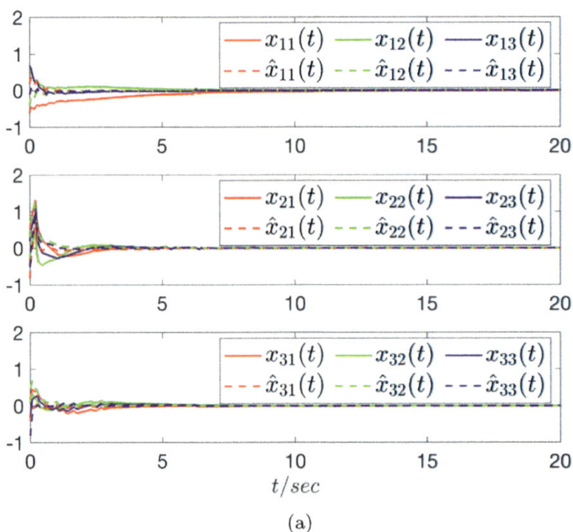

(a)

FIGURE 9.5   Simulation results under CSTC (9.16): (a) state trajectories of (9.34) and observer trajectories of (9.2); (b) $\|\hat{e}_i(t)\|^2$ along with its bound; (c) $|s_{i_e}(t)|$ along with its bound.

while, choosing $\eta = 0.9$, $\psi = 0.1$, $\rho = 0.03$, $\mu = 0.5$, $\phi = 0.01$, and order $\alpha = 0.98$, and then using the Matlab Toolbox, we obtain the 3D image of the motion trajectory of the open-loop system (9.35), which is given in Fig. 9.8. Now, let $x(t) = \begin{bmatrix} v_1(t) \\ v_2(t) \end{bmatrix}$, $A = \begin{bmatrix} 0 & 1 \\ 0 & -0.1 \end{bmatrix}$, $f(x(t)) =$

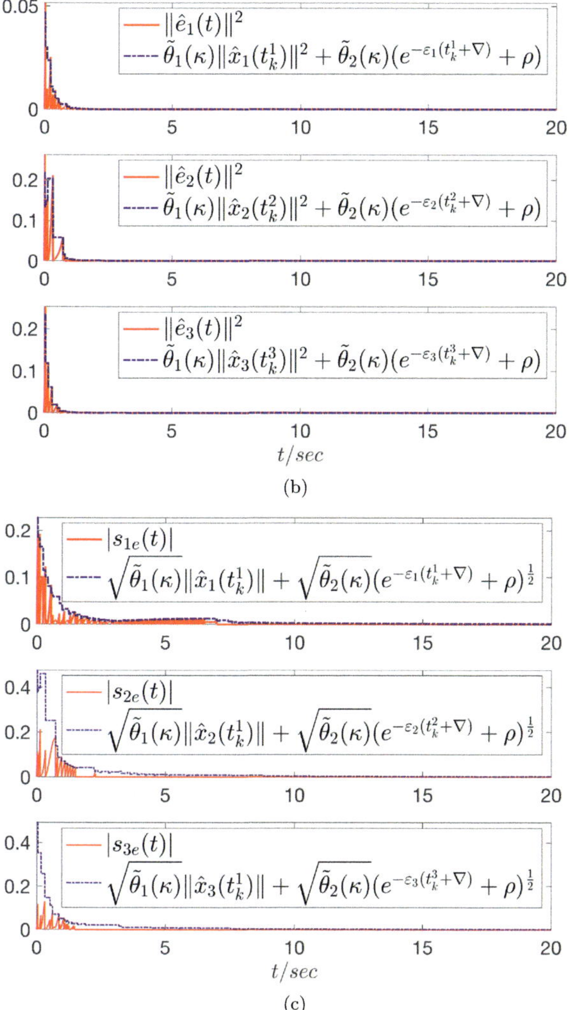

FIGURE 9.5   *(continued)*

$$\begin{bmatrix} 0 \\ -0.9\sin(x_1(t)) + 0.03 + 0.5\cos(0.01t) \end{bmatrix}.$$ Considering the influence of the coupled configuration, control input, and stochastic variables, the dynamical system can be expressed in the following form:

$$\begin{cases} {}^{C}_{t_0}D^{\alpha}_{t}\boldsymbol{x}_i(t) = \boldsymbol{A}\boldsymbol{x}_i(t) + \sum_{j=1}^{4} h_{ij}\boldsymbol{\Gamma}\boldsymbol{x}_j(t) + \beta_i(t)\boldsymbol{f}(\boldsymbol{x}_i(t)) + \boldsymbol{B}\boldsymbol{u}_i(t), \\ \boldsymbol{y}_i(t) = \boldsymbol{C}_i\boldsymbol{x}_i(t), \\ \tilde{\boldsymbol{y}}_i(t) = \boldsymbol{y}_i(t) + \gamma_i(t)\boldsymbol{g}(\boldsymbol{y}_i(t)), \quad i = 1, 2, 3, 4, \end{cases} \tag{9.36}$$

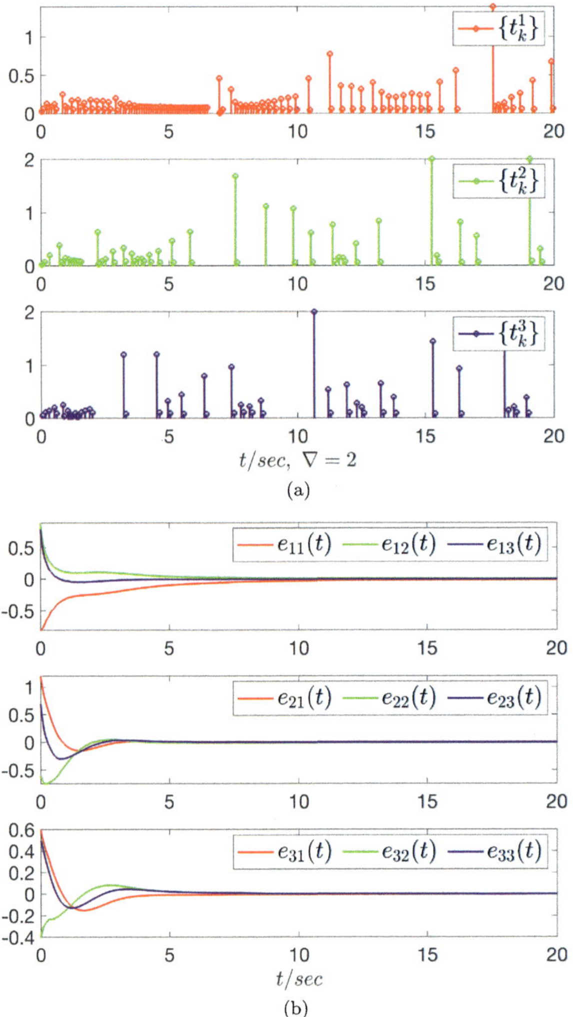

FIGURE 9.6   Simulation results under CSTC (9.16): (a) release instants and release intervals; (b) state estimation error curves; (c) sampled error curves.

where $\alpha = 0.98$, $\boldsymbol{\Gamma} = \begin{bmatrix} 1 & 0 \\ 0 & 1 \end{bmatrix}$, $\boldsymbol{B} = \begin{bmatrix} 0.8 \\ 1 \end{bmatrix}$, $\boldsymbol{C}_1 = \begin{bmatrix} 0 \\ 1 \end{bmatrix}$, $\boldsymbol{C}_2 = \begin{bmatrix} 1 \\ 1 \end{bmatrix}$, $\boldsymbol{C}_3 = \begin{bmatrix} 1 \\ 0 \end{bmatrix}$, $\boldsymbol{C}_4 = \begin{bmatrix} -1 \\ 0 \end{bmatrix}$, and $\boldsymbol{H} = [h_{ij}]_{4\times 4} = \begin{bmatrix} -3 & 1 & 1 & 1 \\ 1 & -2 & 0 & 1 \\ 1 & 0 & -1 & 0 \\ 1 & 1 & 0 & -2 \end{bmatrix}$.

The initial states are chosen as follows:

$$\boldsymbol{x}_1 = [-2.4, \ 2]^{\mathrm{T}}, \ \boldsymbol{x}_2 = [1.6, \ -1.2]^{\mathrm{T}}, \ \boldsymbol{x}_3 = [-0.8, \ 2.4]^{\mathrm{T}}, \ \boldsymbol{x}_4 = [2, \ -3.2]^{\mathrm{T}},$$

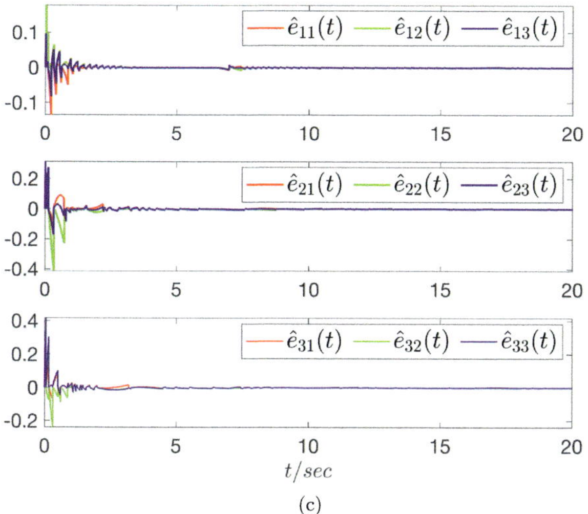

FIGURE 9.6 (*continued*)

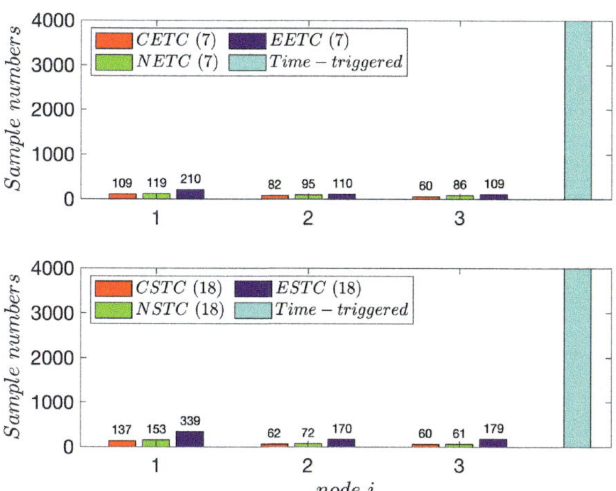

FIGURE 9.7 Sampling numbers under ETC (9.5) and STC (9.16).

$$\hat{x}_1 = \begin{bmatrix} 1.6, & -3.2 \end{bmatrix}^{\mathrm{T}}, \ \hat{x}_2 = \begin{bmatrix} -3.2, & 2 \end{bmatrix}^{\mathrm{T}}, \ \hat{x}_3 = \begin{bmatrix} 3.6, & -3.2 \end{bmatrix}^{\mathrm{T}}, \ \hat{x}_4 = \begin{bmatrix} -2, & 3.2 \end{bmatrix}^{\mathrm{T}}.$$

The nonlinear function is taken as $g(y_i(t)) = 0.2 \sin(y_i(t))$, and the mathematical expectations of networked attacks are chosen as $\beta_1 = 0.3$, $\beta_2 = 0.5$, $\beta_3 = 0.7$, $\beta_4 = 0.9$, while $\gamma_1 = 0.2$, $\gamma_2 = 0.4$, $\gamma_3 = 0.6$, and $\gamma_4 = 0.8$.

According to Assumption 9.1, the Lipschitz matrices of $f(\cdot)$ and $g(\cdot)$ can be selected as $F_1 = \mathrm{diag}\{0, 0.1\}$ and $F_2 = 0.2$, respectively. The parameters of the nonfragile observer are given as follows: $M_1 = \mathrm{diag}\{0.1, 0.2\}$, $M_2 =$

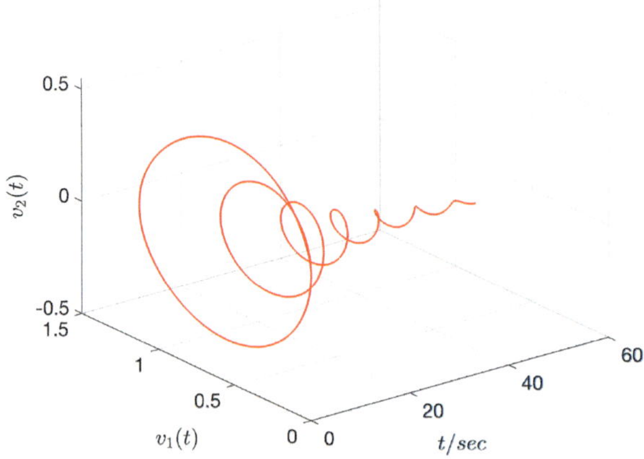

FIGURE 9.8   A three-dimensional graph of the studied power system (9.35).

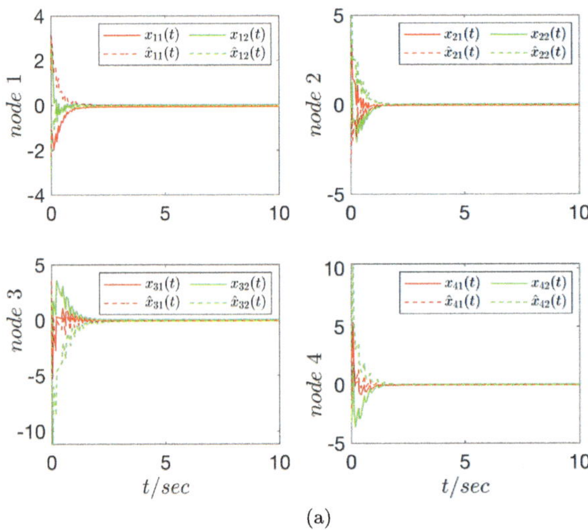

(a)

FIGURE 9.9   Simulation results under CETC (9.5): (a) state trajectories of (9.36) and observer trajectories of (9.2); (b) state estimation error curves; (c) sampled error curves.

diag$\{0.2, 0.1\}$, $\boldsymbol{M}_3 = \mathrm{diag}\{0.3, -0.2\}$, $\boldsymbol{M}_4 = \mathrm{diag}\{-0.1, -0.2\}$, $\boldsymbol{N}_1 = \begin{bmatrix} 1, & 1 \end{bmatrix}^{\mathrm{T}}$, $\boldsymbol{N}_2 = \begin{bmatrix} 1, & -1 \end{bmatrix}^{\mathrm{T}}$, $\boldsymbol{N}_3 = \begin{bmatrix} 1, & 0 \end{bmatrix}^{\mathrm{T}}$, and $\boldsymbol{N}_4 = \begin{bmatrix} -1, & 1 \end{bmatrix}^{\mathrm{T}}$. Meanwhile, choosing $\eta_1 = 0.3481$, $\eta_2 = 0.0469$, $\lambda_1 = 0.2627$, $\lambda_2 = 0.2652$, and $\lambda_3 = 0.4306$, and then solving the LMI (9.11) in Theorem 9.2, the corresponding gain matrices $\boldsymbol{K}_i$ and $\boldsymbol{L}_i$ are obtained as follows:

$$\boldsymbol{K}_1 = \begin{bmatrix} -10.8932, & -20.7334 \end{bmatrix}, \qquad \boldsymbol{K}_2 = \begin{bmatrix} -27.2505, & -20.8893 \end{bmatrix},$$

$$\boldsymbol{K}_3 = \begin{bmatrix} -34.7824, & -0.5032 \end{bmatrix}, \qquad \boldsymbol{K}_4 = \begin{bmatrix} -30.4359, & 1.4324 \end{bmatrix},$$

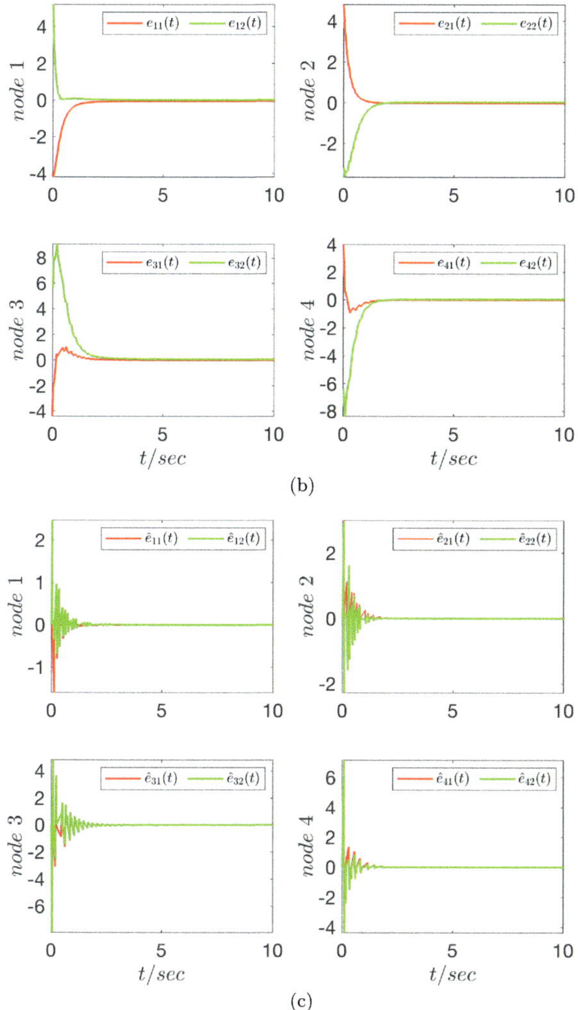

FIGURE 9.9   (*continued*)

$$L_1 = \begin{bmatrix} -4.8185, & -5.9150 \end{bmatrix}^{\mathrm{T}}, \qquad L_2 = \begin{bmatrix} -2.5383, & -3.1644 \end{bmatrix}^{\mathrm{T}},$$

$$L_3 = \begin{bmatrix} -6.5606, & -8.3358 \end{bmatrix}^{\mathrm{T}}, \qquad L_4 = \begin{bmatrix} 5.6888, & 7.2926 \end{bmatrix}^{\mathrm{T}}.$$

And considering the presented CETC (9.5), we choose $\nabla = 2$, $\theta_{1i}(\kappa) = 0.8$, $\theta_{2i}(\kappa) = 0.0001$, $\rho_i = 0.0001$, $\varepsilon_{1i} = 0.725$, $\varepsilon_{2i} = 0.8$, $\varepsilon_{3i} = 0.95$, $\varepsilon_{4i} = 0.98$, and $\xi = 0.25$ in the controllers (9.13). The simulation results are demonstrated in Figs. 9.9 and 9.10. Obviously, from Fig. 9.9(a), it is clear that both system state $x_i(t)$ and observer state $\hat{x}_i(t)$ will tend to zero in a finite-time.

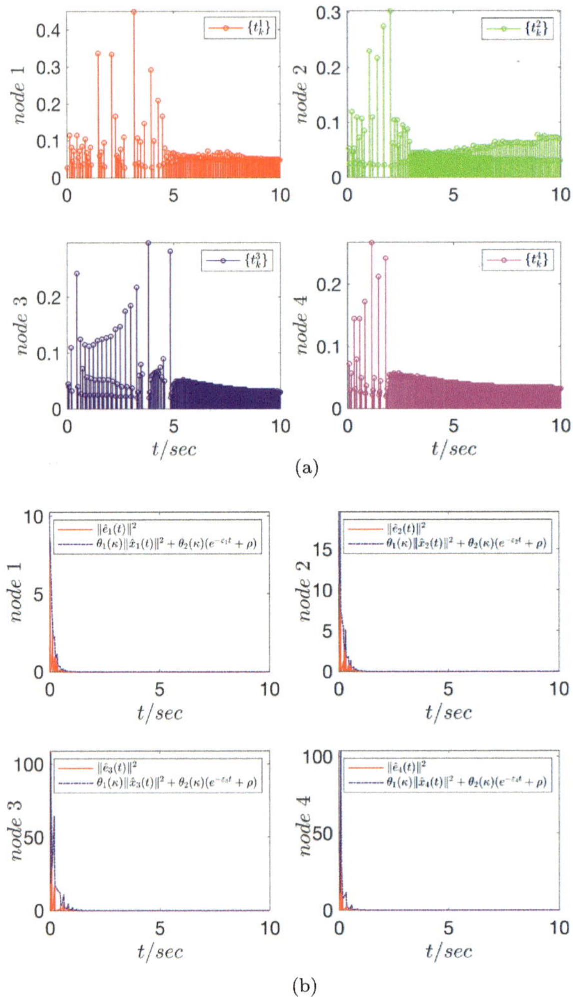

FIGURE 9.10 Simulation results under CETC (9.5): (a) release instants and release intervals, where $t_k^1 = 154$, $t_k^2 = 197$, $t_k^3 = 199$, $t_k^4 = 227$; (b) $\|\hat{\boldsymbol{e}}_i(t)\|^2$ along with its bound; (c) $|s_{ie}(t)|$ along with its bound.

Meanwhile, from Figs. 9.9 and 9.10, we can see that once an event involving $\|\hat{\boldsymbol{e}}_i(t)\|^2$ or $|s_{ie}(t)|$ is triggered, indicating that sampling errors reached their prescribed bound, then these sampling errors will have to be cleared immediately and one should start a new round until they reach the new event-triggered bound again. Therefore, the simulation results demonstrate the effectiveness and feasibility of the proposed control method in this chapter.

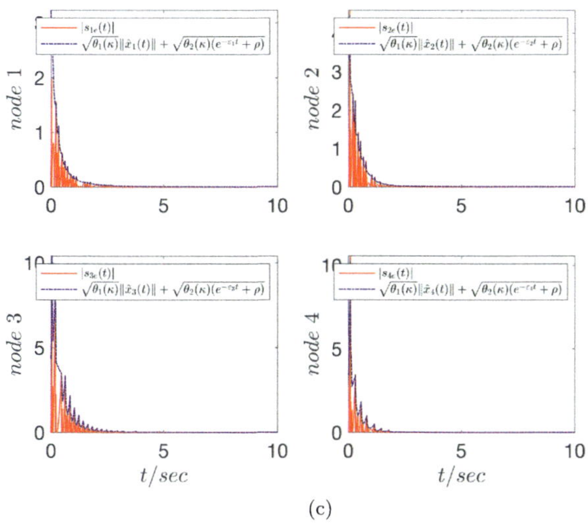

FIGURE 9.10 (*continued*)

9.5 Conclusion

The nonfragile SMC problem of FO-CNSs has been investigated by utilizing the presented combination triggering mechanism in this chapter. First, based on the presented triggered mechanism, a novel CETC that consists of the estimated observer and sliding mode errors has been constructed. Second, a novel event-triggered control strategy has been designed. Then, several sufficient conditions have been derived to guarantee the stability of the augmented closed-loop system by solving the strict LMI. Moreover, to avoid the occurrence of infinitely many triggers in a finite time, this article has also discussed the Zeno behavior through the Gronwall–Bellman inequality. Lastly, two numerical examples have been also provided to illustrate the effectiveness and feasibility of the proposed combination event-triggered SMC strategy.

References

[1] G. Wen, X. Yu, W. Yu, J. Lu, Coordination and control of complex network systems with switching topologies: a survey, IEEE Transactions on Systems, Man, and Cybernetics: Systems 51 (10) (2020) 6342–6357.

[2] X. Liu, Synchronization and control for multiweighted and directed complex networks, IEEE Transactions on Neural Networks and Learning Systems 34 (6) (2023) 3226–3233.

[3] S. Boccaletti, V. Latora, Y. Moreno, M. Chavez, D.-U. Hwang, Complex networks: structure and dynamics, Physics Reports 424 (4–5) (2006) 175–308.

[4] M. Dehmer, Information processing in complex networks: graph entropy and information functionals, Applied Mathematics and Computation 201 (1–2) (2008) 82–94.

[5] Z. Wu, J. Chen, X. Zhang, Z. Xiao, J. Tao, X. Wang, Dynamic event-triggered synchronization of complex networks with switching topologies: asynchronous observer-based case, Applied Mathematics and Computation 435 (2022) 127413.

[6] C. Zhao, S. Zhong, Q. Zhong, K. Shi, Synchronization of Markovian complex networks with input mode delay and Markovian directed communication via distributed dynamic event-triggered control, Nonlinear Analysis: Hybrid Systems 36 (2020) 100883.

[7] H.-S. Hou, H. Zhang, Stability and Hopf bifurcation of fractional complex-valued BAM neural networks with multiple time delays, Applied Mathematics and Computation 450 (2023) 127986.

[8] Y. Yang, Z. Gao, Y. Li, Q. Cai, N. Marwan, J. Kurths, A complex network-based broad learning system for detecting driver fatigue from EEG signals, IEEE Transactions on Systems, Man, and Cybernetics: Systems 51 (9) (2019) 5800–5808.

[9] Y. Liu, B.-Z. Guo, J.H. Park, S.-M. Lee, Nonfragile exponential synchronization of delayed complex dynamical networks with memory sampled-data control, IEEE Transactions on Neural Networks and Learning Systems 29 (1) (2016) 118–128.

[10] W. Yang, Z. Zheng, G. Chen, Y. Tang, X. Wang, Security analysis of a distributed networked system under eavesdropping attacks, IEEE Transactions on Circuits and Systems. II, Express Briefs 67 (7) (2019) 1254–1258.

[11] J. Wu, C. Peng, J. Zhang, B.-L. Zhang, Event-triggered finite-time H_∞ filtering for networked systems under deception attacks, Journal of the Franklin Institute 357 (6) (2020) 3792–3808.

[12] J. Lu, W. Wang, L. Li, Y. Guo, Distributed fusion estimation for non-linear networked systems with random access protocol and cyber attacks, IET Control Theory & Applications 14 (17) (2020) 2491–2498.

[13] X. Meng, B. Jiang, H.R. Karimi, C. Gao, An event-triggered mechanism to observer-based sliding mode control of fractional-order uncertain switched systems, ISA Transactions 135 (2023) 115–129.

[14] I. Podlubny, Fractional Differential Equations: An Introduction to Fractional Derivatives, Fractional Differential Equations, to Methods of Their Solution and Some of Their Applications, Elsevier, 1998.

[15] H.M. Fahad, A. Fernandez, Operational calculus for Caputo fractional calculus with respect to functions and the associated fractional differential equations, Applied Mathematics and Computation 409 (2021) 126400.

[16] F. Wang, Y. Yang, M. Hu, X. Xu, Projective cluster synchronization of fractional-order coupled-delay complex network via adaptive pinning control, Physica A. Statistical Mechanics and Its Applications 434 (2015) 134–143.

[17] L. Zhang, J. Zhong, J. Lu, Intermittent control for finite-time synchronization of fractional-order complex networks, Neural Networks 144 (2021) 11–20.

[18] H. Bao, J.H. Park, J. Cao, Non-fragile state estimation for fractional-order delayed memristive BAM neural networks, Neural Networks 119 (2019) 190–199.

[19] Y. Jia, H. Wu, J. Cao, Non-fragile robust finite-time synchronization for fractional-order discontinuous complex networks with multi-weights and uncertain couplings under asynchronous switching, Applied Mathematics and Computation 370 (2020) 124929.

[20] C. Peng, F. Li, A survey on recent advances in event-triggered communication and control, Information Sciences 457 (2018) 113–125.

[21] X. Ge, Q.-L. Han, X.-M. Zhang, L. Ding, F. Yang, Distributed event-triggered estimation over sensor networks: a survey, IEEE Transactions on Cybernetics 50 (3) (2019) 1306–1320.

[22] L. Zha, R. Liao, J. Liu, X. Xie, E. Tian, J. Cao, Dynamic event-triggered output feedback control for networked systems subject to multiple cyber attacks, IEEE Transactions on Cybernetics 52 (12) (2021) 13800–13808.

[23] M. Xiong, Y. Tan, D. Du, B. Zhang, S. Fei, Observer-based event-triggered output feedback control for fractional-order cyber-physical systems subject to stochastic network attacks, ISA Transactions 104 (2020) 15–25.

[24] S. Zhu, H. Bao, Event-triggered synchronization of coupled memristive neural networks, Applied Mathematics and Computation 415 (2022) 126715.

[25] Q. Sun, J. Ren, Y. Guo, Dynamic event-triggered sliding mode control for the uncertain discrete-time system using the reaching law, Nonlinear Analysis: Hybrid Systems 50 (2023) 101407.

[26] D. Zhang, X. Chen, H. Wang, H. Ren, H. Xu, Event-triggered H_∞ control for networked dynamic system with nonideal interconnection, Applied Mathematics and Computation 463 (2024) 128373.

[27] J. Yuan, T. Chen, Switched fractional order multiagent systems containment control with event-triggered mechanism and input quantization, Fractal and Fractional 6 (2) (2022) 1–29.

[28] X. Meng, B. Jiang, H.R. Karimi, C. Gao, Leader–follower sliding mode formation control of fractional-order multi-agent systems: a dynamic event-triggered mechanism, Neurocomputing 557 (2023) 126691.

[29] Y. Tan, M. Xiong, B. Zhang, S. Fei, Adaptive event-triggered nonfragile state estimation for fractional-order complex networked systems with cyber attacks, IEEE Transactions on Systems, Man, and Cybernetics: Systems 52 (4) (2021) 2121–2133.

[30] Y. Cui, L. Yu, Y. Liu, W. Zhang, F.E. Alsaadi, Dynamic event-based non-fragile state estimation for complex networks via partial nodes information, Journal of the Franklin Institute 358 (18) (2021) 10193–10212.

[31] T. Feng, Y.-E. Wang, L. Liu, B. Wu, Observer-based event-triggered control for uncertain fractional-order systems, Journal of the Franklin Institute 357 (14) (2020) 9423–9441.

[32] M. Li, Y. Chen, Y. Liu, Adaptive disturbance observer-based event-triggered fuzzy control for nonlinear system, Information Sciences 575 (2021) 485–498.

[33] J. Chen, B. Chen, Z. Zeng, P. Jiang, Event-triggered synchronization strategy for multiple neural networks with time delay, IEEE Transactions on Cybernetics 50 (7) (2019) 3271–3280.

[34] Y. Zhang, H. Wu, J. Cao, Global Mittag-Leffler consensus for fractional singularly perturbed multi-agent systems with discontinuous inherent dynamics via event-triggered control strategy, Journal of the Franklin Institute 358 (3) (2021) 2086–2114.

[35] X. Li, Z. Sun, Y. Tang, H.R. Karimi, Adaptive event-triggered consensus of multiagent systems on directed graphs, IEEE Transactions on Automatic Control 66 (4) (2020) 1670–1685.

[36] M. Hymavathi, M. Syed Ali, T.F. Ibrahim, B.A. Younis, K.I. Osman, K. Mukdasai, Synchronization of fractional-order uncertain delayed neural networks with an event-triggered communication scheme, Fractal and Fractional 6 (11) (2022) 1–20.

[37] Q. Li, S. Liu, Y. Chen, Combination event-triggered adaptive networked synchronization communication for nonlinear uncertain fractional-order chaotic systems, Applied Mathematics and Computation 333 (2018) 521–535.

[38] Q. Chang, A. Hu, Y. Yang, L. Li, Pinning exponential boundedness of fractional-order multi-agent systems with intermittent combination event-triggered protocol, International Journal of Systems Science 52 (4) (2021) 874–888.

[39] Y. Xu, H. Wang, J. Yu, W. Li, Synchronization of fractional complex-valued neural networks with pantograph delays and inhibitory factors, Neurocomputing 559 (2023) 126797.

[40] X. Meng, Z. Li, J. Cao, Almost periodic quasi-projective synchronization of delayed fractional-order quaternion-valued neural networks, Neural Networks 169 (2024) 92–107.

[41] Z. Liu, H.R. Karimi, J. Yu, Passivity-based robust sliding mode synthesis for uncertain delayed stochastic systems via state observer, Automatica 111 (2020) 108596.

[42] B. Jiang, C. Gao, Decentralized adaptive sliding mode control of large-scale semi-Markovian jump interconnected systems with dead-zone input, IEEE Transactions on Automatic Control 67 (3) (2021) 1521–1528.

[43] L. Wang, J. Dong, Event-based distributed adaptive fuzzy consensus for nonlinear fractional-order multiagent systems, IEEE Transactions on Systems, Man, and Cybernetics: Systems 52 (9) (2021) 5901–5912.

[44] X. Fan, Z. Wang, Event-triggered sliding mode control for singular systems with disturbance, Nonlinear Analysis: Hybrid Systems 40 (2021) 101011.

[45] W. Qi, G. Zong, W.X. Zheng, Adaptive event-triggered SMC for stochastic switching systems with semi-Markov process and application to boost converter circuit model, IEEE Transactions on Circuits and Systems. I, Regular Papers 68 (2) (2020) 786–796.

[46] X. Fan, Z. Wang, Z. Shi, Event-triggered integral sliding mode control for uncertain fuzzy systems, Fuzzy Sets and Systems 416 (2021) 47–63.

[47] Y. Xu, Y. Li, W. Li, C. Zhang, Synchronization of fractional-order fuzzy complex networks with time-varying couplings and proportional delay, Fuzzy Sets and Systems 478 (2024) 108836.

[48] Z. Yu, Y. Sun, X. Dai, Stability and stabilization of the fractional-order power system with time delay, IEEE Transactions on Circuits and Systems. II, Express Briefs 68 (11) (2021) 3446–3450.

Index

9780443316968